Plant Health Case Studies

Plant Pathology and Crop Diseases

Plant Health Case Studies

Plant Pathology and Crop Diseases

Edited by
Trisna Tungadi and Francis Wamonje

CABI

CABI is a trading name of CAB International

CABI
Nosworthy Way
Wallingford
Oxfordshire OX10 8DE
UK

Tel: +44 (0)1491 832111
E-mail: info@cabi.org
Website: www.cabi.org

CABI
200 Portland Street
Boston
MA 02114
USA

T: +1 (617)682-9015
E-mail: cabi-nao@cabi.org

A catalogue record for this book is available from the British Library, London, UK.

Library of Congress Control Number: 2026948713

ISBN-13: 9781836994411 (hardback)
9781836994428 (ePDF)
9781836994435 (ePub)

DOI: 10.1079/9781836994435.0000

Commissioning Editor: Victoria Taylor
Editorial Assistant: Theresa Regueira
Production Editor: Rosie Hayden

Typeset by Exeter Premedia Services Pvt Ltd, Chennai, India
Printed in the USA

Contents

Original Publication Details

Part I: Viral Diseases in Plant Health

1 **The Banana Aphid and Banana Bunchy Top Disease: Potential Threats to Banana and Plantain Production in Ghana**
Ken Okwae Fening and John P. Carr
DOI: 10.1079/planthealthcases.2025.0024

2 **Pea Necrotic Yellow Dwarf Virus: An Emerging Threat for Legume Production in Europe**
Yahya Z.A. Gaafar and Heiko Ziebell
DOI: 10.1079/planthealthcases.2026.0001

3 **Sugarcane Yellow Leaf Virus – A Silent Threat?**
Andre B. Gama and Madison T. Flasco
DOI: 10.1079/planthealthcases.2024.0020

4 **Soybean Dwarf Virus**
Marta Niedzicka
DOI: 10.1079/planthealthcases.2025.0028

5 **Early Detection of Plant Virus Infection Using Multispectral Imaging and Machine Learning**
Bruce Grieve, Siobain Duffy, Mary M. Dallas, José T. Ascencio-Ibáñez, Vasthi Alonso-Chavez, James Legg, Linda Hanley-Bowdoin and Hujun Yin
DOI: 10.1079/planthealthcases.2024.0010

Contributors

Ali Rhouma, Department of Genetics and Bioengineering, Faculty of Engineering, Yeditepe University, Kayisdagi St., 34755, Istanbul, Turkey; Prima, Nexus 2 Building, Carrer Jordi Girona 29, 2nd floor 2A, Barcelona, Spain

Andre B. Gama, Department of Plant Pathology and Crop Physiology, Louisiana State University Agricultural Center, Baton Rouge, Louisiana

Andrea L. Harper, Department of Biology, University of York, York, UK; Centre for Novel Agricultural Products, University of York, York, UK

Andrea Vadillo-Dieguez, School of Biosciences and the Birmingham Institute of Forest Research, University of Birmingham, Birmingham, UK

Annette Pfordt, Georg-August University Goettingen, Institute of Plant Pathology and Crop Protection, Grisebachstrasse 6, 37077, Goettingen, Germany

Athina Koutouleas, Department of Geosciences and Natural Resource Management, University of Copenhagen, Copenhagen, Denmark

Brent McCallum, Agriculture and Agri-Food Canada, Morden Research and Development Centre, Morden Manitoba, Canada

Bruce Grieve, Department of Electrical and Electronic Engineering, University of Manchester, Manchester, UK

Chapwa Kasoma, CAB International (CABI), Southern Africa Centre, 5834 Mwange Close, Kalundu, Lusaka, Zambia

Colin Hiebert, Agriculture and Agri-Food Canada, Morden Research and Development Centre, Morden Manitoba, Canada

Daniel Narino Rojas, Department of Biology, University of York, York, UK; Centre for Novel Agricultural Products, University of York, York, UK

Davide Danzi, Università degli studi di Verona, Italy

Erica M. Goss, Department of Plant Pathology, University of Florida, Gainesville, FL, 32611, USA; Emerging Pathogens Institute, University of Florida, Gainesville, FL, 32610, USA

Fernanda Batistel, Department of Animal Sciences, University of Florida, Gainesville, FL, 32611, USA

G. Fedele, Department of Sustainable Crop Production (DI.PRO.VES.), Università Cattolica del Sacro Cuore, Via E. Parmense 84, 29122, Piacenza, Italy

Guus Bakkeren, Agriculture and Agri-Food Canada, Summerland Research and Development Centre, Summerland British Columbia, Canada

Heiko Ziebell, Julius Kühn-Institute (JKI) – Federal Research Centre for Cultivated Plants, Institute for Epidemiology and Pathogen Diagnostics, Messeweg 11-12, 38104 Braunschweig, Lower Saxony, Germany

Hujun Yin, Department of Electrical and Electronic Engineering, University of Manchester, Manchester, UK

I. Salotti, Department of Sustainable Crop Production (DI.PRO.VES.), Università Cattolica del Sacro Cuore, Via E. Parmense 84, 29122, Piacenza, Italy

Irda Safni, Department of Agrotechnology, Faculty of Agriculture, Universitas Sumatera Utara, Medan, Indonesia

James G. Menzies, Agriculture and Agri-Food Canada, Morden Research and Development Centre, Morden, Manitoba, Canada

James Legg, International Institute of Tropical Agriculture (IITA), Dar es Salaam, Tanzania

John P. Carr, Department of Plant Sciences, University of Cambridge, Downing Street, Cambridge, CB2 3EA, United Kingdom

Jonathan S. West, Rothamsted Research, Harpenden, AL5 2JQ, UK

José T. Ascencio, Department of Molecular and Structural Biochemistry, North Carolina State University, Raleigh, NC, USA

Karla Cardenas Gomez, Department of Biology, University of York, York, UK; Centre for Novel Agricultural Products, University of York, York, UK

Ken Okwae Fening, Soil and Irrigation Research Centre, School of Agriculture, College of Basic and Applied Sciences, University of Ghana, P.O. Box LG. 68, Accra, Ghana; African Regional Postgraduate Programme in Insect Science (ARPPIS), University of Ghana, Accra, Ghana

Kevin M. King, Rothamsted Research, Harpenden, AL5 2JQ, UK

Linda Hanley-Bowdoin, Department of Plant and Microbial Biology, North Carolina State University, Raleigh, NC, USA

Lisnawita, Department of Agrotechnology, Faculty of Agriculture, Universitas Sumatera Utara, Medan, Indonesia

Lone Buchwaldt, Agriculture and Agri-Food Canada, Saskatoon Research and Development Centre, Saskatoon, Saskatchewan, Canada

Madison T. Flasco, Department of Plant Pathology and Crop Physiology, Louisiana State University Agricultural Center, Baton Rouge, Louisiana

Marco Cesarini, Center Agriculture Food Environment (C3A), University of Trento, San Michele all'Adige, Italy; Research and Innovation Centre, Fondazione Edmund Mach, San Michele all'Adige, Italy

Marta Niedzicka, Crop Science Department, Teagasc, Oak Park, Carlow, R93 XE12, Ireland

Marwa Laribi, Department of Agricultural, Food and Nutritional Science, University of Alberta, Edmonton, Alberta, Canada

Mary M. Dallas, Department of Plant and Microbial Biology, North Carolina State University, Raleigh, NC, USA

Meriam Bouri, Department of Genetics and Bioengineering, Faculty of Engineering, Yeditepe University, Kayisdagi St., 34755, Istanbul, Turkey; Prima, Nexus 2 Building, Carrer Jordi Girona 29, 2nd floor 2A, Barcelona, Spain

Michelle T. Hulin, Department of Plant Soil and Microbial Sciences, Michigan State University, East Lansing, USA

Mohammed Radi, Department of Plant Protection, Phytopathology Unit, Ecole Nationale d'Agriculture de Meknès,Km 10, Rte Haj Kaddour, BPS/40, Meknès, 50001, Morocco; Laboratory of Environment and Valorization of Microbial and Plant Resources, Faculty of Sciences, Moulay Ismail University, PO Box 11201, Zitoune, Meknès, Morocco

My Livbjerg, Department of Agroecology, Faculty of Science and Technology, Aarhus University, Forsøgsvej 1, 4200 Slagelse, Denmark

Oscar Giovannini, Research and Innovation Centre, Fondazione Edmund Mach, San Michele all'Adige, Italy

Puopolo Gerardo, Center Agriculture Food Environment (C3A), University of Trento, San Michele all'Adige, Italy; Research and Innovation Centre, Fondazione Edmund Mach, San Michele all'Adige, Italy

Rachid Lahlali, Department of Plant Protection, Phytopathology Unit, Ecole Nationale d'Agriculture de Meknès,Km 10, Rte Haj Kaddour, BPS/40, Meknès, 50001, Morocco

Reem Aboukhaddour, Agriculture and Agri-Food Canada Lethbridge Research and Development Centre, Lethbridge, Alberta, Canada

Robert W. Jackson, School of Biosciences and the Birmingham Institute of Forest Research, University of Birmingham, Birmingham, UK

Sabrina Sarrocco, Department of Agriculture, Food and Environment, University of Pisa, Pisa, Italy

Samia Berraies, Agriculture and Agri-Food Canada, Swift Current Research and Development Centre, Swift Current, Saskatchewan, Canada

Samuel J. Martins, Department of Plant Pathology, University of Florida, Gainesville, FL, 32611, USA

Sara Franco Ortega, Department of Biology, University of York, York, UK

Sean Walkowiak, Grain Research Laboratory, Canadian Grain Commission, Winnipeg, Canada; Department of Plant Sciences, University of Manitoba, Winnipeg, Canada

Siobain Duffy, Department of Ecology, Evolution, and Natural Resources, Rutgers University, New Brunswick, NJ, USA

Sophie Alice James, Department of Biology, University of York, York, UK

Stephen E. Strelkov, Department of Agricultural, Food and Nutritional Science, University of Alberta, Edmonton, Alberta, Canada

T. Caffi, Department of Sustainable Crop Production (DI.PRO.VES.), Università Cattolica del Sacro Cuore, Via E. Parmense 84, 29122, Piacenza, Italy

Toi Ketehouli, Department of Plant Pathology, University of Florida, Gainesville, FL, 32611, USA

V. Rossi, Department of Sustainable Crop Production (DI.PRO.VES.), Università Cattolica del Sacro Cuore, Via E. Parmense 84, 29122, Piacenza, Italy

Vasthi Alonso-Chavez, Rothamsted Research, Harpenden, UK

Ville-Petri Friman, Department of Microbiology, University of Helsinki, Helsinki, Finland

Xiben Wang, Agriculture and Agri-Food Canada, Morden Research and Development Centre, Morden Manitoba, Canada

Yahya Z.A. Gaafar, Centre for Plant Health – North Saanich, Canadian Food Inspection Agency (CFIA), North Saanich, V8L 1H3, British Columbia, Canada

Zaiton Sapak, Plant Pathology Laboratory, Faculty of Plantation and Agrotechnology, Universiti Teknologi MARA, Jasin Campus, Merlimau 77300, Melaka, Malaysia

Foreword

As the CABI editor-in-chief responsible for plant diseases for *Plant Health Cases*, I wish to endorse the timely publication of this compendium.

Plant Health Case Studies: Plant Pathology and Crop Diseases, edited by Drs Trisna Tungadi and Francis Wamonje, presents 23 of the nearly 100 cases published by CABI since 2023. The majority of the chosen PHCs concern a specific disease threatening crop production, and the choice is designed to present contrasting biology and agricultural methods. A couple of the PHCs take a more general perspective, viz one on diagnostics and another on antibiotics.

The concept of *Plant Heath Cases* lies in the need to have an introduction to a topic that can inspire curiosity and encourage a deeper interest, especially among students advanced in their studies. Through my role as an editor, I have had my eyes opened to the huge diversity of plant diseases affecting diverse crops, production systems and regions. The cases present the programme and many complementary approaches that can be taken to mitigate the negative impact of disease on, especially, food production and security.

The editors' introduction to the compilation emphasizes the challenges facing humanity caused by climate change, urbanization and, not least, population growth. Each case presents new perspectives for solving the problem described and provides inspiration for looking for solutions to similar problems in other crops.

The editors have chosen to group the cases under four themes: "Viral Diseases", "Bacterial Diseases in Plant Health", "Cereals and Oilseeds: Fungal and Oomycete Threats" and "Horticulture and Plantation Diseases". The first two themes are specific to pathogen type, whereas the latter two both concern diseases caused by the filamentous fungal and oomycete pathogens. As the names suggest, the third theme concentrates on "Cereals and Oilseeds: Fungal and Oomycete Threats", in other words, diseases of important primarily temperate bulk field crops, whereas the fourth theme, "Horticulture & Plantation Diseases", concerns tropical and horticultural crops, mostly from a warmer climate perspective.

The editors have faced hard choices in making this compilation. There are many more excellent cases on other diseases, and many more are in the pipeline. I hope that the readers of this tome are inspired to explore this resource further and use these cases in their teaching, as I do. If they do not find the case you need, perhaps you could write one yourself!

Professor David B. Collinge,
University of Copenhagen, Joint Editor-in-Chief, *Plant Health Cases*

Introduction: Plant Diseases in a Changing World

Trisna Tungadi and Francis Wamonje

[1]School of Life Sciences, Keele University, Keele, Staffordshire ST5 5BG, United Kingdom; [2]Niab, Park Farm Campus, Villa Road, Histon, Cambridge, CB24 9AT, United Kingdom

Plant diseases represent one of the most persistent and complex challenges facing global agriculture, threatening food security, economic stability, and environmental sustainability across diverse production systems worldwide. The impact of plant pathogens extends far beyond immediate yield reductions, encompassing grain quality deterioration, market access restrictions, trade disruptions, and threats to smallholder livelihoods in regions where affected crops serve as dietary staples. As the global population continues to expand toward an estimated 10 billion by 2050, the capacity to protect crop production from bacterial, viral, fungal, and oomycete pathogens becomes increasingly critical for maintaining stable food supplies and supporting sustainable agricultural intensification.

The diversity of plant pathogenic organisms is notable, including obligate biotrophs that obtain nutrients solely from living host tissues, aggressive necrotrophs that actively kill host cells to promote colonization, and hemibiotrophic pathogens that switch between these nutritional strategies during infection. Each pathogen group exhibits distinct epidemiological patterns, transmission methods, and management challenges, requiring tailored approaches based on a thorough understanding of pathogen biology, host–pathogen interactions, and the environmental and anthropogenic factors shaping disease dynamics.

This volume examines plant diseases through four thematic sections focused on key pathogen groups that impact global crop production: bacterial pathogens, viral pathogens, fungal and oomycete pathogens causing diseases of cereals, oilseeds, and horticultural and plantation crops. These sections include case studies from various geographical regions and cropping systems, ranging from annual cereals vital to global food security to perennial plantation crops that bolster regional economies and international trade. The case studies cover both historical diseases, such as coffee leaf rust and wheat rusts, which have influenced agricultural development and societies over centuries, and emerging threats identified through modern molecular and diagnostic techniques, such as kiwifruit bacterial canker and *Trichoderma* ear rot of maize. By exploring pathogen biology, epidemiology, historical impacts, and current management strategies, this volume provides essential insights into the complex challenges of plant diseases and the integrated approaches necessary to manage them effectively in diverse agricultural settings.

A recurring theme in this volume is how human activities greatly influence the emergence, spread, and control of plant diseases. Climate change is significantly changing where diseases occur and how severe they can become, including the spread of tropical pathogens into temperate areas that were once unsuitable and changes in the timing of pathogen and vector activity. Agricultural intensification, marked by less crop diversity, shorter rotation cycles, and a greater dependence on genetically similar cultivars, has fostered environments that support pathogen survival and outbreaks.

The increased globalization of agricultural trade has sped up the spread of plant pathogens across continents, introducing new diseases to previously unaffected populations and disrupting existing farming systems. At the same time, growing societal demands to reduce pesticide use, the adoption of organic farming, and the promotion of environmentally sustainable practices have driven the need for alternative disease management methods. These should reduce chemical inputs while ensuring crop yields. To understand plant diseases in the 21st century, it is essential to combine biological, ecological, social, and economic perspectives. This integrated approach aims to create management strategies that are effective, sustainable, and adaptable to ongoing environmental and agricultural changes.

Bacterial Plant Pathogens: Persistence, Adaptation, and Management Challenges

Bacterial plant pathogens present distinctive challenges that differentiate them from other pathogen groups, particularly their capacity for persistence in diverse environmental reservoirs, rapid genetic exchange facilitating adaptation to control measures, and the limited availability of effective curative treatments once infections become established.

The case studies examining bacterial diseases span diverse pathosystems, including crown gall disease caused by pathogenic agrobacteria, bacterial wilt and brown rot diseases caused by the *Ralstonia solanacearum* species complex, and bacterial cankers of kiwifruit and stone fruits caused by *Pseudomonas syringae* pathovars. These pathosystems collectively illustrate the remarkable host range and adaptive capacity characteristic of bacterial plant pathogens, with some species capable of infecting hundreds of plant species across multiple families, while others demonstrate exquisite host specificity shaped by intricate molecular dialogues between pathogen effectors and host resistance mechanisms.

The case study on crown gall disease, caused by pathogenic agrobacteria after acquisition of the tumor-inducing Ti plasmid, represents a unique pathosystem where the capacity for conjugal plasmid transfer among soil bacteria complicates disease management and necessitates comprehensive analyses of both pathogenic and non-pathogenic agrobacterial populations (Bourin and Rhouma, 2025). The disease has caused serious economic losses across landscapes, nurseries, and orchards worldwide, with recent outbreaks in North African vineyards linked to climate factors. The Ti plasmid-mediated genetic transformation of plant cells that underlies tumor formation has profound implications beyond agricultural pathology, having revolutionized plant biotechnology and our understanding of horizontal gene transfer between kingdoms. However, this same genetic flexibility that enabled biotechnological applications presents management challenges, as the Ti plasmid can spread through agrobacterial populations, rendering soils unsuitable for susceptible crops for extended periods.

The case study on *R. solanacearum* species complex exemplifies the evolutionary and ecological complexity of bacterial plant pathogens, comprising three phylogenetically distinct species (*R. solanacearum*, *Ralstonia pseudosolanacearum*, and *Ralstonia syzygii*) that collectively infect hundreds of plant species and cause economically devastating diseases, including bacterial wilt of solanaceous crops and brown rot of potato (Gomez *et al.*, 2024). In Central and Northern Europe, cold-adapted strains of *R. solanacearum* have caused sporadic potato brown rot outbreaks since the 1970s, while recent detection of warm-adapted *R. pseudosolanacearum* in greenhouse rose production highlights the potential for tropical strains to establish in temperate regions under protected cultivation. This documented range expansion of tropical bacterial species into previously unaffected regions underscores the vulnerability of agricultural systems to emerging bacterial threats, particularly in the context of climate change and the international movement of plant materials. The soilborne nature of *Ralstonia* species, combined with their capacity to survive in surface waters and asymptomatic infection on certain host plants, presents formidable challenges for disease containment and detection, as no single method has proven capable of efficiently limiting pathogen spread or eradicating infections in field settings.

The case studies on bacterial canker diseases of *Prunus* species (stone fruits) and kiwifruit caused by *P. syringae* pathovars illustrate both the remarkable diversification of bacterial pathogens on closely related hosts and the rapid emergence of highly virulent strains that can devastate established production systems (Danzi, 2024; Vadillo-Dieguez *et al.*, 2025). Kiwifruit bacterial canker provides a particularly instructive case study of pathogen evolution and host–microbe coevolution in real time, as kiwifruit cultivation expanded globally only within the past century, and the causal pathogen *P. syringae* pv. *Actinidiae* (Psa) was first isolated merely 40 years ago. Following initial outbreaks in Japan and Italy caused by relatively mild strains, a highly virulent Psa strain emerged from Italy in 2008 that spread rapidly worldwide, causing severe symptoms, substantial yield losses, and even plant death. The emergence of this virulent strain had numerous social and economic consequences, particularly devastating New Zealand's kiwifruit industry, where the pathogen threatened the foundation of a major agricultural sector. Modern genomic approaches have enabled characterization of Psa population structure and principal virulence factors, though molecular mechanisms underlying pathogenicity remain incompletely understood. The recent emergence and rapid global dissemination of this pathogen provide an almost unique opportunity to investigate bacterial pathogen evolution and adaptation to new hosts and environments in contemporary agricultural systems.

A critical consideration in bacterial disease management is the role of antibiotics in agricultural systems and their implications for human and environmental health (Katehouli *et al.*, 2025). While antibiotics are primarily used in animal agriculture, they are used to a limited extent in some countries to control bacterial plant diseases. The continuous use of antibiotics in plants raises serious concerns about the potential transfer of antibiotic resistance genes to pathogens and non-pathogens, as genetic material can be exchanged between organisms from different kingdoms. Furthermore, antibiotic applications can disrupt the balance of plant and soil microbial community, affecting beneficial plant microbes, making plants more susceptible to other diseases, and impacting plant physiological and metabolic processes. The need for careful management of antibiotic use in plant agriculture to mitigate resistance development and environmental impacts underscores the imperative for developing sustainable alternative approaches, including biological control agents, bacteriophages, and enhanced host resistance.

Viral Plant Pathogens: Irreversible Infections and Vector-mediated Spread

Plant viruses present fundamentally different management challenges compared with bacterial and fungal pathogens, as their obligate intracellular parasitism and irreversible nature of infection preclude curative treatments once systemic infection occurs. Management strategies must, therefore, emphasize prevention through vector control, pathogen exclusion, deployment of host resistance, and rigorous certification systems for virus-free planting material, particularly for vegetatively propagated crops where infected propagules perpetuate viral infections across generations. The case studies examining viral diseases span diverse crops and geographical regions, including banana bunchy top disease in West Africa, pea necrotic yellow dwarf virus in European legume production, sugarcane yellow leaf disease globally, cassava brown streak disease in sub-Saharan Africa, and soybean dwarf virus affecting legumes in Asia and Australia. These pathosystems collectively illustrate the critical importance of arthropod vectors, particularly aphids, in viral disease epidemiology and the profound food security implications of viral diseases affecting staple crops in regions where smallholder farmers depend on these crops for subsistence and income.

A unifying aspect across the viral disease case studies is the essential role of vector biology in influencing disease dynamics and control strategies. *Banana bunchy top virus* (BBTV), one of the most damaging viral diseases of cultivated *Musa* species (banana and plantain), is transmitted by the banana aphid *Pentalonia nigronervosa*, which is endemic in Ghana despite BBTV not yet being detected there (Fening and Carr, 2025). The existence of an established vector population without the virus creates a high-risk scenario where the introduction of BBTV would likely lead to rapid spread with severe impacts on both commercial and smallholder growers. This situation exemplifies the vulnerability of agricultural systems to viral disease incursions when competent vectors are already present, underscoring the crucial importance of quarantine protocols, surveillance, and contingency planning for potential disease introductions. Similarly, vector-mediated transmission dynamics are seen in *Pea necrotic yellow dwarf virus* (PNYDV), spread by pea aphids (*Acyrthosiphon pisum*) and black bean aphids (*Aphis fabae*), and *Sugarcane yellow leaf virus*, primarily transmitted by *Melanaphis sacchari* (Gama and Flasco, 2024; Gaafar and Zeibell, 2026).

The genetic complexity of plant viruses adds additional layers of challenge to disease management and diagnostic efforts. *Pea necrotic yellow dwarf virus* is a multipartite virus consisting of at least eight genomic components, each encapsidated within individual virions, with the distribution of segments in host plants and vectors, and the functional roles of some segments remaining poorly understood (Gaafar and Zeibell, 2026). This genomic architecture presents fundamental biological questions about how multipartite viruses maintain genome integrity during transmission and practical challenges for diagnostic approaches and for resistance breeding.

Sugarcane yellow leaf virus similarly exhibits remarkable genetic diversity, with molecular techniques revealing ten major genotypes grouped into three phylogroups, complicating detection efforts and resistance screening (Gama and Flasco, 2024). The variable impact of sugarcane yellow leaf virus on yield across different cultivars, with some varieties showing little to no foliar symptoms yet experiencing severe yield losses, exemplifies the phenomenon of asymptomatic or cryptic viral infections that render traditional visual surveillance ineffective.

Cassava brown streak disease (CBSD) provides a compelling case study of the impacts of viral disease on food security in sub-Saharan Africa, where cassava is a critical climate-resilient crop essential to 21st-century food security. The disease is caused by either *Cassava brown streak virus* (CBSV) or *Ugandan cassava brown streak virus* (UCBSV), both of which typically produce subtle or no symptoms on stems and leaves while destroying root tissue, meaning that farmers are often unaware that their fields are infected until harvest failure occurs (Kasoma, 2023). The invasive nature of CBSD, spreading rapidly through exchange of infected planting material when used to propagate new fields, has resulted in substantial losses in quality, yield, and income, contributing to household food insecurity in affected regions. The absence of certified propagule systems for cassava in countries like Zambia, combined with traditional practices of cutting exchange among farmers, facilitates disease spread and complicates efforts to contain or eradicate infections. Management approaches integrating phytosanitation, recommended cultural practices, and restrictions on cutting sharing show promise, though implementation requires collaborative and coordinated efforts involving research institutions, extension services, and farming communities.

Technological innovations in disease detection offer promising avenues for addressing the diagnostic challenges posed by asymptomatic or subtle viral infections. Multispectral imaging technology coupled with machine-learning algorithms has demonstrated capacity to detect cassava brown streak disease infection with 95% accuracy at 28 days post-infection under laboratory conditions when plants exhibit no visible symptoms (Grieve *et al.*, 2024). This engineering advancement enables rapid determination of plant infection status in field settings, potentially allowing for identification and removal of infected plants before disease spread occurs. The integration of such precision detection technologies with epidemiological modeling frameworks provides opportunities for assessing the efficacy of different intervention strategies and optimizing disease management approaches. Beyond cassava, this technology and modeling framework may prove applicable to diseases of other vegetatively propagated crops, including banana, plantain, potato, sweet potato, and yam, where early detection of asymptomatic infections could substantially improve disease management outcomes.

Fungal and Oomycete Diseases of Cereals and Oilseeds: Diversity in Pathogenic Strategies

Fungal pathogens of cereals and oilseed crops exhibit a remarkable diversity in nutritional strategies, host ranges, and survival mechanisms. This diversity necessitates a range of management approaches tailored to the specific biology and epidemiology of each pathosystem. The case studies on fungal diseases of wheat, maize, and oilseed rape include obligate biotrophic rust fungi that depend on living host tissues, aggressive necrotrophic pathogens that actively kill host cells by deploying specific effector molecules, and facultative saprotrophs capable of surviving on crop residues or alternative hosts for extended periods.

This section also discusses ergot disease caused by *Claviceps purpurea*, providing historical insight into fungal plant diseases through its role in devastating medieval populations via mycotoxin-contaminated grain (Berraies *et al.*, 2023), and highlights the unexpected emergence of *Trichoderma* species as maize pathogens, challenging long-held views about organisms widely used as beneficial biocontrol agents (Pfordt, 2025) .

The biotrophic rust fungi *Puccinia striiformis* (yellow rust) (Livbjerg, 2025) and *Puccinia triticina* (leaf rust) (McCallum *et al.*, 2024) exemplify the evolutionary arms race between host resistance and pathogen virulence that has characterized wheat production for centuries. Yellow rust, once confined to temperate regions with cool, humid climates or high-altitude warmer areas with cold nights, has recently spread to warmer regions where it was previously unseen, raising particular concern amid global climate change (Livbjerg, 2025). The pathogen is now reported in more than 60 countries and appears in all wheat-growing regions worldwide. Both rust species display remarkable genetic diversity and continuously evolve new races capable of overcoming deployed resistance genes, leading to characteristic boom-and-bust cycles where resistant cultivars become susceptible within years as pathogen populations adapt. Managing these dynamic pathogen populations requires continuous monitoring of race structure through systematic virulence surveys, the strategic use of multiple resistance genes that combine race-specific seedling resistance with race-nonspecific adult-plant resistance, and close collaboration between plant breeders and pathologists to ensure that resistance breeding efforts remain ahead of pathogen evolution.

The genetics underlying rust resistance and virulence offer a valuable framework for understanding host–pathogen coevolution and developing durable resistance strategies. Over 80 leaf rust resistance (Lr) genes have been characterized, most of which confer race-specific resistance active from the seedling stage to maturity, while a few provide non-race-specific partial resistance to leaf rust and other diseases at the adult plant stage. Combining Lr genes that confer seedling and adult plant resistance has proven particularly successful and durable, although the ongoing evolution of *P. triticina* populations requires continuous surveillance and adaptive resistance deployment strategies (McCallum *et al.*, 2024). The obligate biotrophic nature of rust fungi, which necessitates living host tissues for reproduction, fundamentally shapes their epidemiology and management options, as these pathogens cannot survive on dead crop residues but instead depend on continuous asexual reproduction cycles on living wheat plants or overwintering on volunteer plants and alternative hosts.

In contrast to biotrophic rusts, necrotrophic fungal pathogens such as *Pyrenophora tritici-repentis*, causing tan spot of wheat (Laribi *et al.*, 2024), and *Plenodomus lingam*, causing phoma stem canker of oilseed rape (West and King, 2024), present distinct management challenges stemming from their capacity to survive saprophytically on crop debris and their deployment of necrotrophic effectors that actively promote host cell death.

Tan spot, caused by the necrotrophic fungus *P. tritici-repentis*, is globally distributed and can cause significant yield losses under conditions that favor disease development (Laribi *et al.*, 2024). The pathogen has a wide host range among grass species and survives on alternative hosts, crop debris, and infected seeds, requiring management strategies such as burning or burying crop debris, rotating with non-cereal crops, and using pathogen-free seed. Although several fungicides are available for controlling tan spot, their use raises economic and environmental concerns, and resistance breeding efforts have largely failed due to limited understanding of host–pathogen interactions and the absence of effective differential sets for identifying races from durum wheat.

Phoma stem canker of oilseed rape, caused by two closely related fungi *P. lingam* and *Plenodomus biglobosus*, illustrates the complexity of managing diseases where systemic pathogen colonization from initial leaf infections to stem tissues occurs through vascular transport (West and King, 2024). Both pathogens survive saprophytically on plant debris and spread via wind-dispersed ascospores, rain-splashed conidia, and infected seeds. Spores germinate on leaves, produce leaf spots, and then release conidia for secondary spread. The critical damage happens when mycelium grows through vascular tissue to the stem, causing stem cankers that reduce yield by impairing water transport. This systemic colonization pattern makes fungicide-based control difficult, as chemical treatments are ineffective once the pathogen is established in stem tissues. Therefore, preventive applications targeting leaf infections are essential. Disease management must include crop rotation to remove infectious debris, the use of resistant varieties, and strategic fungicide applications combining at least two modes of action to prevent resistance development.

The emergence of *Trichoderma* species as maize pathogens marks a significant shift in understanding fungal–plant interactions, as *Trichoderma* species have long been regarded as beneficial fungi widely used as biocontrol agents and plant growth promoters in sustainable agriculture. Recent findings from maize fields in Germany, France, Italy, and Austria have shown that certain *Trichoderma* isolates can cause ear rot, stalk infections, yield reductions, and considerable tissue damage (Pfordt. 2025). This unexpected pathogenic behavior raises important questions about how damage caused by *Trichoderma* can be distinguished from other diseases, whether commercial biofungicides containing *Trichoderma* as active ingredients may pose hidden risks, and what measures are available to safeguard maize from this unforeseen threat. This case illustrates how agricultural intensification and environmental changes can reveal pathogenic tendencies in previously harmless organisms, emphasizing the importance of comprehensive safety assessments of biological control agents and ongoing surveillance for emerging disease risks.

Diseases of Horticultural and Plantation Crops: Challenges of Perennial Production Systems

Horticultural and plantation crops pose unique disease management challenges due to their perennial nature, high establishment costs, long periods between planting and economic returns, and the continuous availability of hosts that supports pathogen persistence and evolution over decades. The case studies on diseases of grapevine, coffee, oil palm, berry crops, and pineapple cover oomycete pathogens (*Plasmopara viticola*), biotrophic basidiomycetes (*Hemileia vastatrix*, *Ganoderma boninense*), necrotrophic fungi (*Botrytis cinerea*), opportunistic pathogens (*Neoscytalidium dimidiatum*), and bacterial pathogens (*Xylella fastidiosa*). These pathosystems collectively highlight how diseases affecting high-value specialty crops are connected to economic sustainability, market access, consumer preferences for reduced pesticide use, and the socio-economic aspects of agricultural production in both developed and developing regions.

Grapevine downy mildew, caused by the oomycete *P. viticola*, is one of the most important grapevine diseases worldwide, with disease control still largely relying on fungicide applications despite increasing societal demands for agricultural sustainability and reduced chemical inputs (Cesarini et al., 2025). The pathogen's life cycle includes both sexual and asexual reproduction stages, with oospores acting as primary sources of inoculum and asexual sporangia facilitating rapid secondary spread during favorable conditions. The development of partially resistant grapevine varieties through introgression of resistance genes from wild *Vitis* species into the *Vitis vinifera* genome offers promising pathways toward lowering fungicide reliance, although these varieties show partial rather than complete resistance. The pathogen is able to complete its life cycle but causes less severe disease and slows epidemic progression. The durability of resistance in these varieties remains a crucial concern, as intensive use without integrating other management practices could lead to selection of *P. viticola* populations capable of overcoming existing resistance genes. Innovative approaches under development include plant extracts and microbial biocontrol agents as alternatives to traditional fungicides, though issues concerning efficacy consistency and large-scale application remain.

Coffee leaf rust caused by *H. vastatrix* provides historical perspective on how plant diseases have shaped global agricultural development and trade patterns, having devastated early coffee production in Ceylon and fundamentally altered colonial power structures (Koutouleas, 2023). Despite being discovered over 150 years ago and receiving extensive research attention, many unknowns concerning the pathogen's spread and persistence across tropical coffee-growing regions persist. The disease has experienced a resurgence in recent decades, with genetic resistance breaking down in cultivars across the coffee belt. Best practices for disease control remain contentious, particularly regarding compatibility with emerging trends in coffee farming, including organic production, biodiversity-friendly cultivation, and agroforestry systems that integrate shade trees. The complex interactions between shade management, *H. vastatrix*, and natural enemy populations illustrate the multidimensional nature of disease management in perennial tropical cropping systems where pest management decisions intersect with ecosystem services, climate adaptation, and market demands.

Gray mold caused by the necrotrophic fungus *B. cinerea* poses a significant threat to berry crops worldwide, especially strawberries and raspberries, thriving in cool, humid conditions. It infects various plant parts, such as leaves, flowers, and fruits (Radi & Lahlali, 2025). The fungus has a complex disease cycle: it overwinters as sclerotia or mycelia, producing abundant conidia that disperse via wind and water. Infection occurs through direct penetration or wounds. A major challenge in managing gray mold is the pathogen's ability to develop resistance to commonly used chemical fungicides, which requires rotating fungicides with different modes of action and employing cultural practices that enhance air circulation and sanitation. The growing emphasis on biological control agents and biotechnological strategies in berry production reflects regulatory pressures to reduce chemical inputs and consumer demand for sustainably produced fruit. Future research focuses on developing resistant cultivars and improving the efficacy and application methods of biocontrol agents, though the rapid reproductive cycle and genetic diversity of *B. cinerea* continue to challenge long-term disease control.

Basal stem rot of oil palm caused by *G. boninense* demonstrates the devastating impact that soil-borne biotrophic pathogens can have on tropical plantation crops that supply essential oils for food, cosmetics, and biofuel production (Safni & Lisnawita, 2025). Oil palm (*Elaeis guineensis*) is the world's most productive crop for oil, widely cultivated across tropical Asia, Africa, and Latin America, with Malaysia and Indonesia as leading producers. However, basal stem rot disease has become one of the most destructive issues affecting both large plantations and smallholders, with infected trees usually stopping fruit production and eventually collapsing. At least 15 species of biotrophic basidiomycete fungi have been identified worldwide in relation to basal stem rot, with *G. boninense* being the primary causal agent in Southeast Asian production areas. The lack of a single effective control method necessitates integrated disease management strategies focused on reducing incidence after replanting, including sanitation of old plantation materials, soil amendments, biological control methods, and careful selection of planting sites.

Cross-cutting Themes

Several unifying themes emerge across the diverse pathosystems examined in this volume, highlighting commonalities in disease dynamics and management challenges that transcend specific crop–pathogen combinations. The case studies presented in this volume collectively illustrate that effective plant disease management requires integration of diverse knowledge domains spanning molecular biology, genetics, ecology, epidemiology, agronomy, and socio-economics.

No single management tactic provides complete disease control across the diverse pathosystems examined, reinforcing the fundamental importance of integrated disease management approaches that combine multiple complementary tactics tailored to specific crop–pathogen-environment contexts. The challenges are substantial, including pathogen evolution, climate change, globalization facilitating pathogen spread, demands for reduced chemical inputs, and the need to maintain agricultural productivity while enhancing sustainability. However, advances in diagnostic technologies, genomic tools, biological control approaches, precision agriculture technologies, and fundamental understanding of plant–pathogen interactions provide grounds for optimism that innovative solutions can be developed and deployed.

The path forward requires sustained collaboration among researchers, plant breeders, extension specialists, policy makers, and farmers to translate scientific advances into practical management strategies that protect crop production while meeting evolving societal expectations for sustainable agriculture. The comprehensive examination of diverse plant disease systems in this volume provides essential foundational knowledge to inform these efforts and advance the development of effective, sustainable approaches to plant disease management in a changing world.

References

Berraies, S., Walkowiak, S., Buchwaldt, L. and Menzies, J.G. (2023). Ergot (*Claviceps* spp.) of cereals in Western Canada. *CABI Plant Health Cases 2023*. DOI: 10.1079/planthealthcases.2023.0004.

Bourin, M. and Rhouma, A. (2025). Crown Gall Disease: A growing threat to fruit tree health and productivity (a case study from North Africa). *CABI Plant Health Cases 2025*. DOI: 10.1079/planthealthcases.2025.0020.

Cesarini, M., Giovannini, O., Sarroco, S. and Puopolo, G. (2025). Management of *Plasmopara viticola*: From the tradition to the innovation. *CABI Plant Health Cases 2025*. DOI: 10.1079/planthealthcases.2024.0003.

Davide, D. (2025). Kiwifruit bacterial canker: A recent disease for modern plant pathologists. *CABI Plant Health Cases 2025*. DOI: 10.1079/planthealthcases.2024.0022.

Fening, K.O. and Carr, J. P. (2025). The banana aphid and banana bunchy top disease: Potential threats to banana and plantain production in Ghana. *CABI Plant Health Cases 2025*. DOI: 10.1079/planthealthcases.2025.0024.

Gaafar, Y.Z.A. and Ziebell, H. (2025). Pea necrotic yellow dwarf virus: An emerging threat for legume production in Europe. *CABI Plant Health Cases 2025*. DOI: 10.1079/planthealthcases.2026.0001.

Gama, A.B. and Flasco, M.T. (2024). Sugarcane yellow leaf virus - A silent threat? *CABI Plant Health Cases 2024*. DOI: 10.1079/planthealthcases.2024.0020.

Gomez, K.C., Rojas, D.N., James, S.A., Ortega, S.F., Harper, A. l. *et al.* (2024). The epidemiology and management of *Ralstonia solanacearum* and *Ralstonia pseudosolanacearum* in Central and Northern Europe. *CABI Plant Health Cases, 2024*. DOI: 10.1079/planthealthcases.2024.002.

Grieve, B., Duffy, S., Dallas, M.M., Ascencio-Ibáñez, J.T., Alonso-Chavez, V. *et al.* (2024). Early detection of plant virus infection using multispectral imaging and machine learnine. *CABI Plant Health Cases 2024*. DOI: 10.1079/planthealthcases.2024.0010.

Kasoma, C. (2023). Cassava brown streak disease in Zambia: An integrated approach to managing an invasive viral disease of a clonal crop. *CABI Plant Health Cases 2023*. DOI: 10.1079/planthealthcases.2023.0019.

Ketehouli, T., Goss, E.M., Batistel, F. and Martins, S.J. (2025). The benefits and risks of antibiotics use in plant agriculture. *Plant Health Cases, 2025*, Article phcs20250016. DOI: 10.1079/planthealthcases.2025.0016.

Koutouleas, A. (2023). Coffee leaf rust: Wreaking havoc in coffee production areas across the tropics. *CABI Plant Health Cases 2023*. DOI: 10.1079/planthealthcases.2023.0005.

Laribi, M., Aboukhaddour, R. and Strelkov, S.E. (2024). Tan spot (*Pyrenophora tritici-repentis*) of wheat: A minireview. *CABI Plant Health Cases 2024.* DOI: 10.1079/planthealthcases.2024.0017.

Livjberg, M. (2025). A global threat: Epidemiology and management of yellow rush on wheat. *CABI Plant Health Cases 2025.* DOI: 10.1079/planthealthcases.2025.0011.

McCallum, B., Hiebert, C., Wang, X. and Bakkeren, G. (2024). Wheat leaf rust caused by *Puccinia triticana* and mitigation through host genetic resistance. *CABI Plant Health Cases 2024.* DOI: 10.1079/planthealthcases.2024.0023.

Pfordt, A. (2025). Trichoderma ear rot on maize - when friends turn evil. *CABI Plant Health Cases 2025.* DOI: 10.1079/planthealthcases.2025.0030.

Radi, M. and Lahlali, R. (2025). Management strategies of grey mould disease in berries. *CABI Plant Health Cases 2025.* DOI: 10.1079/planthealthcases.2025.0010.

Vadillo-Dieguez, A., Jackson, R.W. and Hulin, M.T. (2025). Global insights into bacterial canker of stone fruits caused by the *Pseudomonas syringae* species complex. *CABI Plant Health Cases 2025.* DOI: 10.1079/planthealthcases.2025.0015.

West, J.S. and King, K.M. (2024). Phoma stem canker (blackleg) of oilseed rape (Canola) and other brassica crops. *CABI Plant Health Cases 2024.* DOI: 10.1079/planthealthcases.2024.0007.

Part I: Viral Diseases in Plant Health

Viral plant pathogens pose a significant threat to global agriculture and food security, characterised by obligate intracellular parasitism, genetic diversity, and reliance on biological vectors for transmission. This section examines viral plant diseases through six case studies covering different crops and regions. A common theme is the crucial role of arthropod vectors, particularly aphids, in virus transmission and disease spread. From Banana bunchy top virus transmitted by *Pentalonia nigronervosa* (Banana aphid) to the multipartite *Pea necrotic yellow dwarf virus* spread by multiple aphid species, these chapters demonstrate how vector biology fundamentally shapes disease patterns. The emergence of *Pea necrotic yellow dwarf virus* as a threat to European legume production and the spread of Cassava brown streak disease across sub-Saharan Africa highlight how viral pathogens rapidly exploit new areas when conducive vector conditions meet vulnerable crop systems. Understanding virus-vector-host interactions is as essential as understanding the pathogens themselves, as illustrated by *Sugarcane yellow leaf virus* and *Soybean dwarf virus*, where vector monitoring is critical for managing diseases.

Many viral diseases cause significant yield losses with minimal or no symptoms, making traditional detection methods ineffective. This is particularly true for *Sugarcane yellow leaf virus* and Cassava brown streak disease, where root necrosis often remains unnoticed until harvest. Innovations such as molecular techniques and multispectral imaging combined with machine learning now enable early detection of infections. Developing sensitive and specific diagnostics is essential for disease monitoring, quarantine, and ensuring virus-free planting material, especially in crops like cassava and banana that propagate infections vegetatively.

The food security implications are clearly evident in the cassava and banana case studies, where millions of smallholder farmers rely on these crops for livelihood and income. Managing viral diseases in vegetatively propagated crops presents unique challenges, as infected planting material enables both long-distance pathogen movement and the persistence of infection within farming systems. The lack of certified propagule systems in many regions, combined with traditional cutting exchange practices, makes eradication difficult and allows rapid disease spread.

The chapters in this section provide comprehensive insights into virus biology, transmission dynamics, diagnostic methods, and management strategies, equipping readers with knowledge to tackle current and emerging viral threats through evidence-based, integrated approaches that balance agricultural productivity with food security priorities.

1 The Banana Aphid and Banana Bunchy Top Disease: Potential Threats to Banana and Plantain Production in Ghana

Ken Okwae Fening[1,2] and John P. Carr[3]

[1]*Soil and Irrigation Research Centre, School of Agriculture, College of Basic and Applied Sciences, University of Ghana, P.O. Box LG. 68, Accra, Ghana*
[2]*African Regional Postgraduate Programme in Insect Science (ARPPIS), University of Ghana, Accra, Ghana*
[3]*Department of Plant Sciences, University of Cambridge, Downing Street, Cambridge, CB2 3EA, United Kingdom*

Introduction: Banana Bunchy Top Disease

Banana bunchy top disease (BBTD) is the most destructive viral disease affecting banana (*Musa* spp.) and plantain (*Musa* × *paradisiaca* L.) cultivation (Bouwmeester *et al.*, 2023). These monocotyledonous crop plants are grown across about 150 countries, where they are extremely important for food security and national income. The dessert (Cavendish) banana is the world's most important fruit in terms of export value (*c.* US$11 billion in 2017). This export trade is dominated by Asian and South American countries. However, in Africa, *c.* 70 million people generate revenue from banana and plantain, which are widely grown for local consumption. Three countries, Côte d'Ivoire, Ghana and Cameroon, are the most significant African exporters. In Ghana, the income from banana has been increasing for farmers and value chain participants, with annual exports to the UK and EU valued at approximately US$53 million (2015 figures: Ghana Export Promotion Authority, 2017). Banana grows well in the temperature range of 15–35°C at a relative humidity of 75–85%, which is characteristic of southern Ghana, where cultivation of the crop is expanding.

In this case study, we discuss the potential threat of BBTD to banana and plantain production in Ghana and what measures can be applied to detect the disease and mitigate its effects. The issue is important since BBTD can cause losses of up to 100%, and if not controlled it can cause significant financial losses to smallholder and commercial farmers and threaten food security.

Furthermore, BBTD is highly invasive, having spread from Fiji, where it was first observed in the late 19th century, to South-east Asia, South Pacific, India, and to Africa, where it was first reported in Egypt in 1900. BBTD has spread to at least 19 African countries, and although at the time of writing this case study the disease has not reached Ghana, there is a high risk of introduction (Bouwmeester *et al.*, 2023; Gaitu *et al.*, 2024).

The causal agent of banana bunchy top bisease

BBTD is caused by banana bunchy top virus (BBTV). BBTV was at one time thought to be a luteovirus, i.e. a phloem-limited RNA virus transmitted by aphids in the persistent mode (see next section on transmission). However, following the successful purification of BBTV particles (virions) in the 1990s, it was found that they contained single-stranded circular DNA. In the latest release from the International Committee on Viral Taxonomy, BBTV is classified as the viral species *Babuvirus musae* in the Genus *Babuvirus* in the Family *Nanoviridae*.

The DNA genome of BBTV is hexapartite, with each of the six circular ssDNA genomic components (DNA-R, DNA-U3, DNA-S, DNA-M, DNA-C, and DNA-N), ranging in length from 1 to 1.1 kb. Each genomic DNA molecule is encapsidated separately in a non-enveloped T = 1 icosahedral particle, assembled from 60 coat protein subunits, that is 18–19 nm in diameter. DNA-C encodes a protein that initiates host cell division by interaction with the banana retinoblastoma-Rb protein orthologue, the DNA-M component encodes the BBTV cell-to-cell movement protein, DNA-N encodes a protein that shuttles viral DNA between the host cell nuclear and cytoplasmic compartments and is implicated in aphid transmission, while the viral coat protein is encoded by DNA-S. DNA-U3 encodes a factor with no currently known function. Nanoviruses and other small single-stranded DNA viruses encode Rep (replication initiator) proteins that facilitate rolling circle DNA replication, which occurs in the nucleus utilising the host's DNA synthesis machinery. In the case of BBTV, the Rep protein is encoded by DNA-R (Guyot *et al.*, 2024). Within the nucleus, the viral mRNAs are transcribed from the BBTV genomic DNAs by host RNA polymerase II.

Some BBTV isolates are associated with alphasatellite DNAs. Alphasatellites are hyperparasitic single stranded circular DNA molecules that depend upon a 'helper' virus (in this case BBTV) for packaging into particles made of the helper virus coat protein, for movement through the plant, and for vector-mediated transmission. Alphasatellites encode their own Rep proteins and initiate replication by host DNA polymerase activity independently of the helper virus (Guyot *et al.*, 2024).

BBTV replication occurs in phloem tissues, where it induces hyperplasia. Cells surrounding the phloem have increased numbers of chloroplasts, giving rise to symptoms of dark green streaks. BBTV can travel down the

Fig. 1.1. Progression of banana bunchy top disease. (a) 'Morse code' streaking (a series of dark green dots and dashes along the veins of a banana leaf, which form 'hooks' at the edge of the midrib). (b) Green J-shaped hooks along the midrib of the leaves. (c) Bunchy top symptom appearance and stunting. (d and e). Bunchy top appearance with yellowed leaf margins. (f) Healthy young banana plants. (g) Diseased inflorescence. (h) Healthy inflorescence. (i) Stunted fruits produced by diseased plants. (j) Healthy young fruit produced by a healthy plant. Photos (f) and (j) by K.O. Fening, University of Ghana. Other photos (a)–(e), and (g)–(i) are courtesy of Hawaii University and the Hawaii Department of Agriculture.

phloem to the vicinity of the meristem, allowing viral spread into developing tissues. Systemic infection with BBTV induces the major symptoms of BBTD, i.e., plant dwarfing, leaf narrowing, leaf margin chlorosis, and induction of dark green discontinuous streaks on the leaves, petioles and pseudotrunks. Additionally, infected plants are unlikely to yield fruit (Halbert and Baker, 2015) (Fig. 1.1). Interestingly, coinfection of plants by BBTV and an alphasatellite may ameliorate symptoms, decrease the titres and dysregulate the ratios of the BBTV DNA components, and inhibit aphid-mediated BBTV transmission. This was recently observed with an alphasatellite associated with BBTV in the Democratic Republic of Congo (Guyot *et al.*, 2022).

Banana Bunchy Top Virus is Vectored by Aphids but Can also Be Disseminated Vegetatively

BBTV is vectored by the banana aphid *Pentalonia nigronervosa* Coquerel (Hemiptera: Aphididae) (Halbert and Baker, 2015). Aphids feed predominantly on pseudostems, unfurled immature leaves and leaf sheaths. Vegetative transmission of BBTV also occurs via infected suckers and planting material, which can act as an inoculum source for other plants if the vector is present (Bouwmeester *et al.*, 2023). *P. nigronervosa* vectors BBTV in a persistent, circulative and non-propagative manner. That is, virus particles enter the body of the aphid and circulate but without virus replication occurring. When BBTV particles reach the salivary glands, the aphid becomes competent to transmit the virus, and the insect remains viruliferous and infectious for life. Both acquisition of viral inoculum and infection require sustained feeding (15 min to 4 h) by aphids from phloem vessels (Halbert and Baker, 2015).

Interestingly, BBTV infection modifies host biochemistry, resulting in the emission of a blend of volatile organic compounds that attract *P. nigronervosa*. However, it was found that once the aphids have acquired BBTV they are less attracted by infected banana plants and will be more likely to migrate to uninfected hosts. It is thought that this apparent manipulation of aphid host selection behaviour will accelerate transmission of BBTV, profound effects on epidemic development and important implications for formulating disease control methods (Murhububa *et al.*, 2024). BBTV also induces changes in the chemical composition and/or nutritional quality of plant tissue so that aphids reared on infected plants grew less well and show morphological differences to those propagated on healthy plants, indicating that infected banana plants are poor hosts for *P. nigronervosa* (Murhububa *et al.*, 2024).

P. nigronervosa was first described in 1859 on the Island of Reunion but is widespread across the world's tropical and subtropical regions (Robson *et al.*, 2007). Although this aphid is a plantain and banana specialist, *P. nigronervosa* can accept a few other host plants, notably *Heliconia*, which is also a host for BBTV. The wide distribution of *P. nigronervosa* means that once BBTV has been introduced into a new territory, the pre-existing vector population will ensure the rapid spread of the virus.

Direct Damage Caused by the Banana Aphid

P. nigronervosa does the greatest harm by vectoring BBTV. Direct damage by the aphid is generally negligible, but heavy infestation weakens the plant, which impedes plant growth and development (CABI, 2021). Additionally, heavy infestations of the fruits will reduce their market value due to the growth of sooty mould fungi fed by honeydew secreted by the aphids (CABI, 2021).

Banana Bunchy Top Disease Symptoms

The initial symptoms of BBTD on banana consist of dark green streaks and J-shaped hooks in the veins of lower portions of the leaf midrib and the leaf stem (Fig. 1.1a and b). As the disease progresses, emerging leaves are progressively stunted and malformed with a more upright bearing than usual and appear to be 'bunched' at the top of the plant; hence, the disease name 'bunchy top' (Hapsari and Masrum, 2012) (Fig. 1.1c–e). Other symptoms include crinkling and mottling of the petioles and overall stunting of infected plants (Fig. 1.1c–e). Banana plants that show BBTD symptoms rarely bear fruit, but when it is produced it is edible, although deformed and decreased in size (Fig. 1.1i and g). Figure 1.1f, h and j represents a healthy plant, inflorescence and fruit.

Disease and Vector Research and Management

Detection of BBTV and identification of *P. nigronervosa*

In addition to the recognition of BBTD or BBTD-like symptoms it is important to be able to definitively identify the causative agent. Serological detection of BBTV with antibodies raised against the coat protein has been possible for many decades and enzyme-linked immunosorbent assay (ELISA) kits are commercially available for BBTV and other viruses affecting banana. Sequence-specific detection of BBTV DNA based on polymerase chain reaction (PCR) and isothermal DNA amplification methods are also well established. DNA amplification methods are generally more sensitive than serological approaches and have the additional benefit that the amplicons can be sequenced, which allows for specific BBTV strains and lineages to be identified. This can be useful in tracking the potential source of an outbreak and the evolution of BBTV as epidemics develop (Wickramaarachchi *et al.*, 2016; Barros *et al.*, 2024). PCR-based DNA detection methods are sufficiently sensitive to identify aphids carrying BBTV (Chen and Hu, 2013) and newer high-throughput sequencing methods allow for simultaneous indexing of plant and insect samples for the presence of BBTV and other banana-infecting pathogens (Rong *et al.*, 2023). Generally, the DNA amplification methods described above could be used to detect BBTV from infected plants and from the viruliferous aphids. However, based on the study by Chen and Hu (2013), high-throughput sequencing using TaqMan®PCR detect rapidly and accurately BBTV in banana plants, whole aphids, aphid honeydew and aphid haemolymph, and requires only a minute initial amount of sample for analysis without virus purification.

Identification of the aphid *P. nigronervosa* is aided by its distinctive morphological features, most notably the alates' black wing venation, which gives the insect its taxonomic name (Fig. 1.2) (CABI, 2021). However, unambiguous identification may be challenging for insects recovered from water or sticky traps, in which case PCR-based 'DNA barcoding' using primers specific for insect genes such as *cytochrome oxidase subunit I* (*CO I*), followed by amplicon sequencing becomes essential (Fening *et al.*, 2020). Indeed, *CO I*-based DNA barcoding is vital in unambiguously distinguishing between *P. nigronervosa* Coquerel (which most typically feeds on banana and plantain and only occasionally on other hosts) and the related *P. caladii* van der Goot that is found only very occasionally on *Musa* spp. (Foottit *et al.*, 2010).

Training personnel in the routine use of these serological and molecular BBTV detection methods and in the trapping and identification of the *P. nigronervosa* vector will be essential to support surveillance, quarantine,

Fig. 1.2. (a) A colony of banana aphids (*Pentalonia nigronervosa*) on a banana petiole. Individual winged (alate) (b) and non-winged (apterous) morphs (c). Adults range between 1.2 and 1.8 mm in length. Image credits Scot Nelson (University of Hawaii), public domain (Available at: https://www.influentialpoints.com/ Gallery/Pentalonia_nigronervosa_banana_aphid.htm, accessed 10 August 2025).

phytosanitary and eradication efforts for BBTD, which are discussed in the following sections.

In Ghana, banana and plantain plants in both households and farms are randomly examined for BBTV symptoms and the presence of the aphid vector. Monitoring of the vector is done by inspecting banana and plantain plants, especially the undersides of leaves, leaf bases, and pseudostems, for the presence of small, reddish-brown to almost black, oval-shaped aphids or using yellow sticky card traps or water bowl traps. Alternative host plants found are also inspected for the presence of the vector. This monitoring is done by Ghana's National Plant Protection Organisation (NPPO), with the support of Academia, and the National Research Institutions and Development Partners (mainly by the Centre for Agriculture and Biosciences International, CABI) as and when resources are available. Leaf samples of suspected (symptomatic) and healthy looking (asymptomatic) plants and aphids are collected into separate vials containing 95% ethanol for identification.

Preventive and active measures

BBTV is persistently transmitted from plant to plant by *P. nigronervosa* and by transporting infected planting materials (Hapsari and Masrum, 2012). It can, however, be limited by culling diseased plants, chemical control of the vector and by using virus-free planting material obtained through *in vitro* meristem tip culture (Ferreira *et al.*, 1997; INIBAP, 2001).

Quarantine measures

If practicable, constant monitoring of national or regional borders to prevent BBTD from being introduced through planting material is an important preventive strategy (Gaitu *et al.*, 2024); where national quarantine measures are less practicable, it is important to discourage farmers from importing planting materials from neighbouring countries where BBTV is present. However, to obtain the cooperation of farmers, it is important to provide training on how to multiply their own healthy and clean planting materials.

Alternative hosts

The broad range of alternative hosts of BBTV seems to be congruent with the host ranges of its oligophagous vectors, *P. nigronervosa* and the taro aphid *P. caladii*, including mainly members of the Musaceae, Cannaceae, Araceae, Zingiberaceae, and Heliconiaceae (Foottit *et al.*, 2010; Bagariang *et al.*, 2019) (Table 1.1). Usually, *P. nigronervosa* favours *Musa* hosts, while *P. caladii* prefers non-*Musa* hosts though there is some overlap in host range (Foottit *et al.*, 2010; Bagariang *et al.*, 2019). Alternative hosts for BBTV may act as reservoirs for infection and could potentially frustrate control programmes, although several listed in Table 1.1 do not occur outside of specific geographic locales, and would not occur as infection reservoirs in Ghana.

Vector management

Since *P. nigronervosa* is BBTV's only effective vector, prompt detection and destruction of this pest should prevent the spread of BBTD. Although BBTV has not yet been reported in Ghana, continual monitoring for its presence is essential. If the virus is detected, an immediate and realistic measure that can be put in place to prevent further spread is prohibition of movement of planting materials from infected areas to other parts of the country. Additionally, the identification, collection and destruction of infected plants (culling, rogueing) has a strong possibility of eradicating it, as was demonstrated in Togo (see below). The success of such a programme pivots on effective legislation and its efficient enforcement at local and national levels.

Table 1.1. Known hosts of BBTV[a].

Genus and species	Family	Common name	Other information
Musa balbisiana	*Musaceae*	plantain	Widely distributed including Africa
Musa acuminata	Musaceae	banana	Ancestral to Cavendish banana, native to Southern Asia, but present in parts of Africa
Musa jackeyi	Musaceae	erect or wild banana	Localised to Queensland, Australia
Musa laterita	Musaceae	bronze banana	Localised to south Asia
Musa textilis	Musaceae	Manila hemp	Endemic to the Philippines
Musa coccinea	Musaceae	Scarlet or red-flowering banana	Localised to China and Vietnam
Musa velutina	Musaceae	Pink or hairy banana	Ornamental, native to Himalayas
Canna indica	Cannaceae[b]	Various including Bandera Espanola	Widely distributed including Africa
Alpinia zerumbet	Zingiberaceae	shell ginger	Native to east Asia
Alpinia galanga	Zingiberaceae	greater galangal	Predominantly south-east Asian distribution
Curcuma longa	Zingiberaceae	turmeric	Predominantly Asian distribution
Heliconia rostrata	Heliconiaceae	lobster claw	South and central America
Colocasia esculenta	Araceae	taro	Widely distributed including Africa

[a]Based on Mendoza *et al.* (2024) and Thomas (2024).
[b]Order: Zingiberales.

Host plant resistance to the virus

Until recently, it was thought that all banana cultivars were susceptible to BBTV. However, work in Australia suggested that the wild species *Musa coccinea* and the banana cultivar Kluai Teparot (ABBB/ABB) possess some degree of resistance. Studies undertaken to evaluate *Musa* genotypes for BBTV resistance found genotypes with tolerance (no symptoms and near normal performance of virus infected plants, e.g., Gros Michel), delayed expression of symptoms (e.g. cv. Dwarf Apple, also known as Santa Catarina) or decreased susceptibility (e.g. cv. Fugamou) (Espino *et al.*, 1993; Niyongere *et al.*, 2011). While wild *Musa* species are known to be resistant to BBTV, and some banana varieties are more readily infected with the virus, no variety of banana is considered truly resistant (Ferreira *et al.*, 1997). A study in East Java in Indonesia indicated that susceptibility is related to genotype. Thus, cultivars in AAB, ABB and BBB genome groups tend to be more tolerant to BBTV than cultivars of AA and AAA, i.e. possession of one or two B genomes enhances tolerance (Hapsari and Masrum, 2012). These genotypes should be evaluated by farmers and consumers in case it becomes necessary to transition to acceptance of these genotypes, to replace current BBTV-susceptible varieties (Gaitu *et al.*, 2024).

Biological control of the vector

In Tonga, the polyphagous aphid parasitoid *Aphidius colemani* Viereck (Hymenoptera: Braconidae) was released in an attempt at biological control of *P. nigronervosa*, but its effects on BBTD incidence were disappointing (Wellings *et al.*, 1994). Another attempt at biological control of *P. nigronervosa* was undertaken in Western Samoa using two coccinellid species, but neither agent could establish breeding populations (Waterhouse and Norris, 1987; Stechmann and Völkl, 1988). In Ghana, biological control options for *P. nigronervosa* have received little attention. Thus, current efforts to document the range of natural enemies (predators and parasitoids) attacking *P. nigronervosa* on banana, plantain and alternative hosts (e.g. *Heliconia*) are important to support development of biological control options for integrated pest management of BBTD (K. O. Fening, unpublished data).

Cultural and chemical control of BBTD

Banana plantations can last for several years in Ghana, but the lifespan of individual plants within a plantation is shorter, ranging from 12 to 16 months, from planting to harvest, depending on the variety and local growing conditions. After a banana plant flowers and fruits, the 'mother' plant dies, and a new 'daughter' plant from the same corm takes over. A plantation can be maintained for many years by cycling through these mother-daughter plants, potentially for 7–10 years or more, depending on the management practices and the lifespans of individual plants. This might result in older plants becoming disease reservoirs.

If this occurs, BBTD can be effectively controlled by culling diseased plants and replacing them with virus-free planting material. Before culling, diseased plants should be sprayed with an appropriate insecticide to kill all potentially viruliferous aphids. To prevent vegetative transmission, the whole stool, including the corm and all associated suckers, must be destroyed by uprooting and chopping into small pieces or by herbicide treatment, as BBTV will ultimately spread to all parts of the mat. Control must be practised across whole production areas to avoid re-infection of planting material (Kumar, 2017).

Insecticides have been used in some countries to control *P. nigronervosa*, and consequent decreases in BBTD incidence have been reported. A 2017–2018 field study in Pakistan evaluated the efficacy of various insecticides and concluded that the highest reduction percentage of *P. nigronervosa* was observed with Spinetoram followed by Methoxyfenozide, a repeat Spinosad treatment, Chlorantraniliprole and Flubendamide on the 1st, 3rd, 5th and

7th days, respectively (Baloch *et al.*, 2023). Injecting plants with herbicides combined with systemic insecticide spraying can kill infected plants and the viruliferous aphids infesting them. However, this is labour intensive, costly for many farmers, and of questionable effectiveness in smallholder farming systems, where it may be difficult to adopt common practices between farms. In addition to prohibitive costs, the development of resistance in the aphids, and concerns with the effects of chemical residues on farmer health and on the environment make it unlikely that vector control using pesticides will gain wide use in Ghana.

Amsavalli *et al.* (2022) applied the entomopathogenic fungi *Lecanicillium lecanii* (at 10^9 colony-forming units/ ml), *Metarhizium anisopliae* (10^8 cfu/ml) and *Beauveria bassiana* (10^8 cfu/ml) to banana plants infested with *P. nigronervosa*, resulting in 63–76% aphid mortality. However, in the same study, two botanical insecticidal preparations, Neem oil (1% effective concentration) and Pongamia oil (2% EC), killed 80–83% and 80–81% aphids, respectively. These higher mortality rates are close in efficacy to the synthetic neonicotinoid insecticides Acetamiprid (82–84%) and Imidacloprid (89–91%) (Amsavalli *et al.*, 2022). Research in Ghana showed that homemade aqueous Neem kernel extract (applied at 50 g/L of water) or a commercial Neem-based formulation (0.3–1% azadirachtin applied at 30–60 ml/15L of water, respectively) were very effective in controlling the aphids *Lipaphis erysimi pseudobrassicae* Davis (Hemiptera: Aphididae) and *Myzus persicae* Sulzer (Hemiptera: Aphididae) on cabbage (Forchibe *et al.*, 2017; Adenka *et al.*, 2021). Taken together, these finding indicate that treatments using homemade Neem extracts or Neem-based products can readily be adapted to control *P. nigronervosa* on banana and plantain in Ghana. The efficacy of these treatments in decreasing *P. nigronervosa* populations can be monitored by sampling alates using yellow water bowl and sticky traps placed near the crop and monitoring the plants for apterous aphids. Trapped aphids can be checked for the presence of BBTV using PCR.

The potential for biotechnological control methods for BBTD

RNA silencing [also known as RNA interference (RNAi) and post-transcriptional gene silencing (PTGS)] is a highly effective means of protecting plants against both RNA and DNA viruses (Jekayinoluwa *et al.*, 2020). RNA silencing is a potent natural antiviral system that is elicited during virus infection by the presence of RNA molecules containing substantial stretches of double-stranded structure, which leads to the generation of small (21–26 nucleotide long) RNAs. These small RNAs direct Argonaute endonucleases to 'slice' homologous sequences in viral RNA. RNA silencing can target the genomic RNAs of RNA viruses and mRNAs produced by RNA and DNA viruses. RNA silencing can be stimulated artificially using double-stranded RNA molecules either through expression in transgenic plants (Cillo and Palukaitis, 2014), or by application of synthetic RNA molecules by spraying leaves of non-transgenic plants (Fletcher *et al.*, 2020). Since RNA silencing in transgenic plants works well against geminiviruses, which like BBTV have single stranded circular DNA genomes, it has been suggested that the approach would be a feasible strategy for control of BBTD (Jekayinoluwa *et al.*, 2020). In other plant systems, both transgenic expression and topical application of double-stranded RNAs with sequence homology to insect transcripts can provide protection against aphids (Fletcher *et al.*, 2020), suggesting a means of controlling BBTV transmission.

Gene editing is a method of introducing targeted mutations into DNA sequences. The most used approaches use adaptations of the prokaryotic antiviral defence system ('CRISPR') comprising a DNA endonuclease, such as Cas9, that can be directed to a specific site in DNA by a guide RNA with homology to the target sequence. CRISPR has been used to generate non-transgenic plants rendered resistant to specific viruses by mutation of plants genes conditioning susceptibility (Chandrasekaran *et al.*, 2016; Pyott *et al.*, 2016). CRISPR has also been shown to be effective when used directly against DNA viruses, including several geminivirus, especially if guide RNAs target the non-coding intergenic regions of the viral genomes (Ali *et al.*, 2016).

RNA silencing (especially using topical double-stranded RNA application) and gene editing have significant potential for control of BBTV. However, the deployment in Ghana of these technologies would depend on regulatory regime, the attitudes of growers, and consumer preferences both within Ghana and in export markets.

The Threat to Ghana and Need for Improved Surveillance for BBTV and Mapping of *Pentalonia nigronervosa* Prevalence

BBTV occurs in countries neighbouring Ghana such as Benin (localised) (Lokossou *et al.*, 2012) and Nigeria (localised) (CABI, 2021), and was present in Togo, where it has been eradicated (Kolombia *et al.*, 2021). Since *P. nigronervosa* is widely distributed in Ghana (CABI, 2021; PPRSD, 2024), there is a very high risk of BBTD introduction. Thus, continuous monitoring for BBTV and vector abundance in banana and plantain plantations is essential to support informed and rapid decision making in the event of an outbreak. Increased surveillance at the borders of Ghana for import of banana and plantain suckers from neighbouring countries (Togo, Burkina Faso and Ivory Coast) will be crucial for interception of infected planting material. Integrating BBTV survey data and reports from neighbouring countries, could also be used for parameterisation of epidemiological models to aid in assessment of the probability of entry of the virus into Ghana and provide an additional means of alerting stakeholders (Retkute and Gilligan, 2025). Early detection of BBTV introduction will allow for its immediate containment and possible eradication. In contrast, failure to detect BBTV introduction early will make containment and eradication difficult since *P. nigronervosa* is widespread. It is, therefore, recommended that Ghana's National Plant Protection Organisation (i.e., the Plant Protection and Regulatory Services

Directorate, PPRSD, of the Ministry of Food and Agriculture) should partner with other border agencies through awareness creation to prevent the entry and spread of this devastating banana and plantain disease. Academic and Research Institutions can offer technical backstopping in the surveillance, detection, accurate identification of both BBTV and its aphid vector and their management for informed decision making. For example, ongoing collaborative research between K.O. Fening (Agricultural Entomologist) from the African Regional Postgraduate Programme in Insect Science (ARPPIS), University of Ghana and others are investigating BBTV/D transmission and the bioecology of the vector *P. nigronervosa* in Ghana.

Discussion Points

1. Discuss how climate change may affect *P. nigronervosa* populations and how this might influence the spread of BBTV in Ghana and elsewhere.
2. What are the reasons biological control of *P. nigronervosa* have not been very successful and what is the best way forward for improving this approach, for example as part of an integrated pest management framework?
3. Discuss how the lessons from the successful eradication of BBTD in Togo might be applied elsewhere.
4. The growth in banana cultivation in southern Ghana for export may inadvertently create an environment conducive for the establishment of BBTD. What combination of preventive and other approaches will have the best chance of avoiding this?
5. Could future technologies based on RNAi or gene editing be useful in controlling either the virus or its vector?Barros *et al.* (2024)

Further Reading

Global Market Reports: Banana (2020) Publisher: The International Institute for Sustainable Development. Available at: https://www.iisd.org/publications/report/global-market-report-banana (accessed 16 April 2025).

International Committee on the Taxonomy of Viruses (n.d.) Family: Nanoviridae. Available at: https://ictv.global/report/chapter/nanoviridae/nanoviridae/babuvirus (accessed 16 April 2025).

Plants of the World Online (n.d.) Musa *L.* Royal Botanic Garden Kew. Available at: https://powo.science.kew.org/taxon/urn:lsid:ipni.org:names:327926-2 (accessed 16 April 2025).

References

Adenka, K.D., Fening, K.O., Afreh-Nuamah, K., Wamonje, F.O. and Carr, J.P. (2021) Susceptibility of five cabbage varieties to attack by aphids (Hemiptera: Aphididae) in the Accra plains of Ghana. *Phytoparasitica* 49, 33–47. DOI: 10.1007/s12600-020-00863-y.

Ali, Z., Ali, S., Tashkandi, M., Zaidi, S.S. and Mahfouz, M.M. (2016) CRISPR/cas9-mediated immunity to geminiviruses: Differential interference and evasion. *Scientific Reports* 6, 26912. DOI: 10.1038/srep26912.

Amsavalli, V., Shinde, B.D., Narangalkar, A.L. and Joshi, S.S.M. (2022) Management of black aphid, *Pentalonia nigronervosa* (Coquerel) (Hemiptera: Aphididae) infesting banana under controlled field conditions. *The Pharma Innovation Journal* 11(11), 1478–1482. Available at: https://www.thepharmajournal.com/archives/2022/vol11issue11/PartR/11-11-125-450.pdf (accessed 16 April 2025).

Bagariang, W., Hidayat, P. and Hidayat, S.H. (2019) Morphometric analysis and host range of the genus *Pentalonia coquerel* (Hemiptera: Aphididae) infesting banana in Java. *Jurnal Perlindungan Tanaman Indonesia* 23, 11–178. DOI: 10.22146/jpti.38220.

Baloch, S.A., Lanjar, A.G., Syed, T.S., Khanzada, M.A., Nahiyoon, S.A. *et al.* (2023) Efficacy of different insecticides and botanicals against banana aphids (*Pentalonia nigronervosa*) under field conditions. *Pure and Applied Biology* 12(2), 1302–1314. Available at: https://thepab.org/index. php/journal/article/view/2662 (accessed 16 April 2025).

Barros, S.C.I.M., Figueira, A.D.R. and Silveira, A.T.L. (2024) Molecular characterization of isolates of the banana bunchy top virus (BBTV) from the district of Chókwè, Mozambique. *Applied Sciences* 14(10), 4291. DOI: 10.3390/app14104291.

Bouwmeester, H., Blomme, G., Omondi, A.B. and Ocimati, W. (2023) Banana bunchy top disease in Africa — Predicting continent-wide disease risks by combining survey data and expert knowledge. *Plant Pathology* 72, 1476–1490. DOI: 10.1111/ppa.13764.

CABI (2021) PlantwisePlus Knowledge Bank – 39598. Available at: https://plantwiseplusknowledgebank.org/doi/10.1079/pwkb.species.39598 (accessed 16 April 2025).

Chandrasekaran, J., Brumin, M., Wolf, D., Leibman, D., Klap, C. *et al.* (2016) Development of broad virus resistance in non-transgenic cucumber using CRISPR/Cas9 technology. *Molecular Plant Pathology* 17(7), 1140–1153. DOI: 10.1111/mpp.12375.

Chen, Y. and Hu, X. (2013) High-throughput detection of banana bunchy top virus in banana plants and aphids using real-time TaqMan® PCR. *Journal of Virological Methods* 193(1), 177–183. DOI: 1016/j.jviromet.2013.06.013.

Cillo, F. and Palukaitis, P. (2014) Transgenic resistance. *Advances in Virus Research* 90, 35–146. DOI: 10.1016/B978-0-12-801246-8.00002-0.

Espino, R.C., Magnaye, L.V., Johns, A.P. and Juanillo, C. (1993) Evaluation of philippine banana cultivars for resistance to bunchy-top and fusarium wilt. In: Valmayor, R.V., Hwang, S.C., Ploetz, R., Lee, S.C. and Roa, N.V. (eds) *Proceedings: International Symposium on Recent Developments in Banana Cultivation Technology, 14-18 December 1992, Chiuju, Pingtung, Taiwan*, Taiwan Banana Research Institute, INIBAP/ASPNET, Los Banos, Philippines, pp. 89–102. Available at: https://www.musalit.org/seeMore.php?id=2226 (accessed 10 August 2025).

Fening, K.O., Wamonje, F.O., Forchibe, E.E., Adama, I., Afreh-Nuamah, K. *et al.* (2020) First report of the Indian mustard aphid, *Lipaphis erysimi pseudobrassicae* (Davis) (Hemiptera: Aphididae), on cabbage (*Brassica oleracea* var *capitata*) in Ghana. *Journal of Economic Entomology* 113, 1363–1372. DOI: 10.1093/jee/toaa057.

Ferreira, S.A., Trujillo, E.E. and Ogata, D.Y. (1997) *Banana Bunchy Top Virus. Plant Disease 12. College of Tropical Agriculture and Human Resources*. University of Hawaii at Manoa, Honolulu, Hawaii. Available at: https://www.ctahr.hawaii.edu/oc/freepubs/pdf/PD-12.pdf (accessed 16 April 2025).

Fletcher, S.J., Reeves, P.T., Hoang, B.T. and Mitter, N. (2020) A perspective on RNAi-based biopesticides. *Frontiers in Plant Science* 11, 51. DOI: 10.3389/fpls.2020.00051.

Foottit, R.G., Maw, H.E.L., Pike, K.S. and Miller, R.H. (2010) The identity of *Pentalonia nigronervosa* Coquerel and *P. Caladii* van der Goot (Hemiptera: Aphididae) based on molecular and morphometric analysis. *Zootaxa* 2358(1), 25–38. DOI: 10.11646/zootaxa.2358.1.2.

Forchibe, E.E., Fening, K.O. and Afreh-Nuamah, K. (2017) Effect of different pesticide management options on the population dynamics of aphids, *Lipaphis erysimi pseudobrassicae* (Davis) and *Myzus persicae* (Sulzer) (Hemiptera: Aphididae), their natural enemies and the yield of cabbage. *Science and Development – Journal of the College of Basic and Applied Sciences* 1(1), 29–44. Available at: https://www.researchgate.net/publication/323870915_Effects_of_different_pesticide_management_options_on_the_population_dynamics_of_aphids_Lipaphis_erysimi_pseudobrassicae_Davis_and_Myzus_persicae_Sulzer_Hemiptera_Aphididae_their_natural_enemies_and_th (accessed 16 April 2025).

Gaitu, C., Agboyi, L.K., Beseh, P., Boafo, H.A., Honger, J.O. *et al.* (2024) *Inspection Guideline for Detection of Banana Bunchy Top Disease and its Vector Pentalonia Nigronervosa. Plant Protection and Regulatory Services Directorate (PPRSD) of the Ministry of Food and Agriculture (MoFA).* Published by CABI West Africa, Accra, Ghana, p. 7.

Ghana Export Promotion Authority (2017) Banana from Ghana. Available at: https://www.gepaghana.org/cms/wp-content/uploads/2017/04/Banana-from-Ghana.pdf (accessed 16 April 2025).

Guyot, V., Rajeswaran, R., Chu, H.C., Karthikeyan, C., Laboureau, N. *et al.* (2022) A newly emerging alphasatellite affects banana bunchy top virus replication, transcription, siRNA production and transmission by aphids. *PLoS Pathogens* 18(4), e1010448. DOI: 10.1371/journal.ppat.1010448.

Guyot, V., Trieu, T.D., Insisiengmay, O., Zhang, T., Iskra-Caruana, M.L. *et al.* (2024) A new genus of alphasatellites associated with banana bunchy top virus in Southeast Asia. *Virus Evolution* 10(1), vead076. DOI: 10.1093/ve/vead076.

Halbert, S.E. and Baker, C.A. (2015) Banana bunchy top virus and its vector. Plant Pathology Circular 417. Florida Department of Agriculture and Consumer Services. Available at: https://www.fdacs.gov/About-Us/Publications/Plant-Industry-Publications/Plant-Pathology-Circulars (accessed 16 April 2025).

Hapsari, L. and Masrum, A. (2012) Preliminary screening resistance of *Musa* germplasms for banana bunchy top disease in Purwodadi botanic garden, Pasuruan, East Java. *Botanic Gardens Bulletin* 15(2), 57–70. Available at: https://www.neliti.com/publications/55107/preliminary-screening-resistance-of-musa-germplasms-for-banana-bunchy-top-diseas (accessed 16 April 2025).

International Network for the Improvement of Banana and Plantain (INIBAP) (2001) *Networking Banana and Plantain: INIBAP Annual Report 2000.* International Network for the Improvement of Banana and Plantain, Montpellier, France, pp. 11–12. Available at: https://cgspace.cgiar.org/items/d0f779d3-c611-45cc-8d06-c633fb7ece9a (accessed 16 April 2025).

Jekayinoluwa, T., Tripathi, L., Tripathi, J.N., Ntui, V.O., Obiero, G. *et al.* (2020) RNAi technology for management of banana bunchy top disease. *Food and Energy Security* 9(4), e247. DOI: 10.1002/fes3.247.

Kolombia, Y.A., Oviasuyi, T., Dzola, A.K., Ale Gonh-Goh, A., Atsu, T. *et al.* (2021) First report of banana bunchy top virus in banana (*Musa* spp.) and its eradication in Togo. *Plant Disease* 105(10), 3312. Available at: https://apsjournals.apsnet.org/doi/full/10.1094/PDIS-03-21-0473-PDN (accessed 10 August 2025).

Kumar, L. (2017) Banana Bunchy Top Virus (Bunchy Top of Banana). CABI Compendium 8161. Available at: https://www.cabidigitallibrary.org/doi/10.1079/cabicompendium.8161 (accessed 16 April 2025).

Lokossou, B., Gnanvossou, D., Ayodeji, O., Akplogan, F., Safiore, A. *et al.* (2012) Occurrence of Banana bunchy top virus in banana and plantain (*Musa* sp.) in Benin. *New Disease Reports* 25, 13. DOI: 10.5197/j.2044-0588.2012.025.013.

Mendoza, N.A.P., Mendoza, J.-V.S., Thomas, J.E. and Dela Cueva, F.M. (2024) Alternative hosts of banana bunchy top virus in the Philippines and the first evidence of seed transmission of BBTV. *Frontiers in Plant Science* 15, 1467331. DOI: 10.3389/fpls.2024.1467331.

Murhububa, I.S., Bragard, C., Tougeron, K. and Hance, T. (2024) Preference of *Pentalonia nigronervosa* for infected Banana plants tends to reverse after *Banana bunchy top virus* acquisition. *Scientific Reports* 14(1), 2993. DOI: 10.1038/s41598-024-53205-x.

Niyongere, C., Ateka, E., Losenge, T., Blomme, G. and Lepoint, P. (2011) Screening *Musa* genotypes for banana bunchy top disease resistance in Burundi. *Acta Horticulturae* 897, 439–447.

Plant Protection and Regulatory Services Directorate (PPRSD) (2024) *Pest List of Ghana*. Ministry of Food and Agriculture, Pokuase, Accra, Ghana. Available at: https://assets.ippc.int/static/media/files/reportingobligation/2020/12/04/GHANA-_QUARANTINE_PEST_LIST_IPPC_doc.pdf (accessed 15 April 2025).

Pyott, D.E., Sheehan, E. and Molnar, A. (2016) Engineering of CRISPR/Cas9-mediated potyvirus resistance in transgene-free *Arabidopsis* plants. *Molecular Plant Pathology* 17(8), 1276–1288. DOI: 10.1111/mpp.12417.

Retkute, R. and Gilligan, C.A. (2025) Developing a spatiotemporal model for Banana bunchy top disease: Leveraging remote sensing and survey data. *Frontiers in Plant Sciences* 16, 1521620. DOI: 10.3389/fpls.2025.1521620.

Robson, J.D., Wright, M.G. and Almeida, R.P. (2007) Biology of *Pentalonia nigronervosa* (Hemiptera, Aphididae) on banana using different rearing methods. *Environmental Entomology* 36(1), 46–52. DOI: 10.1603/0046-225x(2007)36[46,bopnha]2.0.co;2.

Rong, W., Rollin, J., Hanafi, M., Roux, N. and Massart, S. (2023) Validation of high-throughput sequencing as virus indexing test for *Musa* germplasm: Performance criteria evaluation and contamination monitoring using an alien control. *Phyto Frontiers* 3(1), 91–102. DOI: 10.1094/PHYTOFR-03-22-0030-FI.

Stechmann, D.H. and Völkl, W. (1988) Introduction of *Lysi hlebus testacetpes* (Cresson) (Hym:Aphidiidae) into the Kingdom of Tonga. In: Niemczy, E. and Dixon, A.F.G. (eds) *Ecology and Effectiveness of Aphidophaga*. SPB Academic Publishing, The Hague, pp. 271–273.

Thomas, J. (2024) Banana bunchy top virus. In: Drenth, A. and Kema, G. (eds) *Achieving Sustainable Cultivation of Bananas*, Vol. 3. Burleigh Dodds Science Publishing Limited, Cambridge, UK, pp. 315–342. Available at: https://www.taylorfrancis.com/books/edit/10.1201/9781003286752/achieving-sustainable-cultivation-bananas-andr%C3%A9-drenth-gert-kema (accessed 10 August 2025).

Waterhouse, D.F. and Norris, K.R. (1987) *Biological Control: Pacific Prospects*. Inkata Press, Melbourne, p. 454.

Wellings, P.W., Hart, P.J., Kami, V. and Morneau, D.C. (1994) The introduction and establishment of *Aphidius colemani* Viereck (Hym., Aphidiinae) in Tonga. *Journal of Applied Entomology* 118(15), 419–428.

Wickramaarachchi, W.A., Shankarappa, K.S., Rangaswamy, K.T., Maruthi, M.N., Rajapakse, R.G. *et al.* (2016) Molecular characterization of banana bunchy top virus isolate from Sri Lanka and its genetic relationship with other isolates. *Virus Disease* 27(2), 154–e160. DOI: 10.1007/s13337-016-0311-2.

2 Pea Necrotic Yellow Dwarf Virus: An Emerging Threat for Legume Production in Europe

Yahya Z.A. Gaafar[1] and Heiko Ziebell[2]

[1]*Centre for Plant Health – North Saanich, Canadian Food Inspection Agency (CFIA), North Saanich, V8L 1H3, British Columbia, Canada*
[2]*Julius Kühn-Institute (JKI) – Federal Research Centre for Cultivated Plants, Institute for Epidemiology and Pathogen Diagnostics, Messeweg 11-12, 38104 Braunschweig, Lower Saxony, Germany*

Introduction

Legumes play an important role in human nutrition, livestock feed, and soil enrichment through nitrogen fixation. Besides biotic and abiotic factors influencing legume health, viral diseases can cause substantial crop losses, thus threatening sustainable agriculture. Pea necrotic yellow dwarf virus (PNYDV) is an emerging plant virus that poses a significant threat to legume production in Europe, mainly affecting legumes such as peas and faba beans. PNYDV was first identified in Germany in the early 2000s and is now detected in several European countries (Grigoras *et al.*, 2010; Gaafar *et al.*, 2016, 2017, 2018; Fowkes *et al.*, 2023). It is transmitted only by insect vectors, i.e. aphids feeding on legumes such as the pea aphid (*Acyrthosiphon pisum* [Harris]). PNYDV is a nanovirus, with its genome based on eight circular, single-stranded DNA components. In addition, alpha satellite components associated with nanovirus infections have been described. The virus causes severe symptoms, including yellowing, necrosis, and dwarfing, thus reducing crop yield and causing significant economic losses. The severity of yield reduction depends on the time of infection whereas early infections cause more severe symptoms and overall losses. Moreover, the host plant and the variety have an impact on symptom severity. The virus is found often in mixed infection with pea enation mosaic virus 1 and 2 (PEMV-1 and PEMV-2) as they are also vectored by leguminous aphids. Initially considered a minor plant virus with low prevalence and no known epidemic significance, its prevalence has increased due to changing agricultural practices, climate change, and therefore vector dynamics. This case study explores the scientific, economic, and agricultural importance of PNYDV, highlighting its impact on crop yield and food security, understanding its distribution, transmission, and management strategies to mitigate losses and safeguarding legume production.

Taxonomy

PNYDV is a member of the *Nanovirus* genus, *Nanoviridae* family (Thomas *et al.*, 2021). The species name is now called *Nanovirus necropisi*. The genus name "*Nanovirus*" comes from the Greek word nanos which means "dwarf", as these viruses have very small virions and genomic segment sizes; in addition, they can induce dwarfing symptoms in their host plants. The species name "*necropisi*" consists of two parts "necro" from the necrosis caused by the virus and "pisi" from the genus of the pea plant i.e. *Pisum*.

Biology

PNYDV genomic organization and replication

Generally, the genome of PNYDV consists of at least eight genomic segments or components i.e. DNA-C, DNA-M, DNA-N, DNA-R, DNA-S, DNA-U1, DNA-U2 and DNA-U4. In addition, satellite-like DNAs (alphasatellites) can be found associated with the virus. These components are positive-sense single-stranded circular DNA (sscDNA) molecules and are encapsidated separately. The length of the genomic segments ranges from approximately 978 to 1002 nt; the associated alphasatellites can be as long as 1037 nt and DNA-X is 497 nt. They have a similar structural organization, including conserved inverted repeat sequences potentially forming a stem-loop structure that are the origin of replication, a TATA box, an open reading frame (ORF) and a poly-A signal (Fig. 2.1).

Each component encodes a distinct protein involved in various functions. DNA-C encodes the Clink protein which manipulates the plant cell cycle and DNA-M encodes the movement protein (MP). DNA-N produces a protein that is necessary for aphid-mediated transmission and is presumed to be a nuclear shuttle protein (NSP) based on analogy to geminiviruses. DNA-R produces the master replication Initiator Protein (M-Rep) for all the different components and DNA-S produces the capsid protein (CP). A recent study showed that DNA-U2 can act as a viral suppressor of RNA silencing (VSR) (Yan *et al.*, 2024). DNA-U1 and DNA-U4 encode different proteins of unknown functions, and the alphasatellites encode their own replication-associated proteins.

PNYDV replication is believed to occur in the nucleus by rolling-circle replication and is initiated by the M-Rep protein. The virions are uncoated, and dsDNA intermediates are synthesized using host DNA polymerase(s). Afterwards, a host RNA polymerase transcribes the encoded viral mRNAs. Although this has not been studied in PNYDV, research on the eight genomic

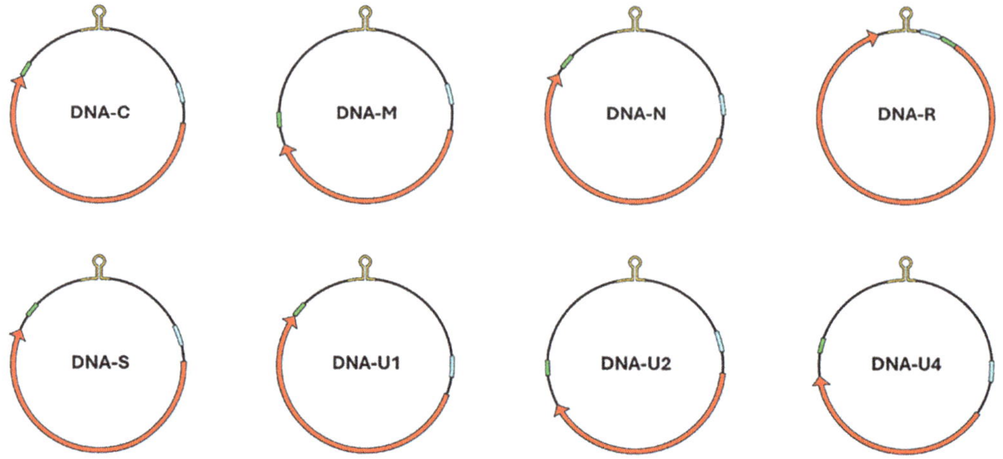

Fig. 2.1. Diagram illustrating the genomic organization of pea necrotic yellow dwarf virus (*Nanovirus necropisi*). Each circle represents a single stranded DNA of ~1 kb and contains (a) open reading frames (ORFs) and the mRNA transcription direction from the dsDNA intermediate (red arrows), (b) the common stem-loop region (yellow), (c) TATA box (light blue) and d) the poly-A signal (green).

components of faba bean necrotic stunt virus (FBNSV), another nanovirus, revealed that some segments are accumulated at low frequencies while others have higher frequencies such as -N and -U4. The relative frequency of the segments can be described as "genome formula" that has an impact on both viral accumulation and symptom expression and varies depending on the host plant (Sicard *et al.*, 2015). The role of the alphasetellites in the replication and severity of the symptoms are unknown. However, it was reported that they might affect the genome formula (Mansourpour *et al.*, 2022).

Research on the cellular localization of PNYDV proteins by fluorescent tagging showed that the M-Rep, CP and Clink proteins are located in the nucleus where PNYDV replication takes place (Krenz *et al.*, 2017). The MP is located at the cell periphery and around the nucleus. NSP shows a nucleo-cytoplasmic localization, whereas U1 appears as small circles or spots inside the nucleus. U2 is localized as spots at the periphery, in the nucleus and U4 at the cellular periphery and around the nucleus, which resembles the localization for endoplasmic reticulum (ER). The MP and U4 complexes appear as vesicle-like spots and are localized at the nuclear envelope and cell periphery and associated with the ER. Bimolecular fluorescence complementation revealed that the M-Rep self-interacts and also interacts with CP and NSP, whereas the MP self-interacts and interacts with U4. In addition, NSP and U4 both exhibit self-interactions. NSP was found to interact with the stress granule component G3BP, which may indicate a possible involvement in modulation of the plant stress response pathway.

PNYDV structure and morphology

PNYDV virions are non-enveloped, icosahedral with T=1 symmetry, composed of capsid proteins of approximately 19 kDa and lack associated lipids or carbohydrates. The virions measure approximately 18 nm in diameter.

Hosts and Symptomatology

So far, only plants from the Fabaceae family have been identified as host of PNYDV. These include peas, faba beans (*Vicia faba*), lentils (*Lens culinaris*), chickpeas (*Cicer arietinum*), and vetches (*V. hirsute*, *V. pannonica* and *V. sativa*). PNYDV infection is associated with significant yield reductions, with early infections potentially resulting in total crop failure. In recent years, substantial losses have been documented in pea, bean, and lentil crops. For example, surveys of five faba bean fields in Denmark, with disease incidences ranging from 2% to 25%, confirmed the presence of PNYDV in all fields (Gaafar *et al.*, 2018). In co-infection with PEMV, yield losses of 9% to 96.6% in faba beans were reported and up to 100% losses in greenhouse-based experiments (Saucke *et al.*, 2019).

Early infection of pea and faba beans with PNYDV results in symptoms of stunted plant growth and dwarfed plants, characterized by smaller leaves, shortened internodes, and reduced root and nodule development (Fig. 2.2) (Saucke *et al.*, 2019). Affected leaves often turn yellow, may curl upward, and in some cases, the shoot tips die. In lentil and chickpea crops,

Fig. 2.2. Symptoms PNYDV (isolate Holtsee) on faba bean infected plants under greenhouse conditions. The symptoms include leaves yellowing and curl upward, and short internodes.

red discoloration may also occur, accompanied by reduced flower and pod set and poor grain development. Late infections of beans may allow plants to reach normal growth height and pod set yet characterized by chlorotic symptoms. The severity of symptoms is variety-dependent (Kartheuse *et al.*, 2025).

Investigation of faba bean infected sites using low altitude true-colour aerial imaging data showed that as the season progressed, yellow patches of infected plants expanded and later developed into a funnel-like shape. The central area comprised of sterile, severely stunted, and senescent plants exhibiting a prematurely blackish coloration. Surrounding the central area, a chlorotic ring was observed, consisting of plants of normal height but showing partial leaf shedding, premature leaflet abortion, and leaf senescence. Another study investigating the impact of PNYDV on faba bean reported a more severe impact of PNYDV on nodulation, leghemoglobin status, N_2 fixation and finally yield with early infections than PEMV (Seeger *et al.*, 2022).

Geographical Distribution of PNYDV

PNYDV was first detected infecting pea plants in Germany in 2009, and in 2010, it was detected in Austria. To date, PNYDV has also been detected in

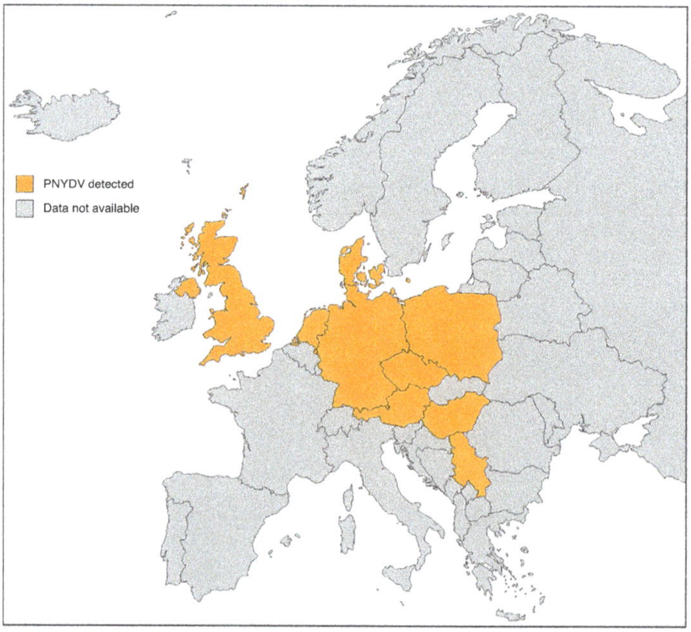

Fig. 2.3. Distribution of pea necrotic yellow dwarf virus in Europe. The virus was reported in nine countries i.e. Germany, Austria, Netherlands, Denmark, Czechia, Hungry, Serbia, Poland and the United Kingdom (in orange).

Denmark, the Netherlands, the Czech Republic, Hungary, Serbia, Poland and the United Kingdom infecting several legume crops (Fig. 2.3).

Transmission

PNYDV is transmitted by a number of aphid species in a persistent, non-propagative manner. It can be transmitted by the leguminous aphids *A. pisum*, *Aphis fabae* [Scopoli], *Megoura viciae* [Buckton], *A. craccivora* [Koch] and the polyphagous aphid *Myzus persicae* [Sulzer]. The virus has also been detected in the cabbage aphid *Brevicoryne brassicae* [Linnaeus]. Among these, *A. pisum*, with its various biotypes, is considered the most important vector of PNYDV. All these aphid species are known to be present in Europe. PNYDV is not known to be transmitted through seeds or by mechanical transmission. Experimentally, nanoviruses can be transmitted by biolistic bombardment using purified virions or cloned DNA components via agroinoculation.

DNA-N was found to be essential for nanovirus aphid transmission as it acts as a transmission helper factor in a study on FBNSV (Grigoras *et al.*, 2018). Purified virions alone could not be transmitted because the preparation lacked the necessary helper protein. It was also reported that DNA-N is required for the internalization and accumulation of FBNSV in the aphid gut cells (Mattia *et al.*, 2020). Figure 2.4 shows a predictive schematic representation of PNYDV cycle in aphid's midgut cell based on studies on nanoviruses. The virions require the presence of the NSP to translocate into and out of the midgut cells in a currently unknown manner. Additionally, the absence of DNA-U1 results in aphid transmission efficiency reduction.

In the body of aphids, it was reported that nanoviruses translocate from the anterior midgut to the principal salivary glands, with no evidence of replication occurring inside the aphid's body. However, the genome formula of FBNSV in the body of aphids were different than in the host plant they fed on (Sicard *et al.*, 2015). These changes were in the early stages of the transmission cycle. The molecular basis of the transmission is still inadequately understood and requires additional investigations. Moreover, the process by which virions enter and exit the gut and the salivary gland cells is unknown.

Detection

Early and accurate detection of PNYDV is important for effective disease management. Currently, traditional diagnostic methods, such as enzyme-linked immunosorbent assay (ELISA) and polymerase chain reaction (PCR), are the main methods used to detect PNYDV.

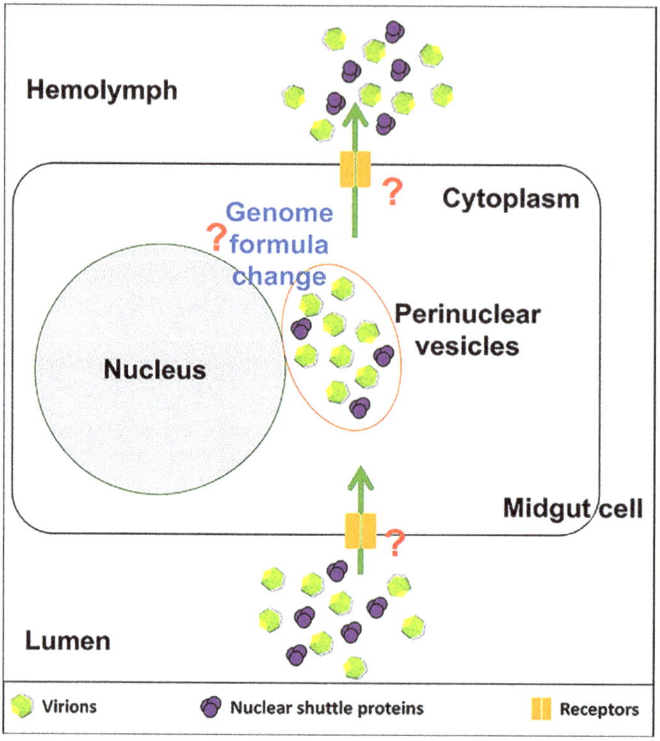

Fig. 2.4. Schematic representation of PNYDV cycle in aphid midgut cells based on studies on nanoviruses. The interaction of the helper factor NSP with virions and/or cell receptors, as well as whether the change in the genome formula occurs in the midgut cells, remains unknown.

Enzyme-linked immunosorbent assay (ELISA)

ELISA has been a standard assay for the detection of PNYDV. Several monoclonal and polyclonal antibodies exist, which can be used for the specific detection of PNYDV or a broadbent detection of various nanoviruses, including FBNSV or faba bean necrotic yellows virus (FBNYV). The serological assays rely on virus-specific antibodies to detect the presence of viral proteins in infected plant material. These assays are relatively inexpensive and easy to perform, making them suitable for routine testing.

Polymerase chain reaction (PCR) and PNYDV-specific primers

PCR-based assays have enhanced the sensitivity and specificity of PNYDV detection by targeting its viral genome. Primers designed against conserved regions of the DNA for each component enabled the detection of low concentrations of the virus in plants. These PCR assays are highly specific and can detect early infections.

Rolling circle amplification (RCA) and restriction fragment length pattern (RFLP) analysis

RCA can be used to amplify circular viral genomes such as PNYDV. The amplified products are then digested using restriction enzyme, such as *Bst*UI, followed by RFLP analysis to identify specific PNYDV signatures. In addition, a PNYDV-PCR can be performed to confirm the presence of the virus.

High-throughput sequencing (HTS)

In addition to ELISA and PCR-based methods, high-throughput sequencing (HTS) has emerged as a powerful tool for the detection and characterization of plant viruses, including PNYDV (Gaafar *et al.*, 2018). For enrichment of viral nucleic acids from host nucleic acid extracts, rolling circle amplification, dsRNA extraction or ribosomal RNA depletion can be used. HTS enables unbiased, genome-wide sequencing of all nucleic acids, allowing for the identification of both known and novel viral sequences without prior knowledge of the pathogen. HTS has been instrumental in detecting the presence of PNYDV in symptomatic plants and studying co-infections with other viruses (Gaafar *et al.*, 2020). Although currently limited by cost and technical complexity, HTS is important for virus discovery, diagnostics, and epidemiological research.

All these diagnostic tools play a fundamental role in integrated PNYDV detection and management strategies and are essential for tracking the spread of the virus, enabling timely responses to protect crops and minimize economic losses.

Challenges in Management and Breeding Strategies

Managing PNYDV presents significant challenges due to the virus nature and complex transmission, and that there are currently no known resistant cultivars. Existing management strategies focus on approaches aimed at reducing infection rates and limiting yield losses. These include monitoring and control of aphid populations, implementing crop rotation, and to adjust sowing dates to avoid peak vector activity. These methods can reduce the incidence of the virus and potentially slow its spread within the fields, although complete control remains difficult. Breeding for resistance is considered a long-term and a more sustainable solution. Efforts are underway to identify PNYDV-resistant pea cultivars. However, it remains an ongoing area of research.

Future Areas of Research to Improve Disease Management

Early detection of PNYDV

Hyperspectral imaging (HIS) and machine learning have shown great potential in the early detection of several plant diseases, including plant viruses. Currently, PNYDV detection is based on symptom assessment done by trained experts followed by molecular confirmation in lab. Drones equipped with hyperspectral cameras can provide an efficient approach for detection of PNYDV-symptomatic plants and the presence of their aphid vectors in the field. This automated technology can be used to increase the throughput of the analysis and early detection of the virus in the field before its establishment and spread.

PNYDV-resistant varieties

As mentioned above, there are currently no known or commercially available PNYDV-resistant plant cultivars. While research is progressing, there is a need for continued research not only in pea but in other important legumes to identify resistant cultivars, including investigating the sources of resistance and markers to help in breeding programs.

PNYDV infection prediction model

PNYDV incidences have increased in Europe since 2009. A reason for that is the spread of their aphid vectors potentially due to the changes in climate in favour of the vector's reproduction. The collection of new data from different fields over time, followed by the analysis of the virus incidences and severity including recording factors such as plant species and variety, weather parameters, vectors presence, and location will help in developing a prediction model for the virus.

Discussion Points

1. List the different symptoms of PNYDV on legume production, and its potential effects on livestock, grain trade and export, producers and consumers in general.
2. List the proteins encoded by PNYDV genes and their roles in the virus infection.
3. How to effectively monitor and control aphids to minimise the risk of PNYDV spread?

4. How can improved knowledge of PNYDV transmission and infection process help manage the disease?
5. Discuss how the use of hyperspectral imaging can be used in PNYDV early detection and disease management.

Further Reading

Available at: https://dafne.at/projekte/nanovir (accessed 22 January 2026).

References

Fowkes, A.R., Duffy, S., Mcgreig, S., Vazquez-Iglesias, I., Alraiss, K. *et al.* (2023) Comparison of sequencing methods for obtaining a whole genome of pea necrotic yellow dwarf virus from U.K. peas. *PhytoFrontiers* 3, 795–801. DOI: 10.1094/PHYTOFR-01-23-0003-R.

Gaafar, Y., Grausgruber-Gröger, S. and Ziebell, H. (2016) *Vicia faba, V. sativa* and *Lens culinaris* as new hosts for pea necrotic yellow dwarf virus in Germany and Austria. *New Disease Reports* 34, 28–28. DOI: 10.5197/j.2044-0588.2016.034.028.

Gaafar, Y., Timchenko, T. and Ziebell, H. (2017) First report of pea necrotic yellow dwarf virus in the Netherlands. *New Disease Reports* 35, 23–23. DOI: 10.5197/j.2044-0588.2017.035.023.

Gaafar, Y., Nielsen, G.C. and Ziebell, H. (2018) Molecular characterisation of the first occurrence of pea necrotic yellow dwarf virus in Denmark. *New Disease Reports* 37, 16–16. DOI: 10.5197/j.2044-0588.2018.037.016.

Gaafar, Y.Z.A., Herz, K., Hartrick, J., Fletcher, J., Blouin, A.G. *et al.* (2020) Investigating the Pea Virome in Germany—Old friends and new players in the field(s). *Frontiers in Microbiology* 11, 2020. DOI: 10.3389/fmicb.2020.583242.

Grigoras, I., Gronenborn, B. and Vetten, H.J. (2010) First report of a nanovirus disease of Pea in Germany. *Plant Disease* 94, 642–642. DOI: 10.1094/PDIS-94-5-0642C.

Grigoras, I., Vetten, H.-J., Commandeur, U., Ziebell, H., Gronenborn, B. *et al.* (2018) Nanovirus DNA-N encodes a protein mandatory for aphid transmission. *Virology* 522, 281–291. DOI: 10.1016/j.virol.2018.07.001.

Kartheuse, C., Tan, S., Bubolz, J., Reckling, M., Menzel, W. *et al.* (2025) First detection of viral pathogens of chickpea in Germany. *Journal of Plant Diseases and Protection* 132, 60. Available at: https://www.bdp-online.de/de/GFPi/Abteilungen_Projekte/Abteilung_Oel-_und_Eiweisspflanzen/EU-Projekt (accessed 22 January 2026).

Krenz, B., Schießl, I., Greiner, E. and Krapp, S. (2017) Analyses of pea necrotic yellow dwarf virus-encoded proteins. *Virus Genes* 53, 454–463. DOI: 10.1007/s41348-024-01048-z.

Mansourpour, M., Gallet, R., Abbasi, A., Blanc, S., Dizadji, A. *et al.* (2022) Effects of an alphasatellite on the life cycle of the Nanovirus Faba bean necrotic yellows virus. *Journal of Virology* 96, e01388–e01321. DOI: 10.1128/JVI.01388-21.

Mattia, J.D., Vernerey, M.-S., Yvon, M., Pirolles, E., Villegas, M. *et al.* (2020) Route of a multipartite nanovirus across the body of its aphid vector. *Journal of Virology* 94. DOI: 10.1128/jvi.01998-19.

Saucke, H., Uteau, D., Brinkmann, K. and Ziebell, H. (2019) Symptomology and yield impact of pea necrotic yellow dwarf virus (PNYDV) in faba bean (*Vicia faba* L. minor). *European Journal of Plant Pathology* 153, 1299–1315. DOI: 10.1007/s10658-018-01643-5.

Seeger, J.N., Ziebell, H. and Saucke, H. (2022) Impact of pea necrotic yellow dwarf virus (PNYDV) on nodulation, N2 fixation and yield in faba bean (*Vicia faba*, L.). *Journal of Plant Diseases and Protection* 129, 1437–1450. DOI: 10.1007/s41348-022-00646-z.

Sicard, A., Zeddam, J.-L., Yvon, M., Michalakis, Y., Gutiérrez, S. *et al.* (2015) Circulative non-propagative aphid transmission of nanoviruses: An oversimplified view. *JournalofVirology* 89, 9719–9726. DOI: 10.1128/JVI.00780-15.

Thomas, J.E., Gronenborn, B., Harding, R.M., Mandal, B., Grigoras, I. *et al.* (2021) ICTV virus taxonomy profile: Nanoviridae. *Journal of General Virology* 102(3). DOI: 10.1099/jgv.0.001544.

Yan, D., Han, K., Lu, Y., Peng, J., Rao, S. *et al.* (2024) The nanovirus U2 protein suppresses RNA silencing via three conserved cysteine residues. *Molecular Plant Pathology* 25, e13394. DOI: 10.1111/mpp.13394.

3 Sugarcane Yellow Leaf Virus – A Silent Threat?

Andre B. Gama and Madison T. Flasco

Department of Plant Pathology and Crop Physiology, Louisiana State University Agricultural Center, Baton Rouge, Louisiana

Introduction

Yellow leaf disease is caused by sugarcane yellow leaf virus (SCYLV), though it was originally reported under different names, including yellow wilt and yellow leaf syndrome (Fig. 3.1). The disease was first identified in Tanzania in 1962 as yellow wilt, and later in 1989, it was recognized as yellow leaf disease by researchers in Hawaii. Nearly a decade later, the aetiology of the disease became clearer when it was associated with a previously undescribed luteovirus, which was subsequently named SCYLV. SCYLV has been taxonomically assigned to the species *Polerovirus SCYLV*. While originally being placed in the *Luteoviridae* family, it has been reclassified and is now a member of the *Polerovirus* genus and *Solemoviridae* family. SCYLV consists of a positive-sense ssRNA [(+)ssRNA] genome that is approximately 5.8–5.9 kilobases in size with six open reading frames. ORF0 encodes an RNA silencing suppressor, ORF1 encodes a multifunctional protein, ORF2 codes for the RNA-dependent RNA polymerase, ORF3 encodes the coat protein, ORF4 codes for the virus movement protein, and ORF5 is hypothesized to encode a protein involved in vector transmission. The genome is packaged in an icosahedral virion with a diameter ranging from 24 to 29 nm.

SCYLV is transmitted by aphid vectors, with the sugarcane aphid (*Melanaphis sacchari* Zehntner) being the most significant. Other insect species, such as *Rhopalosiphum maidis* (Fitch), *R. rufiabdominalis* (Sasaki), and *Ceratovacuna lanigera* (Zehntner), can also transmit SCYLV in a circulative, non-propagative manner, though *M. sacchari* remains the principal vector (Figs 3.2 and 3.3). Upon aphid feeding on the phloem of SCYLV-infected sugarcane, the virus moves through the aphid gut tissue, i.e., acquisition and into the haemolymph. Upon movement to the salivary glands, SCYLV can be transmitted to healthy tissue in subsequent feeding. Transmission is also non-propagative in that while crossing these tissue barriers, the virus does not replicate. Transmission experiments in the greenhouse indicate

Fig. 3.1. Characteristic midvein yellowing (white arrow) due to sugarcane yellow leaf virus infection in an experimental field in Houma, Louisiana.

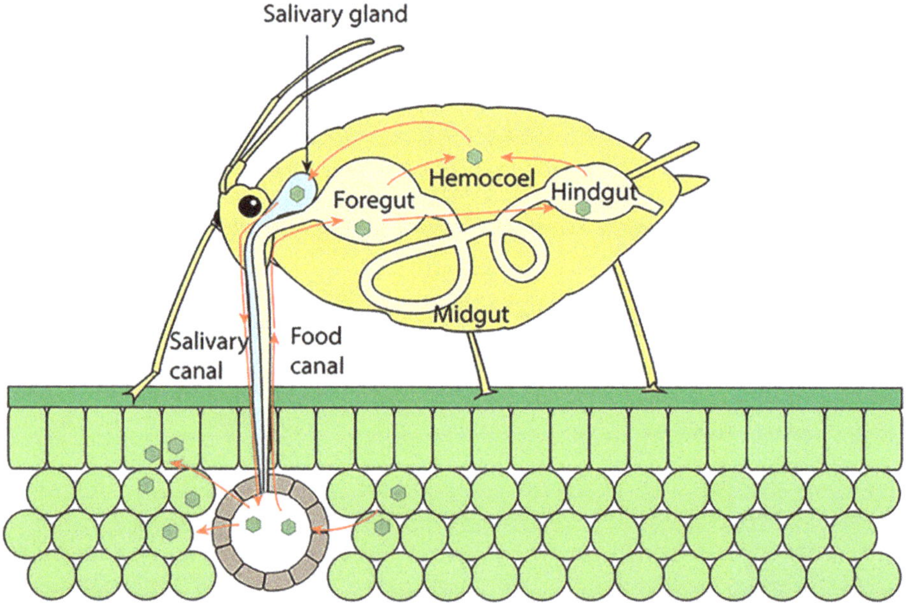

Fig. 3.2. Schematic representation of sugarcane yellow leaf virus transmission in a circulative and non-propagative manner by the sugarcane aphid, *Melanaphis sacchari*. Photo credit: https://viralzone.expasy.org/3738 (image licensed under Creative Commons CC-BY licence https://creativecommons.org/licenses/by/4.0/).

the sugarcane aphid can acquire SCYLV after 4.5h of continuous feeding on infected sugarcane and transmit it to healthy tissue within two days of feeding (Paray *et al.*, 2011). While aphids may acquire SCYLV from infected sugarcane, they could also acquire the virus from non-cultivated host species.

Fig. 3.3. Sugarcane aphid colony on the abaxial surface of a sugarcane leaf in Saint Gabriel, Louisiana, USA.

For example, this virus was found surviving in Columbus grass, a common weed occurring near sugarcane fields in Florida. Despite continued efforts to understand its aetiology, transmission, and impact on sugarcane production, the disease's variable symptom expression across different regions, cultivars, and soil types highlights the need for improved detection methods and a deeper understanding of its epidemiology.

SCYLV Symptomatology

When the disease was first reported in Hawaii, cultivar H 65–7052 showed the typical yellow leaf disease symptoms—yellowing of the mid rib on the abaxial surface of sugarcane leaves, evolving to leaf tip yellowing and necrosis with an eventual appearance of a reddish colour in the otherwise chlorotic mid ribs. However, even though cultivar H 65–7052 was symptomatic, neighbouring sugarcane fields with different cultivars did not show symptoms. It is noteworthy that the lack of reliable detection methods at the time makes it unclear whether those neighbouring sugarcane fields were infected with the virus. The yellowing of the midrib, the most typical symptom of yellow leaf disease, can be confused with abiotic stresses, underscoring the importance of accurate diagnostic techniques. Moreover, the disease can occur asymptomatically in many sugarcane varieties while still negatively impacting yield, making field surveys and assessments of cultivar resistance or tolerance to SCYLV essential.

SCYLV Impact on Yield

The impact of SCYLV infection on sugarcane varies significantly depending on the cultivar. Due to the often-asymptomatic nature of virus infection, accurately assessing its effect on yield can be challenging. The most severe yield loss attributed to yellow leaf disease was reported in Brazil with the cultivar SP 71–6163 in 1993, where sugar yields were reduced by 50%, leading to the cultivar's abandonment by the growers, highlighting the need to understand a cultivar's susceptibility to SCYLV and virus impacts on yield. In Louisiana, sugar yields were reduced by approximately 11% when comparing SCYLV-infected plants with healthy ones of the cultivar HoCP 96–540, a key variety in the state. In Florida, cultivar CP96-1252 experienced a 27% reduction in sugar yield when comparing a field with 100% SCYLV prevalence to another with only 2% prevalence. Importantly, the above yield losses were often recorded on asymptomatic infections, relying on serological and molecular techniques to detect the virus. Since cultivars vary genetically, they have individual yield characteristics and responses to viral infections, therefore, assessing the impact of SCYLV on popular cultivars in each sugarcane-growing region is crucial to ensure that yields are not being compromised by the disease.

Distribution and Diversity of SCYLV

SCYLV is prevalent across sugarcane-producing regions worldwide, including most countries that grow sugarcane in the Americas, Asia, and Africa, such as Argentina, Brazil, Colombia, Cuba, Guatemala, Mexico, China, the Philippines, Senegal, South Africa, and Kenya (Rott *et al.*, 2023). Monitoring yellow leaf disease is challenging due to the variability in symptom expression – symptoms can differ among cultivars, across regions, and between growing seasons. Detection is further complicated by the seasonal nature of symptoms which can appear, disappear, and reappear throughout the growing cycle (Lehrer and Komor, 2008). Following significant outbreaks of yellow leaf disease and the advancement of diagnostic techniques, particularly RT-PCR, SCYLV was detected in numerous sugarcane varieties worldwide, often in asymptomatic plants. Asymptomatic plant material infected with SCYLV was likely propagated by growers unaware of its presence, unintentionally introduced to fields via the planting material, and later exacerbated by aphid-mediated spread (Fig. 3.4). Vegetative propagation of planting material, and subsequently the virus, likely contributed to the virus's widespread distribution and its presence in most sugarcane-producing countries today (Fig. 3.5).

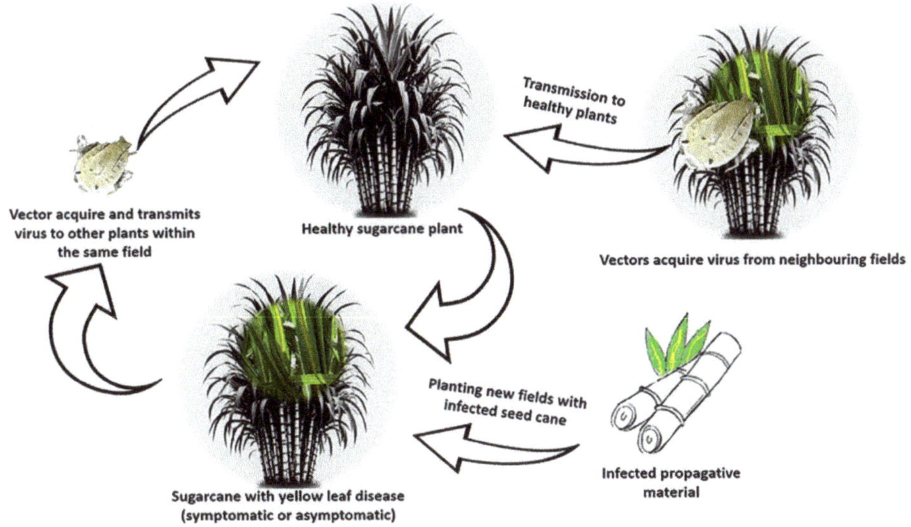

Fig. 3.4. Schematic representation of sugarcane yellow leaf virus (SCYLV) transmission cycle. Aphids may acquire SCYLV from infected plants from neighbouring fields, or within the same field, and transmit the virus to otherwise healthy plants. Infected propagative material may also introduce yellow leaf disease to new areas.

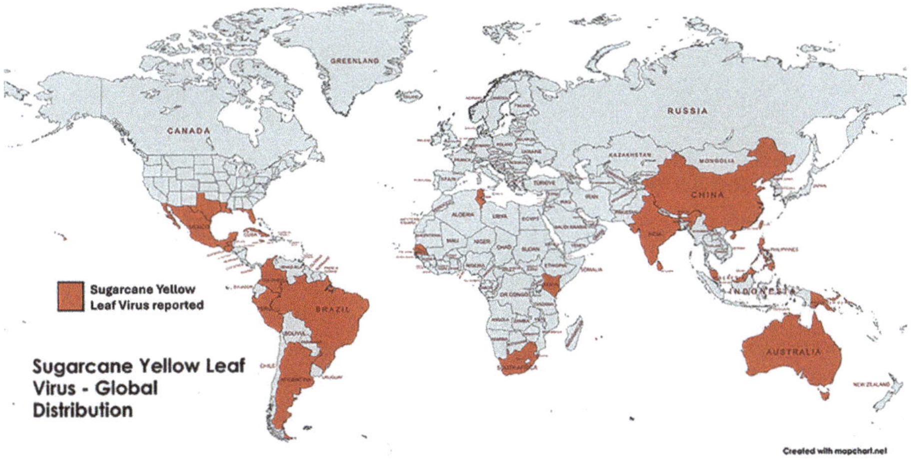

Fig. 3.5. Global distribution of sugarcane yellow leaf virus. Countries shown in red reported the presence of the virus in sugarcane samples.

With advances in molecular techniques, various SCYLV genotypes have been identified. There are 10 major SCYLV genotypes – BRA, HAW, PER, CHN3, REU, FLA3, CHN1, FLA1/2, IND, and CUB – which have been recently grouped into three phylogroups: G1, G2, and G3. Genotypes BRA, HAW, PER, and CHN3 belong to G1; REU is the sole genotype in G2; and FLA3, CHN1, FLA1/2, IND, and CUB are clustered in G3 (Lu *et al.*, 2021).

However, a SCYLV isolate from Guatemala did not fit into any of these major phylogroups, suggesting that SCYLV's genetic diversity may be even broader than previously assumed (Rott *et al.*, 2023). This is especially important considering the global nature of sugarcane may facilitate the sharing of plant material, and subsequently, SCYLV. Regarding the local spread of SCYLV by aphid species, understanding the genotypes present is essential as the sugarcane aphid is reported to transmit different genotypes with different efficiencies (Abu Ahmad *et al.*, 2007). This high genetic diversity further complicates SCYLV detection – primers or antisera may be optimized for some genotypes but fail to detect others, making comprehensive monitoring a challenging task.

Detecting a Barely Visible Threat – A Difficult Challenge

Given the potential of SCYLV to significantly reduce yields in susceptible cultivars, substantial efforts have been directed towards developing reliable diagnostic tools. Yield losses from SCYLV infections can occur even in the absence of visible symptoms, making it difficult for researchers to accurately correlate yield declines with traditional visual disease quantification methods, such as disease scales or severity ratings.

SCYLV can be detected using various serological and molecular techniques, including tissue-blot immunoassays (TBIA), double-antibody sandwich-enzyme linked immunoassays (DAS-ELISA), RT-PCR, and real-time RT-PCR. The sensitivity of these diagnostic methods is crucial in determining the accuracy of SCYLV detection. While TBIA is a quick, inexpensive, and high-throughput option, it is less sensitive than molecular-based assays and may fail to detect low viral titres in plants, especially during milder infections or periods of natural seasonality. Furthermore, such assays may be unable to detect all SCYLV genotypes, resulting in false negatives. Recently, new assays including reverse transcription-loop mediated isothermal amplification (RT-LAMP), reverse transcription-recombinase polymerase amplification (RT-RPA), and reverse transcription-multienzyme isothermal rapid amplification (RT-MIRA) have been developed. These assays show increased sensitivity compared to RT-PCR, though their sensitivity was assessed using a limited genetic diversity. These assays provide a promising future to detection upon optimization with additional genotypes. In Louisiana, current sampling recommendations suggest collecting the top visible dewlap leaf of the plant in question, since previous studies found that the highest proportion of infected vascular bundles was found in the top leaves based on results from tissue immunoblot assays. However, there is limited information on the localization of SCYLV within the host plant throughout the season, although a previous survey indicated that SCYLV incidence tends to rise in late spring to early summer in Louisiana, coinciding with increases in vector populations. This survey, however, did not specifically investigate the seasonal *in planta* distribution of SCYLV within the different parts of the sugarcane plant.

A deeper understanding of SCYLV's movement and localization within the host plant throughout the year is essential for optimizing sampling strategies. Improved knowledge of the virus's epidemiology in the field and distribution within the host will not only enhance detection accuracy for SCYLV but also support research efforts aimed at assessing the impact of SCYLV on sugarcane yields.

Sugarcane Yellow Leaf Disease (SCYLV): Challenges in Management and Breeding Strategies

Like other viral diseases, sugarcane plants cannot be cured of yellow leaf disease using chemical control. Additionally, this virus survives heat treatments effective against other sugarcane diseases, such as ratoon stunting disease. In countries like India, Brazil, and the USA, meristem tissue culture has been employed to produce propagative material free from major sugarcane diseases, including SCYLV. Currently, *M. sacchari* is not a target of pest management programs. In most sugarcane-producing areas, the main pest target is the sugarcane borer, *Diatraea saccharalis* (Fabricius). A previous epidemiological study on SCYLV transmission found that aphid-mediated spread only occurred within short distances, and vector management would not be as effective as the adoption of virus-free seed cane or resistant varieties. No studies measuring the impact of vector management strategies have been performed since then. However, the disease can still spread to multiplication fields from neighbouring areas through vectors, particularly *M. sacchari*. Therefore, varietal resistance is potentially the most impactful management strategy, but breeding SCYLV-resistant varieties is challenging.

The inconspicuous nature of SCYLV infections makes it difficult for plant pathologists and breeders to assess varieties for SCYLV resistance. Typically, viral titre quantification is used to evaluate SCYLV resistance. However, recent studies have found that viral titres do not correlate with either symptom expression or yield losses, as some varieties with high viral titres exhibited lower yield losses than those with lower titres (Nithya *et al.*, 2023). To make matters more complicated, the impact of SCYLV on yield can vary significantly among different cultivars, even in asymptomatic epidemics.

Techniques like TBIA have been used to detect differences in disease incidence in biparental tests involving F1 progeny from highly heterozygotic susceptible cultivars. However, the difficulty in accurately rating SCYLV resistance or tolerance, combined with the variable impact on yield, complicates the decision making process for researchers considering resource allocations to tackle this issue. Understanding the incidence of SCYLV and the susceptibility of current cultivars provides three management strategies of interests:

1. *Assess SCYLV incidence in commercial fields*: If the disease is not widespread or is effectively managed with clean propagative material, an extensive breeding effort may not be justified.

2. *Rate current cultivar tolerance/resistance*: Identify valuable genetic resources for future crossings by comparing yield losses between SCYLV-positive and negative plants under identical environmental conditions. Focus on sugar yields, not just biomass, since viral infections affect the carbohydrate composition within plants. Be mindful of mixed viral infections and how they can influence sugar yields. Keep in mind, cultivars that are asymptomatic with minimal yield loss can serve as inoculum sources to more susceptible neighbouring plantings.
3. *Survey for different SCYLV genotypes*: The impact of infections on yield may vary depending on the SCYLV genotype. Conducting a thorough survey in sugarcane-producing regions to identify the dominant SCYLV genotypes can help in challenging varieties against prevalent strains.

Discussion Points

1. List some reasons why we cannot rely on disease symptoms to assess the resistance of sugarcane varieties to yellow leaf disease.
2. How diverse are isolates of SCYLV? What does that diversity entail for practical purposes? Think of the consequences of diversity in terms of diagnosis and breeding programs. What management strategies are suggested to prevent SCYLV infections? To manage the aphid vector?
3. What should be considered before resources are allocated to breed for SCYLV resistance?

Further Reading

Boukari, W., Kaye, C., Wei, C., Hincapie, M., LaBorde, C. *et al.* (2019) Field infection of virus-free sugarcane by sugarcane yellow leaf virus and effect of yellow leaf on sugarcane grown on organic and on mineral soils in Florida. *Plant Diseases* 103, 2367–2373.

Boukari, W., Wei, C., Tang, L., Hincapie, M., Naranjo, M. *et al.* (2020) Lack of transmission of sugarcane yellow leaf virus in Florida from Columbus grass and sugarcane to sugarcane with aphids or mites. *PLoS One* 15, e0230066.

Espinoza Delgado, H.V., Kaye, C., Hincapie, M., Boukari, W., Wei, C. *et al.* (2016) First report of sugarcane yellow leaf virus infecting columbus grass (*Sorghum almum*) in Florida. *Plant Diseases* 100, 1027.

Gonçalves, M.C., Klerks, M.M., Verbeek, M., Vega, J. and van den Heuvel, J.F.J.M. (2002) The use of molecular beacons combined with NASBA for the sensitive detection of sugarcane yellow leaf virus. *European Journal of Plant Pathology* 108, 401–407.

Gonçalves, C.M., Vega, J., Oliveira, G.J. and Gomes, M.A.M. (2005) Sugarcane yellow leaf virus infection leads to alterations in photosynthetic efficiency and carbohydrate accumulation in sugarcane leaves. *Fitopatologia Brasileira* 30, 10–16.

Gray, S. and Gildow, F.E. (2003) Luteovirus-aphid interactions. *Annual Review of Phytopathology* 41(1), 539–566.

Grisham, M.P., Johnson, R.M. and Zimba, P.V. (2010) Detecting sugarcane yellow leaf virus infection in asymptomatic leaves with hyperspectral remote sensing and associated leaf pigment changes. *Journal of Virological Methods* 167, 140–145.

Islam, M.S., Yang, X., Sood, S., Comstock, J.C., Wang, J. *et al.* (2018) Molecular characterization of genetic basis of sugarcane yellow leaf virus (SCYLV) resistance in *Saccharum* spp. hybrid. *Plant Breeding* 137, 598–604.

Lehrer, A.T. and Komor, E. (2008) Symptom expression of yellow leaf disease in sugarcane cultivars withdifferent degrees of infection by sugarcane yellow leaf virus. *Plant Pathology* 57, 178–189.

McAllister, C.D., Hoy, J.W. and Reagan, T.E. (2008) Temporal increase and spatial distribution of sugarcane yellow leaf and infestation of the aphid vector, *Melanaphis sacchari. Plant Diseases* 92, 607–615.

Species List: Solemoviridae | ICTV (n.d.). Available at: https://ictv.global/report/chapter/solemoviridae/taxonomy/solemoviridae (accessed 4 October 2024).

References

Abu Ahmad, Y., Costet, L., Daugrois, J.-H., Nibouche, S., Letourmy, P. *et al.* (2007) Variation in infection capacity and in virulence exists between genotypes of sugarcane yellow leaf virus. *Plant Diseases* 91, 253–259.

Lehrer, A.T. and Komor, E. (2008) Symptom expression of yellow leaf disease in sugarcane cultivars with different degrees of infection by sugarcane yellow leaf virus. *Plant Pathology* 57, 178–189.

Lu, J.J., He, E.Q., Bao, W.Q., Chen, J.S., Sun, S.R. *et al.* (2021) Comparative genomics reveals insights into genetic variability and molecular evolution among sugarcane yellow leaf virus populations. *Scientific Reports* 11, 7149.

Nithya, K., Parameswari, B., Kumar, S., Annadurai, A., Nithyanantham, R. *et al.* (2023) Prospecting true ScYLV resistance in *Saccharum* hybrid parental population in India by symptom phenotyping and viral titre quantification. *3 Biotech* 13, 125.

Paray, N., Khoodoo, M.H.R., Saumtally, A.S. and Ganeshan, S. (2011) Vector-virus relationship for *Melanaphis sacchari* (Zehnt.) *(Hemiptera: Aphididae)* transmitting *Sugarcane yellow leaf Luteovirus* in Mauritius. *Sugar Technologies* 13, 77–80.

Rott, P., Grinstead, S., Dallot, S., Foster, Z.S.L., Daugrois, J.H. *et al.* (2023) Genetic diversity, evolution, and diagnosis of sugarcane yellow leaf virus from 19 sugarcane-producing locations worldwide. *Plant Diseases* 107, 3437–3447.

4 Soybean Dwarf Virus

Marta Niedzicka

Crop Science Department, Teagasc, Oak Park, Carlow, R93 XE12, Ireland

Introduction

The novel viral disease responsible for dwarfing in soybean was first documented in Japan in 1952 (Tamada *et al.*, 1969). The pathogenic agent responsible for the disease was characterised in 1969 (Tamada *et al.*, 1969) and has since been identified in various countries and across different hosts (Stone *et al.*, 2024). The virus infects many legume species: beans, peas, lentils and clovers, but is symptomless in some of its hosts, for example white clover (Tamada, 1970). Although the virus is globally distributed, its impact on crops is variable, leading to significant yield losses in soybean in Japan, while not being regarded as a threat to soybean in the United States (Hill and Whitham, 2014). Its relevance extends beyond soybean, as it also induces disease in legume crops in other regions worldwide, for example in the Middle East or in Australia (Makkouk *et al.*, 1997; Congdon *et al.*, 2023). Effective pest management requires an understanding of its local hosts, including potential viral reservoirs, and vectors. Development of commercially available resistant varieties of soybean is ongoing.

Luteovirus glycinis

The virions of *L. glycinis* comprise linear, positive sense single-stranded RNA of 5.8 kB, which encodes five open reading frames (ORFs).

Like many other plant viruses, *L. glycinis* is transmitted by insect vectors. The mode of transmission is circulative and non-propagative (Tamada, 1970): upon acquisition from an infected plant during phloem feeding, the virus circulates within the insect's body, including accessory salivary glands, without entering the cells or replicating within the insect. Despite the absence of replication within the insect's body, the virus retains its infectivity within the vector. As a result, when the insect feeds on a new plant, the virus is released through its saliva (Fig. 4.1).

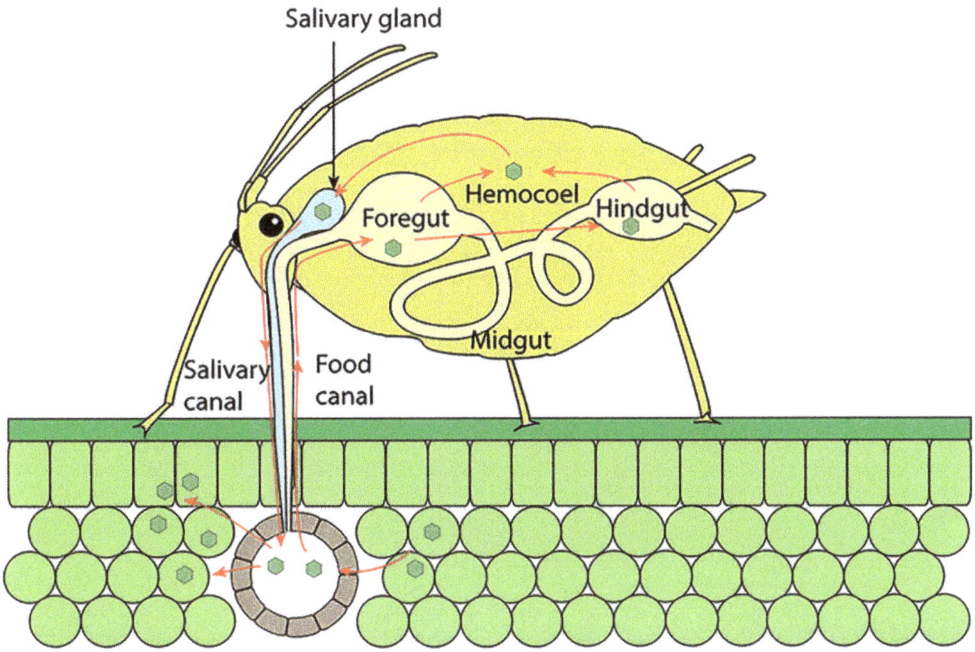

Fig. 4.1. Plant virus circulative route in insects. Courtesy of SwissBioPics (Available at: https://viralzone.expasy.org/3738, accessed 19 February 2025).

The main aphid species serving as vectors worldwide are *Aulacorthum solani* and *Acyrthosiphum pisum* (Terauchi *et al.*, 2001). Other potential vectors have been identified, including *Nearctaphis bakeri* and *Myzus persicae* (Damsteegt *et al.*, 2011; Congdon *et al.*, 2023).

Several strains of the virus have been described:

- Based on the symptoms caused in the soybean: Yellowing (Y) and Dwarfing (D) strain
- Based on the transmission vector – *A. pisum* (P) or *A. solani* (S), with P strain being transmitted also by *N. bakeri*

This classification results in four commonly recognised strains: YP, YS, DP and DS; with natural occurrences of mixed strain infections confirmed. The main strains – Y and D – are distributed worldwide, and subclades of both strains can lead to significant crop diseases. Recent phylogenetic studies have confirmed that yellowing and dwarfing genotypes are separated in distinctive clades, each containing several subclades (Stone *et al.*, 2024). These subclades differ in the severity of disease they cause in soybean. However, there is insufficient information about the strains responsible for diseases in other crops (Fig. 4.2).

The vectors exhibit differing levels of transmission efficiency, with *N. bakeri* demonstrating lower efficacy in transmitting *L. glycinis* from clover to soybean compared to *A. solani*, while *Acyrthosiphon kondoi* is regarded as an inefficient vector (Harrison *et al.*, 2005; Congdon *et al.*, 2023). Furthermore,

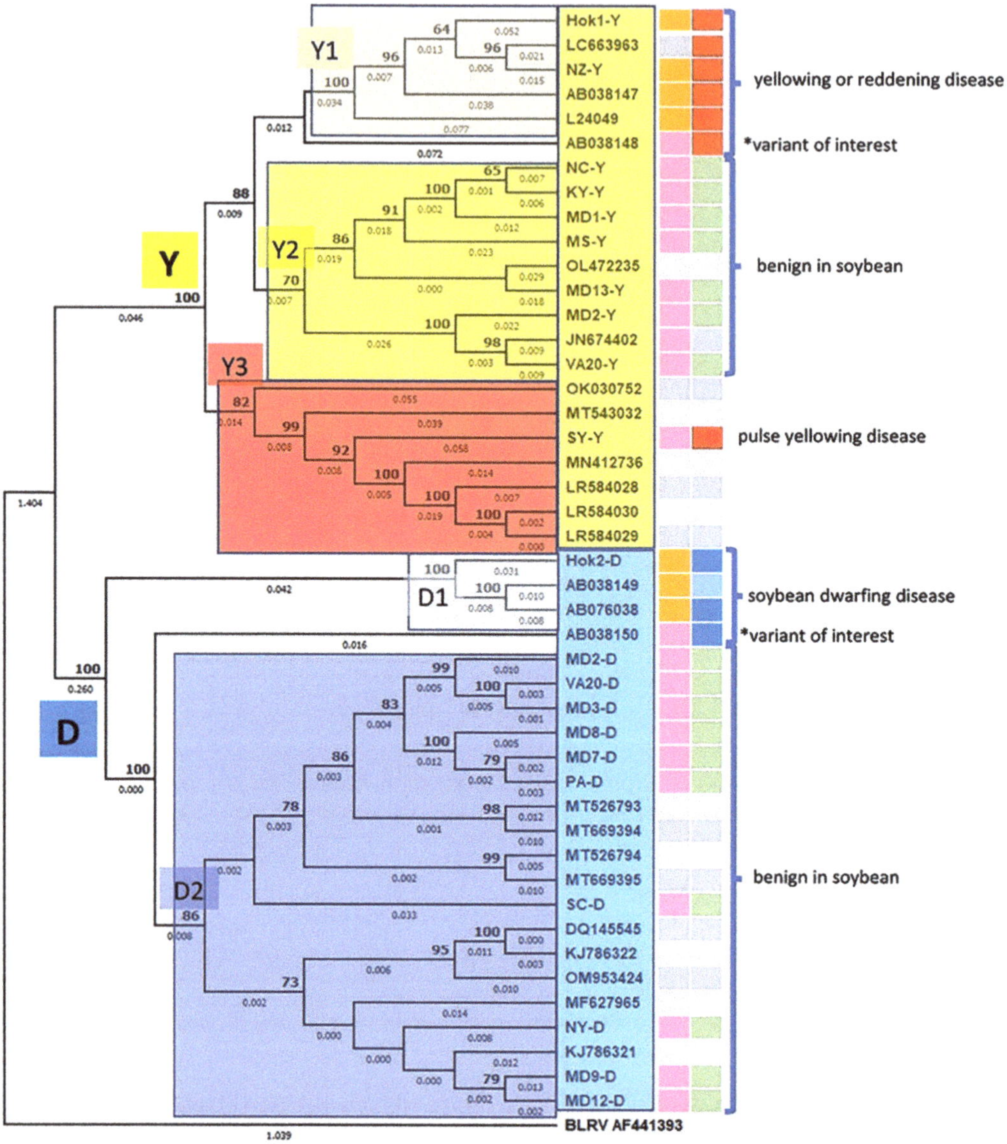

Fig. 4.2. Phylogenetic tree of complete and near-complete *Luteovirus glycinis* genomes. Courtesy of Stone *et al*. (2024).

even within the same vector species, factors such as host preferences, morphological variations and feeding locations on the soybean can contribute to regional differences in the viral transmission efficiency and the perceived threat to crops (Damsteegt and Voegtlin, 1990). For example, *A. solani* populations are limited and do not typically colonise soybean in the USA, unlike the Japanese *A. solani* biotypes.

Hosts Plants and Viral Reservoirs

L. glycinis is distributed worldwide. Since its discovery in Japan in 1952 (Tamada *et al.*, 1969), the presence of the virus has been confirmed in Australia, New Zealand, the USA, Syria and several European countries, including Czech Republic, Germany, UK and Ireland (Harrison *et al.*, 2005; Gaafar and Ziebell, 2020; Fowkes *et al.*, 2021; Luoto *et al.*, 2021; Niedzicka *et al.*, 2025).

The presence of *L. glycinis* has been confirmed in a wide range of legumes—soybean, multiple species of beans, peas, chickpeas, lentils and lupins, and a number of clover species—as well as a few other species, such as *Chenopodium* spp. (Tamada, 1970; Congdon *et al.*, 2023). Although all these species may host the virus, its deleterious effects vary significantly, ranging from severe symptoms, as described in soybean crops in Japan, to an absence of symptoms, commonly noted in white clover (Tamada, 1970). Even within soybean crops, the severity of infection is influenced by regional characteristics: the virus causes significant disease of soybean in Japan, whereas only mild infections are reported in the USA (Damsteegt *et al.*, 2011).

In many regions, clover species and other legume species can act as viral reservoirs, facilitating the transmission of the virus to crops. Clovers from perennial pastures can serve as a primary source of viral infection in nearby crops, making the presence of non-crop legume species a critical factor in risk assessment.

Symptoms and Economic Impact

There are several documented diseases in legume crops caused by *L. glycinis*.

1. Soybean yellowing disease was first documented in Japan during the 1960s. The virus causes severe stunting or yellowing, and yield losses can reach up to 80%. The main symptoms are yellowing of leaves, stunted growth, and reduced seed yield. Mixed infections involving yellowing and dwarfing strains result in more severe symptoms (Hill and Whitham, 2014).
2. Read leaf disease in subterranean clover has been observed in Australia, where the virus causes sporadic outbreaks. Symptoms include reddening of leaves, stunted growth, and in some cases, premature plant death. High virus incidence leads to reduced herbage and growth, losses in seed yield and decreased hay production (Department of Primary Industries and Regional Development, 2018).

The virus causes symptoms in other pulse crops, like pea, lentils, common beans, and faba beans. Field experiments have clearly demonstrated that the virus can cause up to 56% yield loss in faba beans (Makkouk and Kumari, 2001). However, its economic impact in pulse crops others than soybean and subterranean clover remains largely unexplored.

The significance of this threat depends on the region, distribution of legume species in arable margins and pastures, and the presence of vectors. While the soybean dwarfing disease caused significant yield losses in Japan, with all cultivars susceptible to the disease, the same virus found in the USA is not considered a major threat, as plants only exhibit low to moderate symptoms. That discrepancy is likely attributable to variations in the virus strains and transmission vectors, i.e. the main aphid species and its biotype characteristics, such as host preference, transmission efficiency and specificity (Hill and Whitham, 2014). This underscores the importance of monitoring potential threats in the future, as the introduction of efficient transmission vectors could precipitate a disease outbreak.

Detection Methods

The main diagnostic methods of *L. glycinis* are ELISA and RT-PCR (Hill and Whitham, 2014). The monoclonal antibodies have been in use since 1989 (D'Arcy *et al.*, 1989), and accurate virus-specific ELISA tests from plant tissues are commercially available and can be used on material from plants or aphids.

Developed and tested RT-PCR primers are either species or strain specific (Abraham *et al.*, 2007; Damsteegt *et al.*, 2011). For example, strain-specific multiplex PCR assay has been developed by Schneider *et al.* (2011) and is used by other researchers.

Additionally, multiple sequences of virus were assembled via both Sanger and high throughput sequencing and were deposited in GenBank.

Management Practices

To establish management practices, it is essential to understand the interactions between virus strains, vectors and potential viral reservoirs. Viral reservoirs, often asymptomatic and found in pastures or wild legume species, serve as the main sources of infection for crops, as infected vectors spread the virus from these reservoirs. One management practice is to avoid planting soybean next to clover pastures. However, in regions where white and/or red clovers are widespread, preventing the infection of *L. glycinis* in the main crop becomes challenging. The main management approach focuses on early-season insecticides applications, eliminating the vectors (aphids)—for example, in the case of subterranean clover pastures, the insecticide application is recommended at 2 and 6 weeks after seedling emergence. Pest management includes further spraying during the season, aligned with the migration pattern of winged aphids, to prevent the later onset of the disease. Additionally, adjusting planting date may also be considered as a potential management tool, soybean crops planted later in the season, after the main migration of winged aphids, are less likely to be severely impacted

(Hill and Whitham, 2014; Department of Primary Industries and Regional Development, 2018).

At the same time, the virus-resistance Indonesian soybean cultivar "Willis" has been developed, and two different resistance traits are reported and studied. The ongoing soybean breeding program in Japan focuses on introducing the resistance not only in soybean but also common bean (Yamashita *et al.*, 2014). The resistance trait has to be introduced to cultivars that are adapted to specific environmental conditions. However, these solutions are not yet commercially available.

In case of subterranean clover in Australia, the main control measures include the application of insecticides during periods of highest infection risk, manipulation of pasture composition (as grasses do not host *L. glycinis*), or the cultivation of alternative pasture species, as, for example serradella, as it does not develop symptoms. However, serradella species was confirmed to be a viral reservoir, being highly susceptible to infection and serving as an asymptomatic host (Congdon *et al.*, 2023).

Future Research Directions

Ongoing and future research efforts are focused on breeding varieties of soybean with enhanced resistance. Other important research directions include:

1. testing the importance of the disease in other legume crops (e.g. beans and peas);
2. improving aphid monitoring system (testing for virus presence, minimising the usage of pesticide).

Potential future research direction may include better understanding of dependency between aphid biotypes and the genotypes of *L. glycinis* in different crops. Better understanding of the impact of the virus in various legumes will improve our pest management strategies and evaluation of potential risks.

Discussion Points

1. How can the use of insecticides be reduced without diminishing the current management controls of the *L. glycinis*?
2. What is the benefit of studying potential viral reservoirs for the management of plant health?

Funding

HORIZON EUROPE Marie Sklodowska-Curie Actions (Grant/Award Number: '101106728').

Further Reading

Department of Primary Industries and Regional Development (2018) Subterranean Clover Red Leaf Syndrome (SbDV). Available at: https://www.wool.com/globalassets/wool/land/pastures/weed-and-pest-management/accordion/subterranean-clover--red-leaf-syndrome.pdf (accessed 19 February 2025).

Hill, J.H. and Whitham, S.A. (2014) Control of virus diseases in soybeans. In: *Advances in Virus Research*, Vol. 90. Elsevier, pp. 355–390. DOI: 10.1016/B978-0-12-801246-8.00007-X.

Stone, A.L., Damsteegt, V.D., Smith, O.P. and Stewart, L.R. (2024) Global phylogenetic analysis of soybean dwarf virus isolates and their associations with aphid vectors and severe disease in soybeans. *Virology* 591(March), 109984. DOI: 10.1016/j.virol.2024.109984.

References

Abraham, A.D., Menzel, W., Vetten, H.J. and Saucke, H. (2007) First report of soybeandwarf-virus (genus *Luteovirus*) infecting Faba bean and clover in Germany. *Plant Disease* 91(8), 1059–1059. Available at: https://www.wool.com/globalassets/wool/land/pastures/weed-and-pest-management/accordion/subterranean-clover--red-leaf-syndrome.pdf

Congdon, B.S., Baulch, J.R. and Foster, K.J. (2023) Vector species, pasture legume host range, and impact on grain legumes of an Australian soybean dwarf virus isolate. *Archives of Virology* 168(1), 20. DOI: 10.1007/s00705-022-05664-8.

Damsteegt, V.D. and Voegtlin, D.J. (1990) Morphological and biological variation among populations of *Aulacorthum solani* (Homoptera: Aphididae), the vector of soybean dwarf virus. *Annals of the Entomological Society of America* 83(5), 949–955. DOI: 10.1093/aesa/83.5.949.

Damsteegt, V.D., Stone, A.L., Kuhlmann, M., Gildow, F.E., Domier, L.L. *et al.* (2011) Acquisition and transmissibility of U.S. soybean dwarf virus isolates by the soybean aphid, aphis glycines. *Plant Disease* 95(8), 945–950. DOI: 10.1094/pdis-10-10-0726.

D'Arcy, C.J., Torrance, L. and Martin, R.R. (1989) Discrimination among Luteoviruses and their strains by monoclonal antibodies and identification of common epitopes. *Phytopathology* 79(8), 869–873.

Department of Primary Industries and Regional Development (2018) Subterranean Clover Red Leaf Syndrome (SbDV). Available at: https://www.wool.com/globalassets/wool/land/pastures/weed-and-pest-management/accordion/subterranean-clover--red-leaf-syndrome.pdf (accessed 19 February 2025).

Fowkes, A.R., McGreig, S., Pufal, H., Duffy, S., Howard, B. *et al.* (2021) Integrating high throughput sequencing into survey design reveals turnip yellows virus and soybean dwarf virus in pea (*Pisum sativum*) in the United Kingdom. *Viruses* 13(12), 2530. DOI: 10.3390/v13122530.

Gaafar, Y.Z.A. and Ziebell, H. (2020) Complete genome sequence of a soybean dwarf virus isolate from white clover in Germany. *Microbiology Resource Announcements* 9(28), e00637–e00620. DOI: 10.1128/MRA.00637-20.

Harrison, B., Steinlage, T.A., Domier, L.L. and D'Arcy, C.J. (2005) Incidence of soybean dwarf virus and identification of potential vectors in Illinois. *PlantDisease* 89(1), 28–32. DOI: 10.1094/PD-89-0028.

Hill, J.H. and Whitham, S.A. (2014) Control of virus diseases in soybeans. In: *Advances in Virus Research*, Vol. 90. Elsevier, Amsterdam, Netherlands, pp. 355–390. DOI: 10.1016/B978-0-12-801246-8.00007-X.

Luoto, A., Lehtonen, M., Valkonen, J.P.T. and Santala, J. (2021) First report of soybean dwarf virus infecting white clover (*Trifolium repens*) in Finland. *PlantDisease* 105(12), PDIS–04. DOI: 10.1094/PDIS-04-21-0822-PDN.

Makkouk, K.M. and Kumari, S.G. (2001) Reduction of incidence of three persistently transmitted aphid-borne viruses a!Ecting legume crops by seed-treatment with the insecticide Imidacloprid (gaucho). *Crop Protection* 20, 433–437.

Makkouk, K.M., Damsteegt, V., Johnstone, G.R., Katul, L., Lesemann, D.-E. *et al.* (1997) Identification and some properties of soybean dwarf luteovirus affecting lentil in Syria. *Phytopathologia Mediterranea* 36, 135–144.

Niedzicka, M., Ballandras, V., McNamara, L. and Byrne, S. (2025) First report of soybean dwarf virus in clover species in Ireland. *New Disease Reports* 51(1), e70017. DOI: 10.1002/ndr2.70017.

Schneider, W.L., Damsteegt, V.D., Stone, A.L., Kuhlmann, M., Bunyard, B.A. *et al.* (2011) Molecular analysis of soybean dwarf virus isolates in the eastern United States confirms the presence of both D and Y strains and provides evidence of mixed infections and recombination. *Virology* 412(1), 46–54. DOI: 10.1016/j.virol.2011.01.001.

Stone, A.L., Damsteegt, V.D., Smith, O.P. and Stewart, L.R. (2024) Global phylogenetic analysis of soybean dwarf virus isolates and their associations with aphid vectors and severe disease in soybeans. *Virology* 591(March), 109984. DOI: 10.1016/j.virol.2024.109984.

Tamada, T. (1970) Aphid transmission and host range of soybean dwarf virus. *Japanese Journal of Phytopathology* 36(4), 266–274. DOI: 10.3186/jjphytopath.36.266.

Tamada, T., Goto, T., Chiba, I. and Suwa, T. (1969) Soybean dwarf, a new virus disease. *Japanese Journal of Phytopathology* 35(4), 282–285. DOI: 10.3186/jjphytopath.35.282.

Terauchi, H., Kanematsu, S., Honda, K., Mikoshiba, Y., Ishiguro, K. *et al.* (2001) Comparison of complete nucleotide sequences of genomic RNAs of four soybean dwarf virus strains that differ in their vector specificity and symptom production. *Archives of Virology* 146(10), 1885–1898. DOI: 10.1007/s007050170040.

Yamashita, Y., Takeuchi, T., Okuyama, M., Sasaki, J., Onodera, K. *et al.* (2014) Development and validation of DNA markers linked to Sdvy-1, a common bean gene conferring resistance to the yellowing strain of soybean dwarf virus. *Breeding Science* 64(4), 404–408. DOI: 10.1270/jsbbs.64.404.

5 Early Detection of Plant Virus Infection Using Multispectral Imaging and Machine Learning

Bruce Grieve[1], Siobain Duffy[2], Mary M. Dallas[3], José T. Ascencio-Ibáñez[4], Vasthi Alonso-Chavez[5], James Legg[6], Linda Hanley-Bowdoin[3] and Hujun Yin[1]

[1]Department of Electrical and Electronic Engineering, University of Manchester, Manchester, UK
[2]Department of Ecology, Evolution, and Natural Resources, Rutgers University, New Brunswick, NJ, USA
[3]Department of Plant and Microbial Biology, North Carolina State University, Raleigh, NC, USA
[4]Department of Molecular and Structural Biochemistry, North Carolina State University, Raleigh, NC, USA
[5]Rothamsted Research, Harpenden, UK
[6]International Institute of Tropical Agriculture (IITA), Dar es Salaam, Tanzania

Introduction

Assuring food security for 8 billion people is one of the most pressing challenges of the 21st century. Climate change-resilient crops like cassava are projected to play a key role in these efforts. However, cassava production in East Africa is limited by RNA viruses that cause cassava brown streak disease (CBSD). CBSD typically causes subtle to no symptoms on stems and leaves, while destroying the root tissue, which means farmers may be unaware their fields are infected until they have a failed harvest. Distressingly, the visible symptoms are so slight that planting material provided by "clean seed" programs may be infected. While molecular diagnostics like PCR can determine if plants are infected, they are too expensive for testing many plants in a field, and testing needs to be done in laboratories that are distant from affected farms. Since there are no cures for CBSD, farmers can only rogue infected plants (remove them from the field before they can spread infection), but models of roguing based on visible symptoms show the method to be ineffective. Because it is difficult to observe infection prior to harvest, we do not know how quickly CBSD spreads in fields due to transient association with whitefly vectors, which has harmed modeling efforts. While CBSD has been spreading in East Africa since it became a regional epidemic in the 2000s, the development of CBSD-resistant cassava clones has been slow, and there is a high risk of spreading to West Africa.

Recently a multinational (USA, UK, and Tanzania) and multidisciplinary team has come together to research and implement advanced sensor engineering and computer science to rapidly determine the infection status of plants in the field in Tanzania. This partnership has been funded by the National Science Foundation (NSF, USA) and Biological and Bioscience Research Council (BBSRC, UK) and blends engineers, biologists, mathematicians and seed system specialists from across the three nations and five institutions, that host the authors of this article. The handheld sensor technology is based on a low-cost active multispectral imager (MSI). The MSI is used to observe plant leaves at a number of wavebands both in the visible and near-infrared areas of the light spectra. These are then interpreted by machine learning models (artificial intelligence) trained on cassava leaf scans. Under laboratory conditions, the MSI detects CBSD infection with 95% accuracy at 28 days post-infection, when plants have no visible symptoms. The team is studying the spread of CBSD and parameterizing agent-based transmission models to assess the efficacy of roguing with different detection methods in the critical fields with large downstream effects within the clean seed propagation system. Finally, using these parameterized models we are simulating smallholder farms over several harvest seasons, looking at how host susceptibility and environmental factors driving vector abundance affect disease pressure, and modeling interventions by farmers and agricultural extension agents to minimize the local and regional spread of CBSD.

Our study will be the first to accurately model the plant-to-plant spread of this emergent and damaging cassava pathogen, and the first to integrate methods with different limits of detection into models for a vegetatively propagated plant pathosystem. We will create both accurate tools for use in cassava and frameworks for employing our technology and models for other economically important pathosystems in vegetatively propagated crops, such as banana/plantain, potato, sweet potato, and yam.

This research offers the potential to significantly impact the food security of people living in areas affected by CBSD and reduce the likelihood of further spread into other regions of sub-Saharan Africa, where cassava is a staple crop for 800 million people. Our team includes Tanzanian cassava researchers involved in the Tanzanian clean seed program which assures a high likelihood of translation of the conclusions of our work to real applications.

Background

Cassava is the second-most important source of calories across sub-Saharan Africa and is a staple crop for 800 million people (Tomlinson *et al.*, 2018). This insurance crop can withstand drought and grow in nutrient-poor soil

Fig. 5.1. Cassava field inspection to rogue infected plants.

and is predicted to be the most resilient staple crop to climate change in sub-Saharan Africa (Jarvis *et al.*, 2012). However, cassava production is limited in sub-Saharan Africa by two viral diseases: the endemic cassava mosaic disease (CMD) and the emergent CBSD. Our knowledge of CMD is much greater than that of CBSD, both because of its presence in African agriculture, since 1894 (Warburg, 1894), and because CMD has obvious symptoms in most affected plants (Mbewe *et al.*, 2020). Both diseases are propagated not only through infected planting material but also during the cropping period by the whitefly vector, *Bemisia tabaci* (Genn.).

When a farmer detects infection in a cassava plant, the main way they can control the spread of the infection is to rogue the plant – remove it from the field – so that it cannot serve as a source of infection for spread to other plants in the field (Fig. 5.1). The striking visual symptoms of CMD alert farmers to the threat without any modern technology (Rey and Vanderschuren, 2017). In contrast, CBSD is most noticeable post-harvest, when large roots that appear unaffected on the outside are cut open, revealing brown necrotic tissue (Fig. 5.2A). The necrotized tissue is inedible, preventing the sale of whole roots for food.

The two cassava viruses together cause an estimated US$ 1 billion annual losses in cassava production (Legg *et al.*, 2006) and severely impact smallholder farmers, as well as emerging cassava businesses (Hillocks and Maruthi, 2015).

Fig. 5.2. Symptoms of CBSD in roots (a), on stems (b), and on leaves (c).

CBSD symptoms in cassava plants are often very subtle, with brown streaks on the stem (Fig. 5.2B shows strong streak symptoms that are apparent later in the growing season), and faint vein-associated chlorosis on lower leaves (Fig. 5.2C). Infected plants are frequently considered healthy by experienced cassava farmers and they can be difficult to recognize for cassava seed producers (McQuaid *et al.*, 2016). This prevents timely roguing, and modeling predicts roguing based on visual symptoms is ineffective for control of CBSD (McQuaid *et al.*, 2016; Ferris *et al.*, 2020). Fields found to be infected at harvest are often highly infected, with some studies finding an average of 85% of roots affected (McSween *et al.*, 2006) and up to 100% of roots reported (Hillocks and Jennings, 2003). Without a means to control CBSD in the field, farmers in affected regions may be forced to wait until resistant cultivars are identified and developed for commercial use, a lengthy process with no guarantee of success (Mero *et al.*, 2021).

CBSD is caused by two related RNA viral species in the *Potyviridae* family, genus *Ipomovirus: Cassava brown streak virus* (CBSV) and *Ugandan cassava brown streak virus* (UCBSV). Like all potyviruses, they have polyadenylated +ssRNA genomes of ~9000 nt (Yang *et al.*, 2021). Both of these cassava brown streak ipomoviruses (CBSIs) are found throughout the area where

Fig. 5.3. The area affected by cassava brown streak disease.

CBSD occurs: Tanzania, first reported in 1936 (Storey, 1936), Kenya, Malawi, Mozambique, Uganda, Burundi, Rwanda, eastern parts of the Democratic Republic of Congo, and the Comoros Archipelago (Shirima *et al.*, 2022). Prior to 2004, CBSD had been confined to lowland coastal and lakeshore regions of East Africa, leading farmers and researchers to hope that the disease could not thrive at higher altitudes. Since 2004, however, CBSD has spread rapidly through parts of East and Central Africa that are more than 1000 m above sea level, and the disease is no longer thought to be bounded by any environment in sub-Saharan Africa (Jeremiah *et al.*, 2015). This changing pattern of epidemiology has been associated with increased abundance of the whitefly vector of CBSIs—*B. tabaci* (Maruthxi *et al.*, 2005). The new spread westwards and southwards has given rise to significant concern amongst stakeholders in Central and West Africa whose food security depends heavily on cassava (Cameroon, Republic of Congo, Gabon, Central African Republic, and Nigeria—the world's largest cassava producer (Shirima *et al.*, 2020)) (Fig. 5.3).

The Epidemiology and Detection of CBSD

Despite CBSD's increasing threat to cassava in East Africa since the mid-2000s, we do not have a clear understanding of CBSD epidemiology (Gashaka *et al.*, 2021). CBSD is now considered a greater constraint to cassava production in East Africa than CMD, but because of its subtle visual symptoms, it cannot be easily traced within fields. Because whiteflies spread CBSIs in a semi-persistent manner, the whiteflies do not become infected and can only spread the viruses for up to 2 days (Njoroge *et al.*, 2017; Donnelly *et al.*, 2020).

Although it has been assumed that the probability of spread from a point source of infection is directly proportional to the distance from the infected plant, this assumption has not been rigorously tested. One of the most detailed studies that tried to monitor spread only scored visual symptoms, and no new plants were scored as infected until 8 weeks after the experiment began (Maruthi *et al.*, 2017). That study showed that plants closer to the infected source plant are more likely to develop symptoms than those farther away (5/6 plants 2 m away developed symptoms compared to ~25% 6 or 8 m away) over 22 weeks.

The more localized nature of CBSD infections should make it easier to control their spread compared to the persistently transmitted cassava mosaic begomoviruses (CMBs) which cause CMD (Mbewe *et al.*, 2020). Theoretically, this should make CBSD more controllable by roguing than CMD (Lapidot *et al.*, 2014). However, the faint symptoms of CBSD have made roguing virtually impossible to implement in the field without additional technology. Enhanced detection methods and accurate models of spread would make roguing a viable control strategy for CBSD and would alert seed producers and farmers to CBSD infection long before harvest.

CBSD's subtle visual symptoms have prompted researchers to develop better detection methods, including PCR, qPCR (Shirima *et al.*, 2017), LAMP assays (Tomlinson *et al.*, 2013), and AI-trained algorithms using digital images of leaves (PlantVillage Nuru app (Mrisho *et al.*, 2020)). The molecular methods perform well, especially on plants with higher titer infections (Tomlinson *et al.*, 2013; Shirima *et al.*, 2017). These titers build up over many weeks of infection, i.e., PCR is not reliable on cassava plants infected with cassava brown streak virus until 6 weeks post-infection (Duff-Farrier *et al.*, 2019). Thus, these expensive screening methods can provide infection information earlier and more accurately than visual inspection, but they may not be reliable until many weeks after a plant is initially infected when it is already serving as a source plant for further infections. The PlantVillage Nuru smartphone app performs poorly at recognizing CBSD from a photo of a single leaf (≤27% accuracy), requiring the analysis of six photos from a single plant (74% accuracy) to exceed the accuracy of a cassava-trained agricultural extension officer (49%) (Mrisho *et al.*, 2020).

Detection Using Multispectral Imaging

Our team has developed a novel tool for the early detection of CBSD that has been verified under laboratory conditions to offer significantly earlier and more sensitive detection versus existing color image-based technologies, such as the PlantVillage Nuru smartphone app (Mrisho *et al.*, 2020). It has also been shown to detect CBSD infection before CBSIs are detectable by PCR (Peng *et al.*, 2022). The partnership has developed a handheld active multispectral imaging (MSI) device (Fig. 5.4) that takes images with a wide range of wavelengths of light (Owomugisha *et al.*, 2021). The images are declared

Fig. 5.4. The MSI scanning a cassava leaf.

healthy or diseased by a machine learning model that has been trained on images of infected and uninfected cassava leaves.

In our laboratory (NCSU), the model predictions were 95% accurate, with a false positive rate <5%, in a susceptible cultivar at 28 days post-infection (dpi), a time when most plants exhibited no symptoms and PCR failed to detect any of the infected plants (Peng *et al.*, 2022). Previous research noted that whiteflies tend to only pick up CBSIs from more heavily infected, higher titer leaves (Maruthi *et al.*, 2005). Therefore, it is projected that the MSI technology can detect infection when titers are so low that the plant is unlikely to spread viruses to whitefly vectors.

The ability to detect rogue-infected plants very early, even before the plant can spread infection, would be transformative for the management of CBSD in cassava clean seed systems. Importantly, the MSI has a very low cost and time investment per sample and could be deployed affordably in more layers of the clean seed system than molecular testing.

Tanzania's Cassava Clean Seed System

The most important place for early and accurate detection of CBSD is in cassava clean seed systems that sell cuttings to smallholder farmers. Tanzania has recently developed and formalized a countrywide cassava seed system

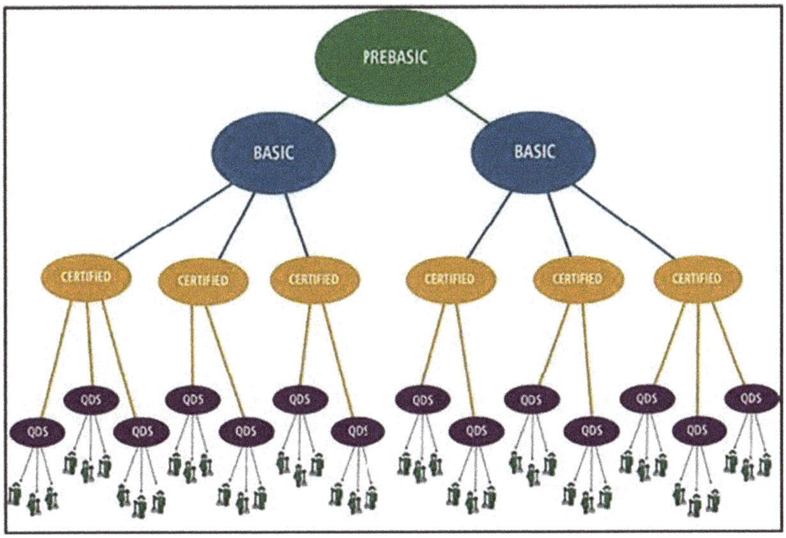

Fig. 5.5. Tanzania's cassava clean seed system has four layers. The Tanzanian Agricultural Research Institute (TARI) manages the pre-basic level, growing cassava from tissue culture to assure it is virus-free before planting in pre-basic screenhouses or fields. Cuttings are distributed to the basic level and cassava seed entrepreneurs (CSEs) propagate those to maximizes the number of stems to be sold to certified-level CSEs. Certified seed CSEs then sell cuttings to quality-declared seed CSEs, who sell cuttings to farmers (Legg *et al.*, 2022).

with a multi-layered structure of seed producers (pre-basic, basic, certified, and quality declared seed, see Fig. 5.5) with legally enforceable certification guidelines for quality standards and inspection by the Tanzania Official Seed Certification Institute (TOSCI).

While TOSCI inspects for several cassava diseases, CBSD is the most significant pathogen of concern for the system and for the farmers who purchase the end products (Tomlinson *et al.*, 2018). Each level provides cassava stems or cuttings to the next layer, where the material is used to establish a new seed field. Certification inspections are conducted by TOSCI twice per season at each of the pre-basic, basic, and certified levels, and during these visit visual assessments are made of the incidences of each of the main cassava diseases and pests, including CBSD and CMD. Molecular testing for CBSIs is only applied at the pre-basic level, as it would be too expensive to do this at the more extensive lower levels of the system. Pre-basic seed represents <1% of the total seed production in the system. Since visual inspection is not a reliable method for detecting CBSD, a proportion of infected plants at a given level in the system will not be identified. That infection will therefore be carried down to the next level of the system and will represent a potential source of further CBSD spread in the newly-established seed production fields. The current guidelines accommodate this reality by allowing cuttings to be propagated to the next level from fields with up to 2% basic and 7% certified cassava plants with CBSD. Quality-declared seed fields can have up to 10% affected plants and still be approved for sale to root-producing farmers (Legg *et al*., 2022). Despite of these allowances, the lack of a fully robust system for detecting CBSD in seed production fields means that seed producers often fail to achieve the minimum standards for CBSD incidence and are therefore prevented from selling their seed.

An increased understanding of the epidemiology of CBSD would benefit this ambitious national system, and the application of field-level technology that increases the accuracy of CBSD detection would have a tremendous impact in increasing the quality of cassava seed, thereby raising overall cassava production in the country and with it, improving the livelihoods of the hundreds of thousands of cassava farmers and their families who depend on cassava for food and income. If the approach can be successfully piloted in Tanzania for cassava, then it will offer enormous potential for wider scaling both to other major-cassava producing countries, as well as for application to other crops.

Developing and Validating MSI for Early CBSD Detection in Cassava Seed Systems

The current position with cassava virus, and other plant disease sensing, using MSI is rapidly developing. During the course of the next 4–5 years, the ongoing NSF/BBSRC research program will address four complementary areas of research. Namely:

- Refine the MSI device and model to accommodate those field conditions where other plant stresses may affect the underlying signals detected by the device.
- Monitor the spread of CBSD in experimental fields with both susceptible and tolerant cassava cultivars.
- Develop epidemiological models describing how different detection methods affect the amount of CBSD in a cassava clean seed system, which could potentially be applied to any vegetatively propagated crop, including the US potato clean seed system. Modelling an improved clean seed system and judicious use of MSI could improve the health of cassava seed produced by smallholder farmers over multiple harvest seasons.
- Produce simple rules of thumb for inspectors, seed producers, and clean seed system stakeholders for the application of MSI diagnostics and the optimal implementation of cassava seed health management strategies.

Conclusion

Our crop sensors and epidemiology research has now reached an exciting milestone, transitioning the technology from the laboratory to practical implementation within an African context. The proposed research has an inherent potential impact, as it focuses on the security of a staple crop for hundreds of millions of people in sub-Saharan Africa. It will provide overdue insight into CBSD epidemiology at the local scale of a devastating pathogen that is moving westward across Africa. The research will also develop models useful for other cassava diseases as well as related pathogens of other vegetatively propagated crops. Our goal is to design and fabricate an in-field MSI system that will ultimately be manufactured at a price point suitable for stakeholders in sub-Saharan Africa, and we will get stakeholder input on design and performance in both Aims 1 and 2.

As the project progresses, the consortium will be disseminating our research results widely, through scientific conferences, regional cassava meetings, and formal publications and through direct interaction with materials developed for Tanzanian CSEs. Our team will also welcome interest from across the community to extend and broaden the research into other geographical regions, disease types and crop species. If that is of interest, please feel free to contact any of the authors.

Discussion Points

1. Compile a table of crops and corresponding viruses that may affect their growth and yields. Include details of the visible and other symptoms of the virus on healthy plants.
2. Study the range of techniques that may be used for detecting crop diseases. These may include invasive laboratory techniques and remote satellite detection methods as well as techniques that may be used in the field. For each technique, consider the potential advantages and disadvantages of their use.
3. Research and write an essay on the possible ways that sensor technologies and epidemiological models may be implemented in sub-Saharan Africa to help control and eradicate the spread of cassava viruses. This could include consideration of who would undertake the measurements, how they would be implemented within the cassava seed system, how this infrastructure could be financed and what opportunities there may be to extend the approach to other similar, virus-affected, crops such as yams and sweet potato.

Further Reading

Legg, J.P., Lava Kumar, P., Makeshkumar, T., Ferguson, M., Kanju, E. *et al.* (2015) Cassava virus diseases: Biology, epidemiology and management. Advances in Virus Research 91, 85–142.

Legg, J.P., Diebiru-Ojo, E., Eagle, D., Friedmann, M., Kanju, E. *et al.* (2022) Commercial sustainable cassava seed systems in Africa. In: Thiele, G., Friedmann, M., Campos, H., Polar, V. and Bentley, J.W. (eds) *Root, Tuber and Banana Food System Innovations*. Springer, Cham, pp. 453–482.

Nichols, R.F.J. (1950) The brown streak disease of cassava: Distribution, climatic effects and diagnostic symptoms. *East African Agricultural Journal* 15, 154–160.

References

Donnelly, R., Sikazwe, G.W. and Gilligan, C.A. (2020) Estimating epidemiological parameters from experiments in vector access to host plants, the method of matching gradients. *PLOS Computational Biology* 16(3), 1–15.

Duff-Farrier, C.R.A., Mbanzibwa, D.R., Nanyiti, S., Bunawan, H., Pablo-Rodriguez, J.L. *et al.* (2019) Strategiesfor the construction of cassava brown streak disease viral infectious clones. *Molecular Biotechnology* 61(2), 93–101.

Ferris, A.C., Stutt, R., Godding, D. and Gilligan, C.A. (2020) Computational models to improve surveillance for cassava brown streak disease and minimize yield loss. *PLOS Computational Biology* 16(7), e1007823.

Gashaka, G., Kanyange, M.C., Mutumwinka, M., Niyonteze, G., Umfuyusoni, J. *et al.* (2021) Changes in the epidemiology of cassava brown streak disease and associated viruses in Rwanda: Occurrence and distribution since 2009. *African Journal of Crop Sciences* 9(10), 1–10.

Hillocks, R. and Jennings, D. (2003) Cassava brown streak disease: A review of present knowledge and research. *International Journal of Pest Management* 49(3), 225–234.

Hillocks, R. and Maruthi, M.N. (2015) Post-harvest impact of cassava brown streak disease in four countries in eastern Africa. *Food Chain* 5(1–2), 116–122.

Jarvis, A., Ramirez-Villegas, J., Herrera Campo, B.V. and Navarro-Racines, C. (2012) Is cassava the answer to African climate change adaptation? *Tropical Plant Biology* 5, 9–29.

Jeremiah, S.C., Ndyetabula, I.L., Mkamilo, G.S., Haji, S., Muhanna, M.M. *et al.* (2015) The dynamics and environmental influence on interactions between cassava brown streak disease and the whitefly, Bemisia tabaci. *Phytopathology* 105(5), 646–655.

Lapidot, M., Legg, J.P., Wintermantel, W.M. and Polston, J.E. (2014) Management of whitefly-transmitted viruses in open-field production systems. *Advances in Virus Research* 90, 147–205.

Legg, J.P., Owor, B., Sseruwagi, P. and Ndunguru, J. (2006) Cassava mosaic virus disease in East and Central Africa: Epidemiology and management of a regional pandemic. *Advances in Virus Research* 67, 355–418.

Legg, J.P., Diebiru-Ojo, E., Eagle, D., Friedmann, M., Kanju, E. *et al.* (2022) Commercial sustainable cassava seed systems in Africa. In: Thiele, G., Friedmann, M., Campos, H., Polar, V. and Bentley, J.W. (eds) *Root, Tuber and Banana Food System Innovations*. Springer, Cham, pp. 453–482.

Maruthi, M.N., Hillocks, R.J., Mtunda, K., Raya, M.D., Muhanna, M. *et al.* (2005) Transmission of *Cassava brown streak* virus by *Bemisia tabaci* (Gennadius). *Journal of Phytopathology* 153, 307–312.

Maruthi, M.N., Jeremiah, S.C., Mohammed, I.U. and Legg, J.P. (2017) The role of the whitefly, *Bemisia tabaci* (Gennadius), and farmer practices in the spread of cassava brown streak ipomoviruses. *Phytopathology* 165(11–12), 707–717.

Mbewe, W., Hanley-Bowdoin, L., Ndunguru, J. and Duffy, S. (2020) Cassava viruses – host jumps, virusrecombination, spread in plant material. In: Records, A. and Ristaino, J. (eds) *Emerging Plant Diseases andGlobal Food Security*. APS Press, St. Paul, Paul, MN, pp. 133–157.

McQuaid, C.F., Sseruwagi, P., Pariyo, A., San, F. and Bosch, F. (2016) Cassava brown streak disease and the sustainability of a clean seed system. *Plant Pathology* 65(2), 299–309.

McSween, S., Walker, T., Salegua, V. and Pitoro, R. (2006) *Economic Impact on Food Security of Varietal Tolerance to Cassava Brown Streak Disease in Coastal Mozambique*. Institute of Agricultural Research of Mozambique, Mozambique.

Mero, H.R., Lyantagaye, S.L. and Bongcam-Rudloff, E. (2021) Why has permanent control of cassava brown streak disease in Sub-Saharan Africa remained a dream since the 1930s? *Infection, Genetics and Evolution* 94, 105001.

Mrisho, L.M., Mbilinyi, N.A., Ndalahwa, M., Ramcharan, A.M., Kehs, A.K. *et al.* (2020) Accuracy of a smartphone-based object detection model, plantvillage nuru, in identifying the foliar symptoms of the viral diseases of Cassava-CMD and CBSD. *Frontiers in Plant Science* 11, 590889.

Njoroge, M.K., Mutisya, D.L., Miano, D.W. and Kilalo, D.C. (2017) Whitefly species efficiency in transmitting cassava mosaic and brown streak virus diseases. *Cogent Biology* 3(1).

Owomugisha, G., Melchert, F., Mwebaze, E., Quinn, J.A. and Biehl, M. (2021) Matrix relevance learning from spectral data for diagnosing cassava diseases. *IEEE Access* 9, 83355–83363.

Peng, Y., Dallas, M.M., Ascencio-Ibanez, J.T., Hoyer, J.S., Legg, J. *et al.* (2022) Early detection of plant virus infection using multispectral imaging and spatial-spectral machine learning. *Scientific Reports-UK* 12(1), 1–15.

Rey, C. and Vanderschuren, H. (2017) Cassava mosaic and brown streak diseases: Current perspectives and beyond. *Annual Review of Virology* 4(1), 429–452.

Shirima, R.R., Maeda, D.G., Kanju, E., Ceasar, G., Tibazarwa, F.I. *et al.* (2017) Absolute quantification of cassava brown streak virus mRNA by real-time qPCR. *Journal of Virological Methods* 245, 5–13.

Shirima, R.R., Legg, J.P., Maeda, D.G., Tumwegamire, S., Mkamilo, G. *et al.* (2020) Genotype by environment cultivar evaluation for cassava brown streak disease resistance in Tanzania. *Virus Research* 286, 198017.

Shirima, R.R., Wosula, E.N., Hamza, A.A., Mohammed, N.A., Mouigni, H. *et al.* (2022) Epidemiological analysis of cassava mosaic and brown streak diseases, and bemisia tabaci in the Comoros Islands. *Viruses* 14(10), 2165.

Storey, H.H. (1936) Virus diseases of East African plants VI. A progress report studies of the diseases of Cassava. *East African Agricultural Journal* 2, 34–39.

Tomlinson, K.R., Bailey, A.M., Alicai, T., Seal, S. and Foster, G.D. (2018) Cassava brown streak disease: Historical timeline, current knowledge and future prospects. *Molecular Plant Pathology* 19(5), 1282–1294.

Tomlinson, J.A., Ostoja-Starzewska, S., Adams, I.P., Miano, D.W., Abidrabo, P. *et al.* (2013) Loop-mediated isothermal amplification for rapid detection of the causal agents of cassava brown streak disease. *Journal of Virological Methods* 191(2), 148–154.

Warburg, O. (1894) *Die Kulturpflanzen Usambaras*. ES Mittler & Sohn, Berlin.

Yang, X., Li, Y. and Wang, A. (2021) Research advances in potyviruses: From the laboratory bench to thefield. *Annual Review of Phytopathology* 59, 1–29.

6 Cassava Brown Streak Disease in Zambia: An Integrated Approach to Managing an Invasive Viral Disease of a Clonal Crop

Chapwa Kasoma

CAB International (CABI), Southern Africa Centre, 5834 Mwange Close, Kalundu, Lusaka, Zambia

Introduction

Cassava (*Manihot esculenta* Crantz family Euphorbiaceae) is Zambia's second most important crop after maize. It is traditionally grown in four of Zambia's 10 provinces and provides dietary caloric needs for an estimated 30% of the country's population (Chikoti *et al.*, 2019). Unfortunately, since 2018, the arrival of the cassava brown streak disease (CBSD), an invasive pest threatens the production of cassava in the country. CBSD is a viral disease of cassava, which is currently considered the most threatening constraint to cassava production in Africa (Muhindo *et al.*, 2020; Ano *et al.*, 2021). It manifests as hard rots in cassava roots that are often accompanied by brown streaks on stems and discolouration of leaves. CBSD has caused substantial yield losses in East, Central and Southern Africa, where over 200 million people depend on the crop for their livelihoods (Elegba *et al.*, 2020). In the wake of climate change, characterized by recurring crop failures, cassava is regarded as a drought insurance crop because of its capacity to withstand sub-optimal agronomic practices including a limited supply of irrigation. Additionally, modelling studies have shown the resilience of cassava over other crops like maize making it ideal for climate adaptation. However, CBSD poses a major challenge to harnessing the potential of the cassava subsector in enhancing food security in Africa.

Origin and Distribution of CBSD

The first report of CBSD was in the early 1930s in the low-altitude regions of Tanzania (Chikoti and Tembo, 2022). It remained latent and restricted to the coastal areas of East Africa for many decades until early 2000 when it emerged more aggressively in the higher altitude inland regions of Uganda, Kenya and Tanzania (Alicai *et al.*, 2007; Legg *et al.*, 2011; Masinde *et al.*, 2016). It has since spread further in East Africa and has extended to several countries in Central and southern Africa. Presently, the countries with confirmed CBSD presence in Africa include Tanzania, Kenya, Uganda, Rwanda, Congo, Madagascar, Gabon, South Sudan, Burundi, Angola, Mozambique, Democratic Republic of Congo, Zimbabwe, Malawi and Zambia (Legg *et al.*, 2011; Masinde *et al.*, 2016; Mulenga *et al.*, 2018; Munguti *et al.*, 2023). In West Africa, various pre-emptive measures including screening for resistance to CBSD in order to prevent its entry and subsequent spread in the world's largest-producing cassava region are ongoing (Elegba *et al.*, 2020; Ano *et al.*, 2021).

In Zambia, CBSD was first reported in 2018 in Kaputa and Chienge districts located in Northern and Luapula provinces, respectively (Mulenga *et al.*, 2018) (Fig. 6.1). In 5 years, the disease has spread to further districts within the two provinces. There are also unconfirmed reports of CBSD-like symptoms in some selected districts of north-western province such as Kapombo.

Damage caused by CBSD and its impacts

The typical symptoms of CBSD on cassava can be identified in three main plant parts namely; the leaves, stems and roots (Fig. 6.2). In some cases, CBSD symptoms have been observed on cassava fruits as well as the insect vector *Bemisa tabaci* on cassava leaves (Fig. 6.3).

CBSD infection on cassava leaves manifests as chlorotic patches or blotches sometimes referred to as mottling. The patches form a characteristic viral mosaic that runs along the primary and secondary leaf veins culminating in a feathery chlorotic pattern. As the name of the disease suggests, CBSD is marked by the appearance of brown streaks on the plant and these typically occur on the stem. The brown streaks resemble lesions that can be more prominently seen on younger cassava stems because of the contrast between the green colour of the stems and that of the brown streaks (Muhindo *et al.*, 2020). Cassava roots are the major edible portion although the leaves are also consumed as a vegetable in selected parts of Africa including Zambia, Malawi and the Democratic Republic of Congo (Achidi *et al.*, 2005). CBSD causes hard, odourless rots in cassava roots known as necrotic rots that render the roots inedible by both humans and animals. This damage can translate into significant losses in yield and quality of produce, which may threaten food security, especially in traditional cassava-growing regions where the crop

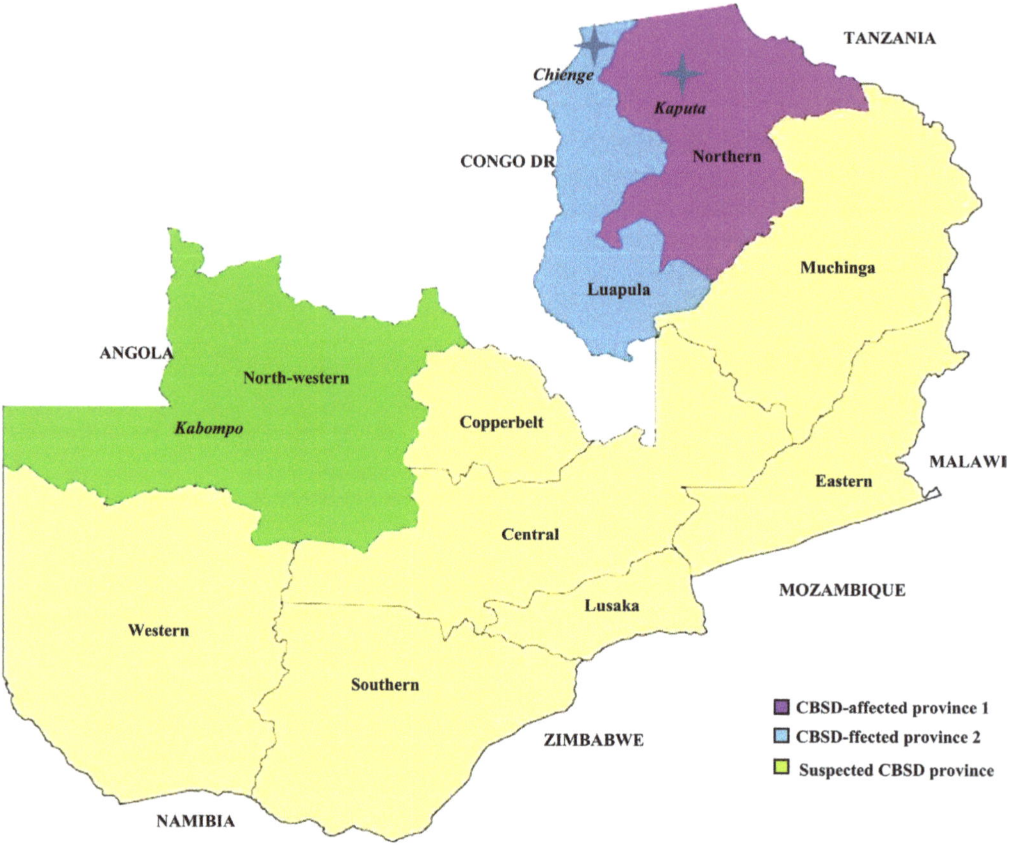

Fig. 6.1. Regions where CBSD was first detected and suspected to be present in Zambia. denotes the districts within the affected provinces where CBSD was first reported.

is the main staple food. CBSD is capable of causing 100% yield losses when susceptible cassava varieties are used coupled with poor agronomic practices.

Presently, the average global yield is 11 t/ha (Ritchie *et al.*, 2022). In Zambia, the average yield for the crop is even lower at 6 t/ha (Kaluba *et al.*, 2022). With the arrival of CBSD, cassava yields are further threatened as both the incidence and severity of the disease continue to increase in the country. As the second most important staple crop after maize in Zambia, cassava production is being promoted for crop diversification and climate change mitigation. Particularly in the past 10 years, there have been marked developments within the cassava subsector of Zambia in an effort to promote production and diverse utilization of the crop to enhance the agricultural sector and drive economic growth. These developments and other sectoral plans are documented in the country's National Cassava Sector Development Strategy of 2020–2024 (Bundi *et al.*, 2022). The presence of CBSD offsets some of the progress made to promote cassava and its utilization as a crop of strategic importance for the country.

Fig. 6.2. Symptoms of cassava brown streak disease of varying severity on cassava leaves, stem and root.

Fig. 6.3. (a) Cassava brown streak disease symptoms on cassava fruits, (b) *Bemisia tabaci* on cassava leaves.

Biological and Ecological Factors Influencing the Invasiveness of CBSD

The causal viruses of CBSD known as cassava brown streak viruse (CBSV) and Ugandan cassava brown streak virus (UCBSV) belong to the family Potyviridae and the genus Ipomovirus. Both viruses are transmitted through the vector *Bemisia tabaci* Gennadius (Hemiptera: Aleyrodidae), over short distances such as within a cassava field or across fields that are in close proximity with each other (Maruthi *et al.*, 2005). Transmission of both CBSV and UCBSV through *B. tabaci* is in a semi-persistent manner unlike the persistent transmission of the CMD causative virus by the same vector (Legg *et al.*, 2011). Across longer distances, CBSD spread happens when cassava cuttings that are infected with either or both causal viruses are used to propagate new fields of the crop, thereby transferring viral inoculum from one field to the next (Patil *et al.*, 2015). Within field transmission of CBSV through aphids is also reported (Ateka *et al.*, 2017). Studies suggest that UCBSV is more prevalent than CBSV in Africa, although the two viruses cause the same characteristic symptoms in cassava. CBSV is also more genetically diverse and is believed to generally cause more conspicuous foliar symptoms on cassava than UCBSV (Mulenga *et al.*, 2018). Further studies are required to characterize the nature of damage caused by the two CBSVs to inform improved management methods. Efficient and cost-effective methods of inoculation for the transmission of CBSD causal viruses using chip bud grafting have been demonstrated (Wagaba *et al.*, 2013). This will enable improved methods of laboratory and greenhouse studies on CBSD aetiology and biology.

Current Efforts to Manage CBSD in Zambia

In Zambia, CBSD is fairly new compared to some of her neighbouring countries that have had the disease for much longer such as Malawi, Tanzania and the Democratic Republic of Congo. For most farmers that have experienced the disease on their crop, CBSD has elicited uncertainty of what is damaging their cassava crop and how to manage it. Consequently, efforts to manage the disease at the farmer level have been limited. Additionally, some methods for managing CBSD in Zambia are either not yet available or are unfamiliar to farmers in the current CBSD-affected regions of the country. However, in the 5 years that CBSD has been present in Zambia, there have been efforts to mitigate the spread of the disease that have also considered lessons from other countries grappling with the problem.

Following the first report of CBSD presence in Zambia by Mulenga *et al.* (2018), pre-breeding efforts to identify suitable cassava varieties with appreciable levels of CBSD tolerance were among the first national responses to the disease. The Zambia Agricultural Research Institute (ZARI) in collaboration with the International Institute for Tropical Agriculture (IITA) embarked on germplasm screening aimed at identifying genotypes with tolerance to the disease. Owing to the complex nature of the genetics that govern resistance to CBSD in cassava and the sole reliance on conventional breeding techniques, these efforts are still ongoing, although with slow progress. Learning from the experience of Uganda, breeding for resistance to CBSD can take many years before a fairly acceptable variety can be achieved. It took over a decade to develop the first variety with appreciable levels of CBSD tolerance in Uganda (Kawuki *et al.*, 2016).

Besides the ongoing pre-breeding and breeding efforts, the need for additional approaches to curb the spread of CBSD in Zambia became more apparent as the disease began to spread rapidly to new districts. In 2021, when nearly half of the farming camps in Kaputa and Chienge, the districts in which the disease was first detected had reported the presence of CBSD, CABI embarked on a CBSD awareness campaign. The campaign aimed to stimulate a nation-wide response to mitigate the spread of CBSD and its impacts in the country. Prior to the implementation of the campaign, a survey targeting the most affected districts and one with reported CBSD-like symptoms was conducted to determine the prevalence of CBSD and farmers' levels of awareness about the disease. The survey provided valuable baseline information on CBSD, cassava production practices and farmers' circumstances in these areas (Bundi *et al.*, 2022). A focus group discussion component was also incorporated in the survey and this provided gender-responsive insights into farmers' cassava production and resources management profiles in Kaputa, Chienge and Kabombo districts. Notably, women and youth in these communities were more involved in field-level cassava production activities. For the activities related to the management of resources, men's participation was found to increase compared to their participation in production-level activities (Figs 6.4 and 6.5). These factors were important when designing the awareness campaign on CBSD to make it gender-inclusive by soliciting the participation of women in gatherings on CBSD awareness. This follows

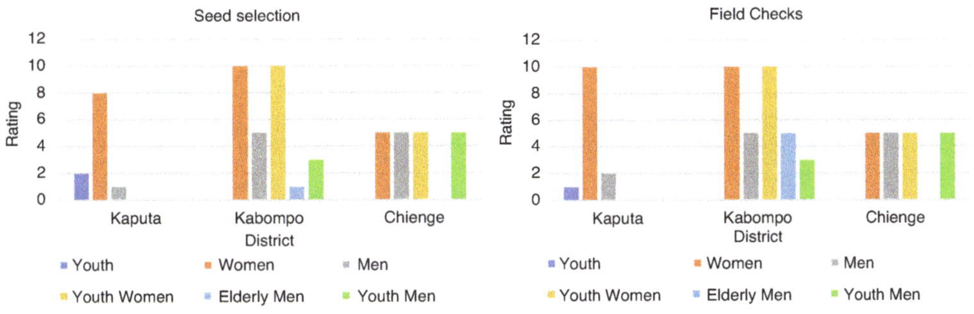

Fig. 6.4. Production profile—Participation of various gender groups of Kaputa, Kabompo and Chienge communities in seed selection and field checks in cassava fields.

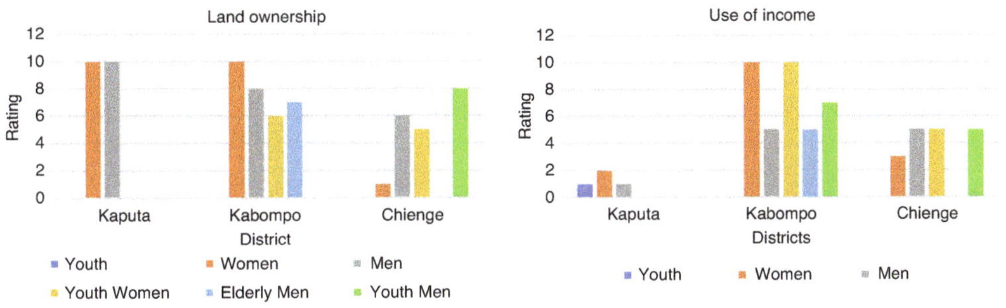

Fig. 6.5. Resources profile—Participation of various gender groups of Kaputa, Kabompo and Chienge communities in land ownership and use of income from cassava produce.

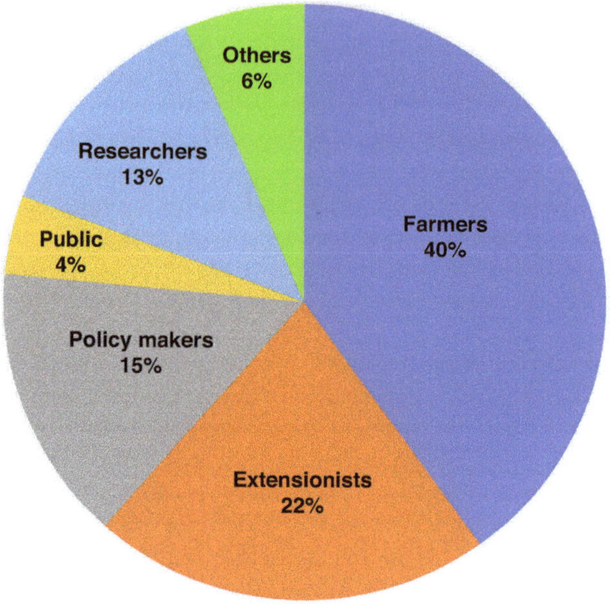

Fig. 6.6. Prioritization of recipient target groups for focussing key messages on mitigation of CBSD spread in Zambia.

from a low participation of women in community gatherings and meetings in general. These factors are also important considerations in the design of a gender-responsive and certified propagule system for cassava.

Stakeholder consultation was also conducted and it provided insights on the design and implementation of the awareness campaign, which was actively launched in November 2021 and is currently ongoing. Outputs of the stakeholder meeting helped to prioritize the target recipients of important messages on mitigating CBSD spread and to identify the appropriate modes of communication to recipients (Fig. 6.6). So far, the awareness campaign has covered some of the most affected districts in Zambia's traditional cassava-growing provinces including northern, western, north-western and western provinces. Key messages on methods to mitigate the spread and impact of CBSD have been shared with farmers, extensionists, policy makers, regulators, media and the general public. Modes of information sharing included meetings, radio programmes, plant health rallies, trainings, short messaging system (SMS), short videos, jingles, television adverts and distribution of various information, education and communication (IEC) materials such as posters and flyers. The content in the key messages was prepared in English and translated into four local languages commonly used in the four target provinces including Tabwa, Kaonde, Lozi, and Luvale. Additionally, local radio stations were used to host the radio programmes and phone in sessions in the target provinces.

Sharing of key messages through the awareness campaign on CBSD revealed the need for the provision of disease-free cassava cuttings to farmers in the CBSD-affected regions. This arose when farmers provided feedback following the key messages shared with them on curbing CBSD spread. Particularly, the key message on ensuring the use of disease-free cassava planting material in new cassava fields was met with questions on where to source such planting material. Consequently, CABI embarked on a collaborative intervention involving the Disaster Management and Mitigation Unit (DMMU) under the Office of the Vice President of the Republic of Zambia and other private and public sector organizations to procure and distribute disease-free cuttings to farmers in five of the most affected northern districts. In addition, CABI enrolled 16 lead farmers in an official cassava propagule production training to initiate local production of disease-free cassava cuttings and promote local access by other farmers in the CBSD-affected areas.

Prospects for sustainable management of CBSD

Considering that CBSD has been present for a long time in several eastern and southern African countries, Zambia can learn valuable lessons and build on the efforts of these countries in the management of CBSD. In east Africa for instance, there are notable efforts in the area of breeding for tolerance against CBSD involving diverse sources of germplasm. Extensive screening to identify CBSD-resistant germplasm and development of CBSD breeding populations derived from exotic and local sources has been conducted (Kawuki *et al.*, 2016). This has enabled the identification of promising breeding material that countries like Zambia with a limited spread can exploit to develop their own local breeding programmes. Currently in Zambia, there is no cassava variety that has been validated as resistant or tolerant to CBSD that farmers can readily use (Chikoti and Tembo, 2022). Conventional breeding often can take a long time for appreciable gains to manifest and therefore building on the breeding efforts in Uganda of more than 17 years suggests a logical approach to breeding against a new disease. Further notable breeding efforts have assessed the response of advanced cassava genetics to field infection by CBSD (Mukiibi *et al.*, 2019). This led to the identification of promising superior cassava clones for downstream breeding against CBSD, which can be combined with locally adapted cassava germplasm and integrated into ongoing CBSD-breeding efforts in Zambia.

Another promising approach to alleviate the spread of CBSD in Zambia is community phytosanitation, which was used to eradicate the presence of the disease in Uganda when it first appeared in the country (Jameson, 1964). Under localized and varying levels of disease incidence, phytosanaitation can be applied effectively to protect disease-free regions, prevent a spike in low disease incidence areas and reduce incidence in the medium- to high-CSBD affected areas. Out of four traditional cassava-growing provinces in Zambia, CBSD's presence is officially confirmed in the Northern and Luapula provinces. It is suspected that the disease is present in selected areas of north-western and western provinces. This localized and sporadic prevalence of CBSD within the affected provinces creates an opportunity to design a tailored community phytosanitation response for mitigating disease impacts. Success of community phytosanitation, however, requires multi-stakeholder awareness that should enhance coordinated action.

Presently in Zambia, there is still a general paucity of information on CBSD despite the awareness efforts so far implemented. This is because of the limited span of time within which the awareness campaign has been implemented as well the targeted nature of most activities, which focussed primarily on the most affected districts and value-chain actors. Thus, CBSD awareness-raising campaigns are still required both to reach a wider audience and to reinforce key messages that will elicit meaningful action from stakeholders. An assessment is underway to determine the level of impact of the CBSD awareness campaign and inform the next steps that should ensure increased impact.

The absence of a formal, functional, accessible and certified propagule system for cassava was foreseen as one of the major impediments to the use of disease-free planting materials by farmers. This implies that prospects exist for mitigating CBSD impacts through a regulated cassava propagule system such as the one in operation for seed maize in Zambia. A reliable propagule system would provide a credible source for recommended cassava planting material for farmers. Establishing a certified propagule system involves various activities and the participation of various multi-sectoral partners including farmers, extensionists, researchers, government regulatory institutions and private sector institutions. Collaborative action from identified stakeholders is possible when there is a unified purpose underlined by a good understanding of the devastating nature of CBSD to the cassava sub-sector as evidenced by the experience in many affected countries.

The application of advanced molecular tools in agriculture presents future prospects for dealing with CBSD, although biotechnology is still a subject of considerable contention in Zambia and other countries in Africa. Among the promising transgenic approaches, host-induced gene silencing (HIGS), is favoured because the resulting plants do not produce a new protein but resistance is specific to the targeted pathogen (Collinge and Sarrocco, 2021). In the particular case of CBSD, post-translational gene silencing (PTGS) has been demonstrated to successfully enhance disease resistance in cassava grown in field and greenhouse trials (Ogwok *et al.*, 2012; Vanderschuren *et al.*, 2012).

Unfortunately, biotechnology techniques in their entirety have limited application in Zambia as regulated by the Biosafety Act No. 10 of 2007 (UNEP-LEAP, 2007). Currently, this Act prohibits the application of transgenesis for crops of strategic importance and this includes cassava, which is Zambia's second most important staple (Chikoti *et al.*, 2019). However, aspects of biotechnology that are permissible by the act, such as viral strain diagnosis, marker-assisted selection in breeding and investigations into the nature of disease resistance that do not result in genetically modified crops can still be applied to fast-track progress in research.

Overall, more successful management of CBSD in Zambia and similar agro-ecologies entails a multi-faceted, multi-sectoral, collaborative and well-coordinated approach. This should integrate various methods of disease control preferably in the order of cultural, host-resistance and biocontrol as suggested by Collinge and Sarrocco (2021). This should enable an effective, sustainable, inclusive and nature positive response to the CBSD challenge and the growth of the cassava subsector for improved food security.

Conclusion

The invasive CBSD is among the most serious biotic challenges faced by cassava farmers in sub-Saharan Africa (SSA). In Zambia, the spread of the disease is exacerbated by many factors. Notably, the traditional farming systems that promote the exchange of planting material and limited application of recommended agronomic practices by farmers owing to the informal status of the cassava value chain. There are prospects for effective management of CBSD to mitigate spread and offset substantial yield losses. However, there are prerequisites to be met for the identified management options to effectively address the management of the disease. Key among them is the need to develop suitable tolerant cassava cultivars for use by farmers and to implement a collaborative and coordinated community phytosanitation plan that is anchored on adequate information dissemination to stakeholders. Overall, CBSD presents an opportunity to establish a functional, inclusive, sustainable and certified propagule system for cassava that will integrate modern technology in agriculture to ensure the viability of the cassava sub-sector.

Discussion Points

1. List the factors that contribute to the rapid spread of CBSD under tropical and sub-tropical conditions.
2. Describe the damage caused by CBSD to cassava and its impacts on the cassava subsector, food security and trade.
3. What are the characteristics of a functional propagule system for cassava and how can it be used to mitigate the impacts of CBSD in an SSA country like Zambia?
4. Using some of the CBSD management methods discussed in this plant health case and others from literature, describe an integrated approach for effectively managing an invasive pest of a clonal crop of strategic importance.

Further Reading

Evidence note highlights impacts and management strategies for cassava brown streak disease in Zambia. Available at: https://blog.plantwise.org/2022/07/27/evidence-note-highlights-impacts-and-managementstrategies-for-cassava-brown-streak-disease-in-zambia/ (accessed 16 January 2023).

CABI joins forces with the Zambia government to help curb the spread of cassava brown streak. Available at: https://www.cabi.org/news-article/cabi-joins-forces-with-zambian-government-to-help-curb-spread-ofdevastating-cassava-brown-streak-disease/ (accessed 5 March 2023).

CBSD campaign takes hold. Available at: https://blog.invasive-species.org/2021/11/22/campaignto-tackle-cassava-brown-streak-disease-in-zambia-takes-hold-after-workshop/ (accessed 21 April 2023).

References

Achidi, A.U., Ajayi, O.A., Bokanga, M. and Maziya-Dixon, B. (2005) The use of cassava leaves as food in Africa. *Ecology of Food and Nutrition* 44(6), 423–435. DOI: 10.1080/036702405003487711.

Ano, C.U., Ochwo-Ssemakula, M., Ibanda, A., Ozimati, A., Gibson, P. *et al.* (2021) Cassava brown streakdisease response and association with agronomic traits in elite Nigerian cassava cultivars. *Frontiers in Plant Science* 12. DOI: 10.3389/fpls.2021.720532.

Ateka, E., Alicai, T., Ndunguru, J., Tairo, F., Sseruwagi, P. *et al.* (2017) Unusual occurrence of a DAG motif in the Ipomovirus cassava brown streak virus and implications for its vector transmission. *PLoS ONE* 12(11), e0187883. DOI: 10.1371/journal.pone.0187883.

Alicai, T., Omongo, C.A., Maruthi, M.N., Hillocks, R.J., Baguma, Y. *et al.* (2007) Re-emergence of cassava brown streak disease in Uganda. *Plant Diseases* 91(1), 24–29. DOI: 10.1094/PD-91-0024.

Bundi, M., Kasoma, C., Mbugua, F., Williams, F., Tambo, J. *et al.* (2022) Cassava Brown Streak Disease: An Evidence Note on Impacts and Management Strategies for Zambia. Available at: https://www.cabi.org/wp-content/uploads/Cassava-Brown-Streak-Disease-Evidence-Note-2022.pdf (accessed 21 May 2023).

Chikoti, P.C. and Tembo, M. (2022) Expansion and impact of cassava brown streak and cassava mosaic diseases in Africa: A review. *Frontiers in Sustainable Food Systems* 6, 2022. DOI: 10.3389/fsufs.2022.1076364.

Chikoti, P.C., Mulenga, R.M., Tembo, M. and Sseruwagi, P. (2019) Cassava mosaic disease: A review of a threat to cassava production in Zambia. *Journal of Plant Pathology* 101, 467–477. DOI: 10.1007/s42161-019-00255-0.

Collinge, D.B. and Sarrocco, S. (2021) Transgenic approaches for plant disease control: Status and prospects 2021. *Plant Pathology* 71, 207–235. DOI: 10.1111/ppa.13443.

Elegba, W., Gruissem, W. and Vanderschuren, H. (2020) Screening for resistance in farmer-preferred cassava cultivars from Ghana to a mixed infection of CBSV and UCBSV. *Plants* 9(8), 1026. DOI: 10.3390/plants9081026.

Jameson, J.D. (1964) Cassava mosaic disease in Uganda. *East African Agricultural and Forestry Journal* 30(3), 208–213. DOI: 10.1080/00128325.1964.11661927.

Kaluba, P., Mwamba, S., Moualeu-Ngangue, D.P., Chiona, M., Munyinda, K. *et al.* (2022) Performance of cassava under lime, fertilizer, and legume intercropping on exhausted land in northern Zambia. *International Journal of Agronomy* 1–10, 1–17. DOI: 10.1155/2022/3649355.

Kawuki, R.S., Kaweesi, T., Esuma, W., Pariyo, A., Kayondo, I.S. *et al.* (2016) Eleven years of breeding efforts to combat cassava brown streak disease. *Breeding Science* 66(4), 560–571. DOI: 10.1270/jsbbs.16005.

Legg, J.P., Jeremiah, S.C., Obiero, H.M., Maruthi, M.N., Ndyetabula, I. *et al.* (2011) Comparing the regional epidemiology of the cassava mosaic and cassava brown streak virus pandemics in Africa. *Virus Research* 159(2), 161–170. DOI: 10.1016/j.virusres.2011.04.018.

Maruthi, M.N., Hillocks, R.J., Mtunda, K., Raya, M.D., Muhanna, M. *et al.* (2005) Transmission of cassava brown streak virus by *Bemisia tabaci* (Gennadius). *Journal of Phytopathology* 153(5), 307–321. DOI: 10.1111/j.1439-0434.2005.00974.x.

Masinde, E.A., Ogendo, J.O., Maruthi, M.N., Hillocks, R., Mulwa, R.M.S. *et al.* (2016) Occurrence and estimated losses caused by cassava viruses in Migori County, Kenya. *African Journal of Agricultural Research* 11(24), 2064–2074. DOI: 10.5897/AJAR2016.107863.

Muhindo, H., Yasenge, S., Casinga, C., Songbo, M., Dhed'a, B. *et al.* (2020) Incidence, severity and distribution of Cassava brown streak disease in north-eastern Democratic Republic of Congo. *Cogent Food and Agriculture* 6(1), 1789422. DOI: 10.1080/23311932.2020.1789422.

Mukiibi, D.R., Alicai, T., Kawuki, R., Okao-Okuja, G., Tairo, F. *et al.* (2019) Resistance of advanced cassava breeding clones to infection by major viruses in Uganda. *Crop Protection* 115, 104–112. DOI: 10.1016/jcropro.2018.09.015.

Mulenga, R.M., Boykin, L.M., Chikoti, P.C., Sichilima, S., Ng'Uni, D. *et al.* (2018) Cassava brown streak disease and Ugandan cassava brown streak virus reported for the first time in Zambia. *Plant Disease* 102(7), 1410–1418.

Munguti, F.M., Nyaboga, E.N., Kilalo, D.C., Yegon, H.K., Macharia, I. *et al.* (2023) Survey of cassava brown streak disease and association of factors influencing its epidemics in smallholder cassava cropping systems of coastal Kenya. *Frontiers in Sustainable Food Systems* 6, 1015315. DOI: 10.3389/fsufs.2022.1015315.

Ogwok, E., Ilyas, M., Alicai, T., Rey, M.E.C. and Taylor, N.J. (2012) Comparative analysis of virus-derived small RNAs within cassava (Manihot esculenta Crantz) infected with cassava brown streak viruses. *Virus Research* 215, 1–11. DOI: 10.1016/j.virusres.2016.01.015.

Patil, B.L., Legg, J.P., Kanju, E. and Fauquet, C.M. (2015) Cassava brown streak disease: A threat to food security in Africa. *Journal of General Virology* 96(5), 956–968. DOI: 10.1099/vir.0.000014.

Ritchie, H., Rosado, P. and Roser, M. (2022) *Crop Yields*. OurWorldInData.Org. Available at: https://ourworldindata.org/crop-yields (accessed 29 July 2023).

UNEP-LEAP (United Nations Environment Programme-Law and Environment Assistance Platform) (2007) Biosafety Act, 2007 (No.10 of 2007). Available at: https://leap.unep.org/countries/zm/national-legislation/biosafety-act-2007-no-10-2007 (accessed 26 July 2023).

Vanderschuren, H., Moreno, I., Anjanappa, R.B., Zainuddin, I.M. and Gruissem, W. (2012) Exploiting the combination of natural and genetically engineered resistance to cassava mosaic and cassava brown streak viruses impacting cassava production in Africa. *PLoS One* 7(9), e45277. DOI: 10.1371/journal.pone.0045277.

Wagaba, H., Beyene, G., Trembley, C., Alicai, T., Fauquet, C.M. *et al.* (2013) Efficient transmission of cassava brown streak disease viral pathogens by chip bud grafting. *BMC Research Notes* 6, 516. DOI: 10.1186/1756-0500-6-516.

Part II: Bacterial Diseases in Plant Health

Bacterial pathogens pose an ongoing and evolving threat to global plant health, agriculture, and food security. This section highlights their remarkable diversity and adaptability through five detailed case studies covering various crops, regions, and pathogen strategies. The chapters examine both longstanding diseases that continue to challenge growers and newly identified threats discovered through modern molecular techniques. From the tumour-causing *Agrobacterium* species, which have revolutionized our understanding of horizontal gene transfer, to the highly destructive *Ralstonia solanacearum* species complex with its broad host range, these cases show how bacterial pathogens exploit plant vulnerabilities and environmental factors to infect. The recent appearance of *Pseudomonas syringae* pv. actinidiae on kiwifruit offers a rare chance to observe pathogen evolution and host adaptation in real time, illustrating how bacterial populations can quickly diversify into different biovars with varying levels of virulence.

A common theme across these chapters is how climate change and globalisation significantly influence disease patterns. Variations in crown gall outbreaks in North African vineyards correspond with climate changes, while the introduction of warm-adapted *R. pseudosolanacearum* strains into Northern and Central Europe highlights the vulnerability of temperate farming systems to tropical pathogens. Additionally, bacterial canker diseases affecting stone fruits and kiwifruit have expanded geographically, facilitated by international plant trade and environmental conditions that support pathogen persistence. These trends emphasize the importance of improved surveillance, diagnostics, and flexible strategies to combat the evolving risks posed by bacterial diseases.

Managing bacterial plant diseases involves unique challenges that set them apart from other pathogen groups. The ability of bacteria to persist in soil, water, and asymptomatic host tissues, along with their capacity for rapid genetic exchange and adaptation, reduces the effectiveness of traditional control methods. Although antibiotics have been used in some agricultural systems to control bacterial diseases, their application raises serious concerns about the development and spread of antibiotic resistance genes, potential disruption of beneficial microbial communities, and environmental contamination. The final chapter of this section critically evaluates both the advantages and risks of using antibiotics in plant cultivation, emphasising the need for sustainable alternatives that do not endanger environmental or human health.

This section's case studies offer vital insights into these complex systems, giving readers the scientific basis to tackle current and emerging bacterial threats to plant health with evidence-based, sustainable strategies that integrate agricultural productivity and environmental care.

7 Crown Gall Disease: A Growing Threat to Fruit Tree Health and Productivity (A Case Study from North Africa)

MERIAM BOURI[1,2] AND ALI RHOUMA[1,2]

[1]*Department of Genetics and Bioengineering, Faculty of Engineering, Yeditepe University, Kayisdagi St., 34755, Istanbul, Turkey*
[2]*Prima, Nexus 2 Building, Carrer Jordi Girona 29, 2nd floor 2A, Barcelona, Spain*

Introduction

Agriculture and food security are highly vulnerable to climate change. Unpredictable variations in weather conditions such as changing rainfall patterns, heat waves, droughts, and floods affect crop yields dramatically.

Scientific evidence has particularly identified plant health issues related to climate change. As abiotic conditions undergo changes, pathogens respond with evolutionary adaptation to novel environments that may lead to the emergence of new diseases or recurrent outbreaks of previously controlled ones. Furthermore, variation in weather conditions impacts plant immune system, and therefore, affects host susceptibility to pathogens and pests (Roussin-Léveillée *et al.*, 2024). Mediterranean agriculture is not exempt from the impact of climate change. In contrast, agricultural systems in the region are continuously experiencing challenging phytosanitary concerns related to climate change. For instance, *Xylella fastidiosa*, historically known as an endophytic bacterium native to Central America, has been detected recently in Italy, where it was responsible for the rapid decline in olive tree plantations (Morelli *et al.*, 2021). Since then, *X fastidiosa* has officially been described as a source of worldwide concern due to global warming (Godefroid *et al.*, 2019). Although intensive national and international actions have been undertaken by various organizations, the disease has, at least in part, been prevented from establishing in the Near East and North Africa (NENA) countries—despite favorable weather conditions, frequent trade, and the high prevalence of olive orchards and insect vectors (Bouri *et al.*, 2023). Several serious diseases are threatening olive trees and other fruit trees in North Africa. Crown gall is one of the most serious diseases affecting nursery production of fruit trees in the region. Once infected, the plant continues to develop galls although the pathogen is no longer present. Therefore, prophylactic solutions, based mainly on early detection of the disease for pathogen-free nurseries and pathogen-free rootstocks, are highly recommended for successful crown gall control. Efficient diagnosis may require very specific and accurate identification tools at the molecular level. This can be achieved only in the presence of a solid knowledge about the genetic diversity of the pathogen communities in relevant biotopes. The relevance of studying crown gall in North Africa is twofold. First, the disease has shown a troubling recent emergence in this region, especially in crops like grapevines and stone fruit trees. Second, climate change may be altering the distribution and aggressiveness of the pathogen, as warmer temperatures, increased humidity in nurseries, and shifting agricultural zones create more favorable environments for disease establishment. Despite the global significance of crown gall disease, the North African context presents distinct challenges—such as fragmented surveillance systems, underreported outbreaks, and limited region-specific data—that demand focused investigation. This document presents an overview for past and current situation for crown gall in North Africa. Insightful information are collected from several studies that have been conducted in the region over the last few decades in order to discuss the state of the art about this disease in the context of climate change, identify critical knowledge gaps, and highlight urgent research priorities for sustainable disease control.

Rootstocks Production Worldwide and Crown Galls

Crown gall causes important economic losses both in nurseries and orchards. Therefore, crown gall disease is challenging for both growers and nursery owners worldwide. The disease is of a particular economic importance in pome, stone fruits, and grapes sectors.

Global stone fruits production has almost reached 615.7 thousand metric tons in 2022 (Shahbandeh, 2024). The worldwide total value of pome fruit production amounts to more than 83.7 million tons, while the almond production was estimated at 1.45 Million (USDA, 2024). The disease threat to stone and pome fruit trees is particularly observed on their corresponding rootstocks: GF677, Myrobolan and Garnem. Garnem was the most susceptible, Myrobolan was moderately susceptible and GF 677 was the least susceptible to crown gall disease infection (Horuz *et al.*, 2018).

Global table grape production was estimated at 28.09 million metric tons this year. Among North African countries, Egypt, with 6% of the total production value, is ranked sixth world producer (USDA, 2024).

These crops – particularly stone fruits and grapes – play a critical role in the agricultural economies of several North African countries, contributing substantially to export earnings, employment, and rural livelihoods. For instance, grapes are the second largest fruit crop in Egypt, with a volume of

73.67 thousand tons as an average of exportation for the period 2008–2018 (Abd El Menem *et al.*, 2022). Consequently, any factor threatening the viability of these sectors, such as crown gall disease, has direct implications for food security, economic stability, and farmer resilience in the region.

History of the Disease

Although the isolation of the crown gall disease was first reported from grapevines in Naples, Italy (1897) by Fridiano Cavara and was named *Bacterium tumefaciens* (Bergey *et al.*, 1923), the first scientific description of galls on grapevines is attributed to Fabre and Dunal. The origin of the disease is believed to date to biblical times on trees and grapevines galls or nodules (Lacroix and Citovsky, 2013). In 1907, the name was changed to *Phytomonas tumefaciens* (Smith & Townsend). Later on, the crown gall disease has been described on fruit trees (mainly apple and stone fruit trees) and the name of the causal agent was changed to *Agrobacterium tumefaciens* (Smith and Townsend, 1907). Several other names have been applied to this phytopathogen that caused the disease wherever grapevines and woody crops were cultivated. By the late 1940s, the discovery of the tumor-inducing factor (extrachromosomal element termed the Ti plasmid) launched plant genetic engineering with *A. tumefaciens* (Kado, 2014). This inter-kingdom transfer from bacteria to plants occurs via a type IV secretion system (T4SS) (Gelvin, 2010; Lacroix and Citovsky, 2013). Agrobacteria are also using some other genes which are involved in the catabolism of opines, to assist the infection by the T-DNA and bacterium/bacterium conjugation-mediated plasmid transfer (Bhatia and Bera, 2015). Once the T-DNA is integrated into the plant genome, an overproduction of the plant hormones auxin and cytokinin is lunched and overgrowths on the infection site are resulted. Although artificial inoculations could reproduce galls on at least 643 different plan species (De Cleene and De Ley, 1976), the number of naturally susceptible plants to crown gall agents is still much lower (Fig. 7.1).

Looking at the chronological history of this disease, from the finding and isolation to the revolutionary research in modern technology of plant transformation, agrobacteria responsible for crown gall are unusual pathogens. They could develop methods enhancing its pathogenicity to overcome host's defences and microbial competitors along millions years of co-evolutionary process. Although pioneering molecular research has turned crown gall pathogens into a useful tool in plant transformation, the disease continues to be responsible for serious economic losses around the world.

Biology of Crown Gall Disease

Crown gall disease, caused primarily by *A. tumefaciens* (now reclassified into several species within the *Rhizobium* genus), follows a unique disease cycle that involves both biological and environmental factors. The cycle includes

Fig. 7.1. Gall formations induced after artificial inoculation with A. fabrum C58 on a carrot disc (a) and a stem of almond (b).

several key stages (Fig. 7.2). The cycle begins when the bacterium enters the plant through fresh wounds caused by grafting, pruning, insect feeding, or mechanical injury. Upon entry, the bacterium transfers a segment of its tumor-inducing (Ti) plasmid DNA (T-DNA) into the genome of the plant cell, triggering uncontrolled cell division through the production of auxins and cytokinins (Gelvin, 2010). This results in the formation of characteristic galls, typically at the crown or root zone of the plant. These galls interfere with water and nutrient uptake, reducing plant vigor and productivity. As the gall tissues decay, agrobacteria can be released into the surrounding soil, where it can survive saprophytically for many years, particularly in moist nursery environments (Krimi *et al.*, 2002). Dissemination of the pathogen occurs via infected planting material, contaminated tools, soil, irrigation water, or even through insect vectors that cause fresh wounds. The disease is often difficult to eradicate because of the pathogen's long-term survival in soil and the asymptomatic nature of infected nursery stock, making clean propagation material and resistant rootstocks essential for effective management.

Host susceptibility, particularly at the rootstock level, plays a key role in the epidemiology of crown gall. Susceptibility varies widely across genotypes, influencing the severity of outbreaks, especially in nursery settings where infection conditions are optimal. Grape rootstocks varieties such as "779P", "99R", "1447P", "Rupestris du lot", "110R", "Freedom", and "41B" are reported to be susceptible to crown gall. Other varieties including "1103P", "Teleki 5C", "420A", "Golia", and "5BB" exhibited moderate susceptibility.

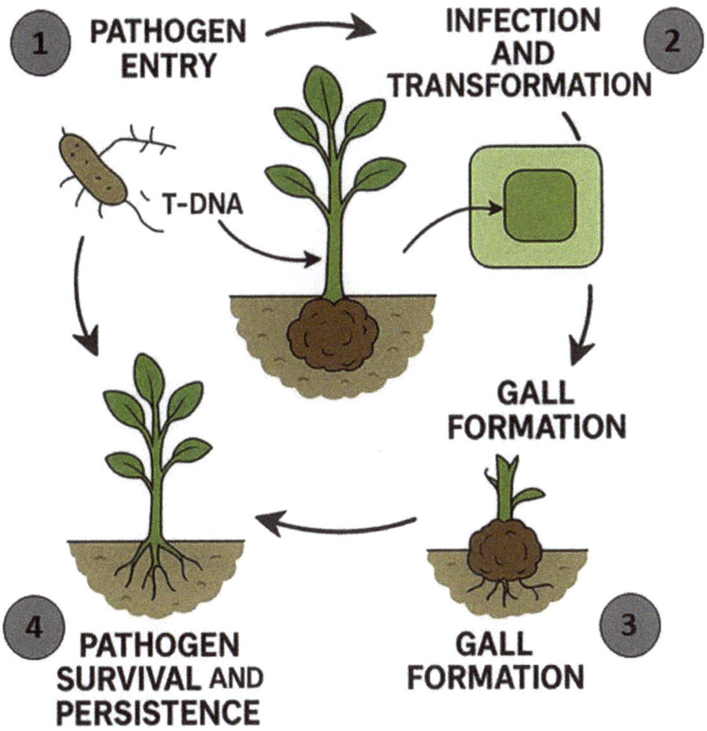

Fig. 7.2. Crown gall disease cycle.

Where Does Crown Gall Disease Occur in North Africa?

Crown gall disease is believed to be particularly serious in Mediterranean countries due mainly to climate conditions. In North Africa, the presence of crown gall disease is reported with restricted distribution according to the EPPO (2024). The most affected crops are stone fruit trees, grapevines and roses. Although tumorigenic agrobacteria were first isolated from grapevine galls in 1897 in Europe (Cavara, 1897a,b), the disease on grapevines was only recently observed in the region; Morocco in 2008 (Habbadi *et al.*, 2019); Egypt in 2011 (Tolba and Zaki, 2011) Tunisia in 2011, (Abdellatif *et al.*, 2013) (Fig. 7.3).

This relatively recent emergence suggests a shifting dynamic in disease epidemiology, possibly influenced by climate change and increased trade of propagation material, which may have introduced the pathogen into new environments or amplified its impact in previously low-risk areas.

Although the disease is typically detected in vineyards and nurseries, its incidence has been rarely observed in these countries. In fact, culture conditions of temperature and humidity in nurseries are known to be favorable for the development of the pathogen. Moreover, agricultural intensification and high-density nursery production create microenvironments that

Fig. 7.3. Symptoms of Crown gall on grapevine detected in the region of Regueb, Tunisia: a – on young grapevine, b – on old grapevine.

promote infection and pathogen persistence. Typically, crown gall disease occurs sporadically in every province or state where susceptible crops are grown. Additionally, some soils can facilitate the persistence of tumorigenic agrobacteria for decades and so that they become unsuitable for susceptible cultures (Krimi *et al.*, 2002). For instance, in Morocco, the area of Meknes where the majority of stone fruit nurseries are located, the crown gall disease is considered a serious threat (Boulif, 2016). In Tunisia, nurseries in the region of Ouerdanine were characterized by recurrent attacks of the disease which made the production of the rootstock GF677 impossible (Rhouma *et al.*, 2005).

While empirical evidence specifically linking climate change to crown gall in North Africa is still limited, studies have shown that climate factors such as temperature and water stress can influence the incidence and severity of various plant diseases. We believe that shifts in rainfall and soil moisture levels may influence the availability of water in the soil, affecting plant susceptibility to crown gall disease and bacterial infection rates. Additionally, changing climatic conditions can lead to the disruption of ecosystems, altering the types of plants that are prevalent in certain regions. This could affect the host range for pathogenic agrobacteria and lead to the emergence of crown gall in new geographic areas. For example, crops that are more tolerant to drought may become more dominant in regions that experience reduced rainfall, potentially creating environments where crown gall is more likely to take hold. Given the sensitivity of the North African climate and the potential for these changes to affect crown gall disease dynamics, further research is needed to fully understand the relationship between climate change and this disease in the region.

A systematic review exploring different sources as Google scholar, Web of Science, and PubMed, has yielded 44 published papers on crown gall disease in North Africa from 2000 to 2024 (Table 7.1). This literature analysis has considered "crown gall", "agrobacterium", and "North Africa" as research key words, and excluded review publications and research articles related to genetic transformation mediated by *Agrobacterium*.

Although serious initiatives have been conducted in several countries of the region for the control of the disease as an ultimate objective, scientific studies on crown gall are still considered fragmentary and far from reflecting the importance of the disease in North Africa. According to this systematic review, there is a serious lack of data from Libya has been recorded. Except some recent works from Morocco and Tunisia, the current situation of the crown gall disease in North African countries is still far from reporting accurate information related to climate change (Fig. 7.4). This contrasts with the more extensive literature available from Europe, North America, and parts of Asia. Stone fruits and grapevines are economically critical crops in North Africa, often cultivated in areas already constrained by limited arable land, water scarcity, and energy challenges. In contrast to areas where crown gall has been long-established, well-documented, and effectively managed, the relative lack of regional research and knowledge in North Africa highlights a critical gap. In this context, the economic consequences of crown gall disease may be disproportionately severe compared to regions with more diversified or industrialized agricultural systems. This underscores the urgent need for a focused, locally-informed understanding of the disease to develop prevention and control strategies that are both regionally relevant and responsive to the unique environmental and agricultural constraints of the area.

Impact of Crown Gall

Agrobacteria commonly inhabit soils and in their avirulent form can exhibit beneficial rhizospheric relationship with plants mainly through their capacity to produce phytohormones as noted by Hao *et al.* (2012). However, in the presence of the tumor-inducing (Ti) plasmid, agrobacteria become tumorigenic and can induce tumor growth once they precede successful transfer of their specific DNA fragment (T-DNA) to the host-plant. Crown gall disease occurs worldwide and affects a large number of both woody and herbaceous plants. Infected soils with pathogenic bacteria can rend nurseries inadequate for susceptible plantations. Consequently, maintaining plant stock nurseries free of crown gall-affected plants and prophylactic culturing practices is of ultimate importance. Members of the rose family are considered to be the most susceptible host. Almond, apple, apricot, euonymus, fig, grape, peach, pecan, plum, walnut, and willow are the common woody hosts of this disease. However, almost 140 genera of woody and herbaceous plants belonging to more than 60 families are reported to be susceptible to crown gall disease.

Pathogenic agrobacteria are the most interesting gall-inducing disease as they have the widest host range of any phytopathogen. The disease is responsible of important annual losses due to unmarketable symptomatic

Table 7.1. Overview of research on crown gall disease and biocontrol in North African crops from 2000 to 2024 according to Google Scholar database: Summary of studies by crop, country, keywords, and source.

No	Crops	Country	Keywords	DOI/Source
1	Fruit trees	Morocco	*Agrobacterium*, biovar, inhibition, antagonism, temperature, time	https://www.inra.org.ma/sites/default/files/10309.pdf
2	Fruit rootstocks, Rose	Morocco, Tunisia, Algeria	Not available	10.1046/j.1365–2338.2002.00589.x
3	Fruit trees, Rose	Morocco	Not available	10.1046/j.1365–2338.2002.00606.x
4	Not specified	Algeria	*Agrobacterium* spp., soil populations, seasonal fluctuations, persistence	10.1128/AEM.68.7.3358–3365.2002
5	Not specified	Morocco	Not available	doi/full/10.1111/j.1365–2338.2004.00769.x
6	Not specified	Tunisia	*Agrobacterium tumefaciens*, biocontrol, agrocin 84, ALS 84, rootstock	https://www.jstor.org/stable/26456700
7	Stone and pome fruits	Tunisia	*Agrobacterium tumefaciens*, crown gall, susceptibility, rootstocks	https://www.jstor.org/stable/26463214
8	Not specified	Tunisia	Plant extracts, antitumorigenic activity, crown gall	PMID: 18390447.
9	Bitter almond	Tunisia	Rhizobacteria, K84, K1026, *A. tumefaciens*, *A. radiobacter*, *Bacillus, Pseudomonas,* crown gall, control	https://www.jstor.org/stable/41998547
10	Bitter almond	Tunisia	Olive mill wastewater, crown gall, bitter almond, suppression, biological control	10.1007/s10658-008-9317-y

11	Tomato	Tunisia	*Agrobacterium, Bacillus subtilis*, bacteriocin, biocontrol, optimization	10.1111/j.1472–765X.2008.02524.x
12	Not specified	Egypt	quaternary N-alkyl chitosans; IR and NMR spectroscopy; antibacterial activity; antifungal activity; fungal spore germination	https://doi.org/10.1002/app.31492
13	Stone fruit rootstocks	Tunisia	Not available	10.1007/s00248-010-9685-7
14	Olive tree	Tunisia	Not available	10.1080/07060661.2011.606430
15	Not specified	Tunisia	*Bacillus thuringiensis*, Bacthuricin F103, Chromatographic methods, Antimicrobial activity, N-terminal sequencing	10.1007/s12010-011-9252-9
16	Grapevine	Egypt	*Agrobacterium vitis*, grapevine, crown gall, characterization	https://doi.org/10.1016/j.aoas.2011.06.001
17	Not specified	Egypt	Not available	http://lis.cl.cu.edu.eg/cgi-bin/koha/opac-detail.pl?biblionumber=35277&shelfbrowse_itemnumber=53018
18	Squash fruits, kalanchoe and rose	Egypt	Not available	10.1016/j.aoas.2013.01.007
19	Tomato	Tunisia	*Lawsonia inermis*, phenolic compounds, olive knot, crown gall, biological control	10.1016/j.cropro.2012.11.014

Continued

Table 7.1. Continued

No	Crops	Country	Keywords	DOI/Source
20	Grapevine	Tunisia	Crown gall, grapevine, pathogenicity, *Agrobacterium vitis*, *A. tumefaciens*.	https://www.cabidigitallibrary.org/doi/pdf/10.5555/20133138147
21	Not specified	Morocco	Antibacterial, bacteria, plant extracts, *Cotaria myrtifolia*, *Agrobacterium tumefaciens*.	10.5897/AJMR2013.6284
23	Almond	Algeria	Not available	https://doi.org/10.19044/esj.2013.v9n30p%p
24	*Medicago ciliaris*	Algeria	Not available	10.1353/bae.2015.0019
25	Not specified	Tunisia	*Agrobacterium tumefaciens*, *Bacillus amyloliquefaciens*, biological control, crown gall disease, lipopeptides	10.1111/jam.12797
26	Not specified	Tunisia	*Brazilian peppertree*, Leaf extracts, the crown gall disease, Antibacterial activity	20.1001.1.22519041.2015.4.1.7.9
27	Not specified	Tunisia	*Ruta montana*, antifungal activity, antibacterial activity, essential oil, *in vivo* assay	10.17179/excli2014-655
28	*Retama monosperma* (L.) Boiss.	Algeria	*Retama monosperma* (L.) Boiss., *Bradyrhizobium*, symbiosis, root nodules, nitrogen fixation, endemic legumes, Algeria.	https://www.ajol.info/index.php/ajb/article/view/137196

29	Guava Plants	Egypt	*Agrobacterium tumefaciens*; *Meloidogyne incognita*; *Pisidum guajava*; Ozone; Ozonated water; Hydrogen peroxide; SEM.	http://agr.p.alexu.edu.eg/Data/Sites/1/%D9%85%D8%AC%D9%84%D8%A9/06/3341%D8%A7%D9%84%D9%85%D8%A7%D9%84%D8%A3%D9%88%D9%84.pdf
30	Grapevine and stone fruits	Tunisia	reck, BOX-PCR, Ti plasmid, strain K84, L-tartrate utilization.	https://www.jstor.org/stable/44280445
31	Tomato plants	Tunisia	*Bacillus methylotrophicus*; bacterial plant disease; biological control agent; mass spectrometry.	10.1002/ps.4331
32	Not specified	Egypt	Bacterial pathogens; Bioactive molecules; Crown gall; Leaf extracts; Soft rot; Woody plants.	10.1016/j.micpath.2018.05.032
33	Not specified	Tunisia	*Bacillus amyloliquefaciens*, lipopeptides, biocontrol, Agrobacterium tumefaciens	10.1007/s11356-018-3570-1
34	Not specified	Tunisia	*Bacillus amyloliquefaciens* subsp. plantarum, Rhizosphere adaptation, Rhizosphere competence, Plant growth-promotion, Biocontrol.	10.1016/j.biocontrol.2018.01.013
35	*Genista microcephala*, *Argyrolobium uniflorum*	Algeria	Rhizobia, *Genista microcephala*, *Argyrolobium uniflorum*, SDS-PAGE, *Gammaproteobacteria*, Heavy metal tolerance	10.1016/j.jare.2018.06.001
36	Tomato	Tunisia	Not available	10.1007/s10658-018-1441-8

Continued

Table 7.1. Continued

No	Crops	Country	Keywords	DOI/Source
37	Grapevine	Northern Hemisphere	Not available	10.3389/fmicb.2019.01896
38	Tomato	Tunisia	*Agrobacterium tumefaciens*; *Mentha piperita*; antibacterial activity; biological control; essential oil.	10.5650/jos.ess18261
39	Tomato	Algeria	Not available	10.1007/s10658-020-01975-1
40	Almond	Tunisia	*Bacillus velezensis*, biocontrol, crown gall, *Agrobacterium fabrum*	10.1007/s13199-021-00791-y
41	*Anthyllis henoniana*	Tunisia	Rhizobia, nitrogen fixation, arid soils, *Anthyllis*.	10.1007/s00203-021-02153-6
42	Not specified	Morocco	Medicinal plants, phytochemistry, antibacterial, toxicity	10.1016/j.sajb.2022.08.016
43	Not specified	Morocco	Essential oils, *Allorhizobium vitis*, antibacterial, phytochemicals	10.1016/j.sajb.2022.03.038
44	Almond	Morocco	Almond tree, crown gall, *Agrobacterium tumefaciens*, *Agrobacterium radiobacter*, K1026, agrocin, biological control	10.34874/IMIST.PRSM/afrimed-i140.43733
45	Eucalyptus occidentalis	Algeria	ITS – Intergenic transcribed spacer; MG – Mannitol glutamate; MS – Murashige and Skoog; tms – tumor morphology shoot; vir – virulence.	10.1007/s10658-006-9055-y

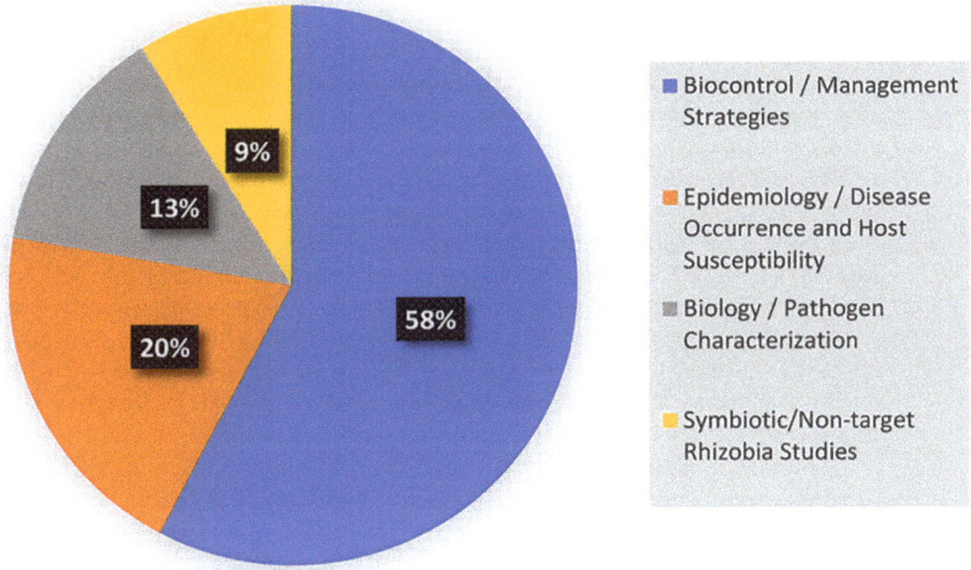

Fig. 7.4. Research focus areas on crown gall disease in North Africa (2000–2024), based on Google Scholar Publications.

rootstocks, low productivity of galled trees, and vulnerability of infected plants to biotic and abiotic stresses (Bliss *et al.*, 1999). With time, tumors enlarge, become convoluted, and dark brown colored (Agrios, 2005). Therefore, the disease is more fatal on young trees. Severe infection can cause plant death. Poor growth and reduced yield are the main consequences of the disease as galls can obstruct the water and nutrients uptake. For instance, infected grape plants show vigor and yield reduction by up to 40% depending on the infection degree (Schroth *et al.*, 1988).

Crown gall disease is believed to be a growing concern, particularly in the context of nursery production and climate change, in Mediterranean countries and North Africa. Although the disease is still making losses in crop productions, there is a very little information about the real damage caused by the crown gall in the Maghreb region. According to previous studies: in Algeria, symptoms have been detected in 99% of nurseries, whereas in Morocco crown gall affected 15–20% of young plants, with 80% in some regions (Benjama *et al.*, 2002), and in Tunisia, the infection rate reached 30% (Boubaker, 1999). Quarantine restrictions have been applied to crown gall disease in North African countries for years. The previously called *A. tumefaciens* and *R. rhizogenes* have figured in the quarantine A2 list before to be removed recently, given the ubiquitous nature of the disease throughout the world (EPPO, 2024).

Causal Agent and Its Vectors

Although the causal agent of the crown gall is traditionally identified as *A. tumefaciens*, there is a large disagreement about the pathogen name. The nomenclature of *Agrobacterium* species is still controversial even at the genus level. For instance, *A. tumefaciens* (earlier designated Biovar 1) is now recognized as species complex, consisting of several genomospecies. To avoid ambiguity in agrobacteria's taxonomy, we will refer, in this chapter, to the new genus classification proposed by Lindström and Young (2011) and Mousavi *et al.* (2014, 2015), which excluded both Biovar 2 and Biovar 3 from the genus *Agrobacterium* and delineated them to *Rhizobium rhizogenes* and *Allorhizobium vitis*, respectively. Moreover, we will focus on the most reported species in North Africa, which are: *A. tumefaciens* complex species, *Al. vitis* and *A. rubi* main phytopathogenic agents of stone fruit trees, grapevine and roses, respectively.

The tumor inducing plasmid Ti is transferred by pathogenic agrobacteria to plants via a type IV secretion system (T4SS) (Gelvin, 2010; Lacroix and Citovsky, 2013). Plant hormones auxin and cytokinin are overproduced once the T-DNA is integrated into the plant genome. Subsequently, overgrowths on the infection site appear. Additionally, secondary tumors at different distance from the primary tumor can be observed once expanding tissues are infected. Therefore, infected plant tissues are the primarily vectors of crown gall. Crown gall bacteria enter plant roots through wounds. Wounds and injuries are the first infiltration point of pathogenic bacteria to plant tissues. Plant tissues damages can be created by biotic factors such as soil insect feeding or by physical factors such as rain splashed. Some cultural practices, like root pruning, grafting and excavation, can also spread the disease. Although root pruning of rootstocks is commonly used prior to transplantation, the resulted wounds from this routine practice facilitate infection with tumorigenic agrobacteria inhabiting the soil (Lacroix and Citovsky, 2013). Bacteria can return to the soil, the main natural reservoir of the disease, from contaminated areal parts by water or equipment. Pathogenic bacteria can survive in soil for years, infect other roots and spread the plasmid Ti into other telluric agrobacteria.

Management Strategies and Control Methods

The infection of plants by pathogenic agrobacteria is irreversible. Once the T-DNA is transferred to the plant genome, the gall growth will continue even if agrobacteria are removed. Therefore, prevention is the foundation for crown gall control (Fig. 7.5).

Detection techniques

An early detection of pathogenic agrobacteria in plants or soil of nurseries and fields is very critical to avoid the propagation of the disease and the contamination of new areas. Small galls require careful diagnosis because they may be confused with excessive wound callus. Nonpathogenic Agrobacterium cells are often prevalent in soil and their presence in high population in galls can make diagnosis difficult (Bouri *et al.*, 2016). Nowadays, Thanks to new developed molecular methods specific to plasmids and different species of agrobacteria, the detection of crown gall disease is made easier and more reliable even in symptomless samples (Shams *et al.*, 2013).

Cultural practices

Good sanitation and cultural practices are very important to prevent the establishment of crown gall disease by avoiding the introduction of

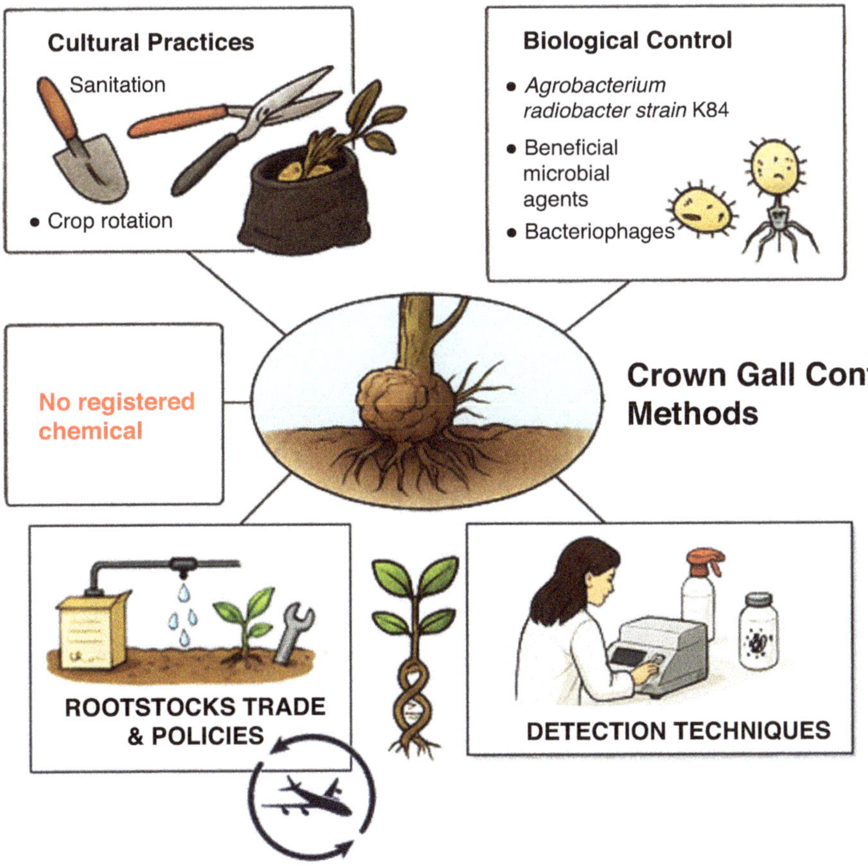

Fig. 7.5. Crown gall management strategy.

pathogenic agrobacteria at all production stages of plantations. In the case of crown gall disease detection, unfortunately the best action is to remove and destroy the infected plants to avoid contaminating healthy ones. Since pathogenic agrobacteria can persist in the soil for years after the plant is gone, it is recommended to avoid planting susceptible species to deprive pathogens from their host plants. Moreover, control of root-attacking insects and nematodes should be performed and, hygiene is needed for pruning equipment, and wounding of plants should be avoided in infected fields.

Even though to be promising, heat therapy soil fumigation and solarization (Fig. 7.6) are not commonly used due to many difficulties (Pu and Goodman, 1993; Pinkerton *et al.*, 1996).

Biological control

A. radiobacter K84 and its genetically engineered strain called K1026 are currently commercialized for the biological control of the disease. These two strains are recognized as very effective agents against crown gall on a number of hosts, with some limitations. In fact, not all strains of tumorigenic agrobacteria are sensitive to agrocin (the toxin of K84). Moreover, transfer of the agrocin plasmid from K48 to pathogenic agrobacteria can make them in turn resistant. Therefore, K1026 strain was engineered to minimize the breakdown of the K84 control. Although K1026 is not able to transfer the agrocin plasmid to other agrobacteria, the use of this genetic modified strain is still rejected by GMO's (genetically modified organism) protesters. It is very important to note here that biological control can only be used as

Fig. 7.6. Assay of solarization technique to reduce pathogenic population of agrobacteria in the soil. Experimental station of Olive institute, Sfax, Tunisia (2018).

preventive and not curative alternative, and the timing of the application is critical to efficiently protect wounded plants during harvest or pruning. Obviously K84 and K1026 are the most common strains used in the control of the crown gall disease, but many studies reported several *Bacillus* strains that have controlled the disease as effectively as the commercial strains (Rhouma *et al.*, 2008; Ben Gharsa *et al.*, 2021). Several studies have been reported the efficiency of some biological alternatives for the control of the disease in North Africa (Table 7.1). Recent studies have explored the use of bacteriophages as a promising biocontrol method against *Rhizobium radiobacter* (formerly *A. tumefaciens*) (Ramadhan *et al.*, 2022; Chantapakul *et al.*, 2024). These findings highlight bacteriophages as an environmentally friendly and targeted alternative to commercial treatments.

Chemical control

Currently, no chemical that can effectively control crown gall has been registered.

Rootstocks trade and policies

With the increase in commercial exchanges of nursery plant material, the spread of crown gall disease started to take an epidemic aspect. Although tumorigenic agrobacteria are considered as quality pathogens according to the European Union regulations, in some North African countries they are vigorously controlled by national regulations. In Tunisia, the importation of rootstocks is under strict sanitary controls: visible tumors must not exceed 1% in nursery plant material for trading authorization (Anon, 1981). Accordingly, the sanitary status of Tunisian nurseries was likely improved 20 years after the decree (Rhouma *et al.*, 2008). In Morocco, due to recent detection of the crown gall disease in vineyards, the importation of grapevine rootstocks was officially regulated by several phytosanitary requirements; the plant material must (i) show no symptoms, (ii) be produced in crown gall free regions, and (iii) have an analysis certification according to standard methods (Anon, 2018).

Future Research Areas to Improve Crown Gall Disease Control

An epidemiological survey of crown gall requires a specific determination of the *Agrobacterium* taxonomy. Although the causal agent of the crown gall is traditionally identified as *Agrobacterium tumefaciens*, there is a large disagreement about the pathogen name. The nomenclature of *Agrobacterium* species is still controversial even at the genus level. A set of biochemical and molecular methods has been set up to facilitate the taxonomic assessment of agrobacteria either in pure culture or directly from soils, rhizospheres or/and tumors. The INCO-DC European program ERBIC18CT970198, "Integrated Control of Crown Gall in Mediterranean Countries" presented a simple and efficient identification scheme for agrobacteria based on simple and fast urease and/or esculinase biochemical tests. Nevertheless, the detection of pathogenic *Agrobacterium* in complex environments without prior strain isolation can still be easily done by PCR using primers designed in the conserved vir region. The identification of the agrobacteria's species and genomospecies present in the soil or galls is also possible directly by specific primers (Shams *et al.*, 2012). A better disease management of crown gall can be achieved only with exhaustive investigations to afford large data about the genetic diversity of the agrobacteria populations upon huge variety of soils and plants. Non-pathogenic agrobacteria should not be excluded from any epidemiological study as they can become pathogenic after Ti plasmids transfer and subsequently can induce tumors and participate in the outbreak of the disease. Moreover, future research could explore the potential of biotechnological tools such as RNA interference (RNAi), including host-induced gene silencing (HIGS), to suppress key virulence genes in *Agrobacterium*, such as those involved in hormone synthesis or opine catabolism. Early studies have shown promising results using RNAi constructs targeting oncogenes like iaaM and ipt (Escobar *et al.*, 2001), offering a potential path toward durable, genetically based resistance strategies.

In summary, understanding the interplay between the crown gall disease and shifting climate dynamics is crucial for sustainable crop production in North Africa, especially given the region's growing agricultural vulnerability and increasingly unpredictable weather patterns. Therefore, it is imperative to expand monitoring and surveillance efforts, and to develop region-specific predictive models that integrate climateand disease data. Strengthening collaboration across the region is essential for integrated disease management. Additionally, investments in the promotion of biocontrol and soil management strategies that enhance microbial diversity antagonistic to *Agrobacterium*, are key to sustainable crown gall control.

Discussion Points

1. Although agrobacteria are ubiquitous in soils, what measures should be implemented in North African countries to prevent new outbreaks in vineyards?
2. How can controversies about the systematic of agrobacteria impact management strategies and efforts to control crown gall diseases?
3. What are the current prospects for developing resistant or tolerant rootstocks varieties to mitigate crown gall, and how effective are these strategies?

4. Why has crown gall been more effectively controlled in some regions compared to others, and how climate change contributes to the variability in its impact across different areas?

Useful Websites

1. Available at: https://depts.washington.edu/agro/ (accessed 27 December 2024).
2. Available at: http://agro.vbi.vt.edu/ (accessed 27 December 2024).

Further Reading

To gain a deeper understanding of the significant impacts of climate change on agriculture in North Africa, especially in relation to plant diseases and pest dynamics, the following resources are recommended:

1. Has North Africa turned too warm for a Mediterranean forest pest because of climate change? Available at: https://link.springer.com/article/10.1007/s10584-021-03077-1 (accessed 27 December 2024).
2. Climate Change Impact on Agriculture in Africa. Available at: https://toxigon.com/climate-change-impact-on-agriculture-in-africa (accessed 27 December 2024).
3. Adapting agriculture and pesticide use in Mediterranean regions under climate change scenarios: A comprehensive review. Available at: https://www.sciencedirect.com/science/article/pii/S1161030124002582 (accessed 27 December 2024).

References

Abdellatif, E., Valentini, F., Janse, J.D., Bouri, M. and Rhouma, A. (2013) Occurrence of crown gall of the grapevine in Tunisia and characterization of *Tunisian agrobacterium* vitis and *A. Tumefaciens* strains. *Journal of Plant Pathology* 95, 115–126.

Abd El Menem, A., Mohamed, A. and Enany, D. (2022) Analytical study of the economic efficiency of Egyptian grapes in the most important global markets. *Nature Science* 18(11), 24–38.

Agrios, G.N. (2005) Plant diseases caused by prokaryotes: Bacteria and mollicutes. *Plant Pathology* 615–703. DOI: 10.1016/b978-0-08-047378-9.50018-x.

Anon (1981) Arrêté du 31/12/1980, fixant les conditions et les normes de contrôle de la production et dela commercialisation des plants des espèces fruitières. Ministère de F Agriculture. *Journal Officiel de laRépublique Tunisienne (JORT) 002 du 09/01/1981 pp. 0035-0037. Imprimerie Officielle de la RépubliqueTunisienne, Tunis, Tunisie.*

Anon (2018). Available at: https://apps.fas.usda.gov/newgainapi/api/report/downloadreportbyfile name?filename=Import%20Requirements%20for%20Grapevine%20Propagation%20Material_Rabat_ Morocco_12-13-2018.pdf

Ben Gharsa, H., Bouri, M., Mougou Hamdane, A., Schuster, C., Leclerque, A. *et al.* (2021) *Bacillus velezensis* strain MBY2, a potential agent for the management of crown gall disease. *PLoS ONE* 16, e0252823. DOI: 10.1371/journal.pone.0252823.

Benjama, A., Boubaker, A., Khlief, H., Krimi, Z., Nesme, X. *et al.* (2002) Susceptibility of stone-fruit rootstocks, rose and grapevine to agrobacterium radiobacter var. tumefaciens in Arab Mediterranean countries. *EPPO Bulletin* 32(3), 463–466. DOI: 10.1046/j.1365-2338.2002.00589.x.

Bergey, D.H., Harrison, F.C., Breed, R.S., Hammer, B.W. and Huntoon, F.M. (1923) *Bergey's Manual of Determinative Bacteriology*, 1st edn. Baltimore, The Williams and Wilkins Co, pp. 1–442.

Bhatia, S. and Bera, T. (2015) Classical and nonclassical techniques for secondary metabolite production in plant cell culture. In: Bhatia, S., Sharma, K., Dahiya, R. and Bera, T. (eds) *Modern Applications of Plant Biotechnology in Pharmaceutical Sciences*. Academic Press, San Diego, CA, pp. 231–291. DOI: 10.1016/B978-0-12-802221-4.00007-8.

Bliss, F.A., Almedhi, A.A., Dandekar, A.M., Schuerman, P.L. and Bellaloui, N.I. (1999) Crown gall resistance in accessions of Prunus species. *HortScience* 34, 326–330.

Boubaker, A. (1999) *Second Scientific Progress Report. INCO- CG No. ERBIC18CT970198*. Tunis, Tunisia.

Boulif, M.A. (2016) Review of research on diseases and pests of almonds in Morocco. In: Kodad, O., López-Francos, A., Rovira, M. and Socias i Company, R. (eds) *XVI GREMPA Meeting on Almonds and Pistachios*. CIHEAM, Zaragoza, pp. 199–204.

Bouri, M., Arslan, K.S. and Şahin, F. (2023) Climate-smart pest management in sustainable agriculture: Promises and challenges. *Sustainability* 15, 4592. DOI: 10.3390/su15054592.

Bouri, M., Chattaoui, M., Gharsa, H.B., McClean, A., Kluepfel, D. *et al.* (2016) Analysis of agrobacterium populations isolated from Tunisian soils: Genetic structure, avirulent-virulent ratios and characterization of tumorigenic strains. *Journal of Plant Pathology* 98, 265–274.

Cavara, F. (1897a) Eziologia di alcune malattie di piante cultivate. *Le Stazioni Sper. Agraric Itliana* 30, 482–509.

Cavara, F. (1897b) Tuberculosi della vite.Intorno alla eziologia di alcune malattie di piante cultivate. *Le Stazioni Sper. Agraric Itliana* 30, 483–487.

Chantapakul, B., Sabaratnam, S. and Wang, S. (2024) Isolation and characterization of bacteriophages for controlling rhizobium radiobacter – causing stem and crown gall of highbush blueberry. *Frontiers in Microbiology* 15, 1437536.

De Cleene, M. and De Ley, J. (1976) The host range of crown gall. *The Botanical Review* 42, 390–466.

EPPO (2024). Available at: https://gd.eppo.int/taxon/AGRBTU/distribution (accessed 18 October 2024).

Escobar, M.A., Civerolo, E.L., Summerfelt, K.R. and Dandekar, A.M. (2001) RNAi-mediated oncogene silencing confers resistance to crown gall tumorigenesis. *Proceedings of the National Academy of Sciences* 98(23), 13437–13442.

Gelvin, S.B. (2010) Plant proteins involved in agrobacterium-mediated genetic transformation. *Annual Review of Phytopathology* 48, 45–68.

Godefroid, M., Cruaud, A., Streito, J.C., Rasplus, J.Y. and Rossi, J.P. (2019) Xylella Fastidiosa: Climate suitability of European continent. *Scientific Reports* 9, 1–10.

Habbadi, K., Duplay, Q., Chapulliot, D., Kerzaon, I., Benkirane, R. *et al.* (2019) Characterization and phylogenetic diversity of *Allorhizobium vitis* isolated from grapevine in Morocco. *Journal of Applied Microbiology* jam.14523. DOI: 10.1111/jam.14523.

Hao, X., Xie, P., Johnstone, L., Miller, S.J., Rensing, C. *et al.* (2012) Genome sequence and mutational analysis of plant-growth-promoting bacterium *agrobacterium tumefaciens*

CCNWGS0286 isolated from a zinc-Lead mine tailing. *Applied and Environmental Microbiology* 78(15), 5384–5394. DOI: 10.1128/AEM.01200-12.

Horuz, S., Çaglar, B.K., Kusek, M. and Aysan, Y. (2018) Crown gall disease susceptibility of some stone fruit rootstocks in Turkey. *TarımBilimleri Dergisi* 24, 439–444.

Kado, C.I. (2014) Historical account on gaining insights on the mechanism of crown gall tumorigenesis induced by *agrobacterium tumefaciens*. *Frontiers in Microbiology* 7(5), 340. DOI: 10.3389/fmicb.2014.00340.

Krimi, Z., Petit, A., Mougel, C., Dessaux, Y. and Nesme, X. (2002) Seasonal fluctuations and long-term persistence of pathogenic populations of agrobacterium spp. *Applied and Environmental Microbiology* 68(7), 3358–3365.

Lacroix, B. and Citovsky, V. (2013) The roles of bacterial and host plant factors in agrobacterium-mediated genetic transformation. *The International Journal of Developmental Biology* 57(6–8), 467–481. DOI: 10.1387/ijdb.130199bl.

Lindström, K. and Young, J.P.W. (2011) International committee on systematics of prokaryotes. Subcommittee on the taxonomy of agrobacterium and rhizobium: Minutes of the meeting, 7 September 2010, Geneva, Switzerland. *International Journal of Systematic and Evolutionary Microbiolology* 61(12), 3089–3309.

Morelli, M., García-Madero, J.M., Jos, Á., Saldarelli, P., Dongiovanni, C. *et al.* (2021) *Xylella Fastidiosa* in olive: A review of control attempts and current management. *Microorganisms* 9, 1771.

Mousavi, S.A., Österman, J., Wahlberg, N., Nesme, X., Lavire, C. *et al.* (2014) Phylogeny of the rhizobium-Allorhizobium-agrobacterium clade supports the delineation of Neorhizobium gen. nov. *Systematic and Applied Microbiology* 37(3), 208–215. DOI: 10.1016/j.syapm.2013.12.007.

Mousavi, S.A., Willems, A., Nesme, X., de, P. and Lindström, K. (2015) Revised phylogeny of Rhizobiaceae: Proposal of the delineation of Pararhizobium gen. Nov., and 13 new species combinations. *Systematic and Applied Microbiology* 38(2), 84–90. DOI: 10.1016/j.syapm.2014.12.003.

Pinkerton, J.N., Canfield, M.L., Ivors, K. and Moore, L.W. (1996) Effect of soil solarization and cover crops on populations of selected soilborne plant pathogens. *Phytopathology* 86(supplement), S98.

Pu, X.A. and Goodman, R.N. (1993) Effects of fumigation and biological control on infection of indexed crown-gall-free grape plants. *American Journal of Enology and Viticulture* 44, 241–248.

Ramadhan, F., Alfiko, Y., Purwantomo, S., Mubarok, A.F., Budinarta, W. *et al.* (2022) A new approach for controlling agrobacterium tumefaciens post transformation using lytic bacteriophage. *Plants* 11(22), 3124.

Rhouma, A., Boubaker, A., Nesme, X. and Dessaux, Y. (2005) Susceptibility of some stone and pome rootstocks to crown gall. *Phytopathologia Mediterranea* 44, 275–284.

Rhouma, A., Bouri, M., Boubaker, A. and Nesme, X. (2008) Potential effect of rhizobacteria in the management of crown gall disease caused by *agrobacterium tumefaciens* Biovar 1. *Journal of Plant Pathology* 90(3), 517–526.

Roussin-Léveillée, C., Rossi, C.A.M., Castroverde, C.D.M. and Moffett, P. (2024) The plant disease triangle facing climate change: A molecular perspective. *Trends in Plant Science* 29(8), 895–914.

Schroth, M.N., McCain, A.H., Foott, J.H. and Huisman, O.C. (1988) Reduction in yield and vigor of grapevine caused by crown gall disease. *Plant Disease* 72, 241–246.

Shahbandeh, M. (2024) Stone fruit production worldwide from 2000 to 2022. Available at: https://www.statista.com/statistics/577598/world-stone-fruit-production/ (accessed 23 September 2024).

Shams, M., Campillo, T., Lavire, C., Muller, D., Nesme, X. *et al.* (2012) Rapid and efficient methods toisolate, type strains and determine species of agrobacterium spp. in pure culture and complex environments. In: Jimenez-Lopez, J.C. (ed.) *Biochemical Testing 4*. Poland, pp. 3–20.
Shams, M., Vial, L., Chapulliot, D., Nesme, X. and Lavire, C. (2013) Rapid and accurate species identificationand exhaustive population diversity assessment of agrobacterium spp. using recA- based PCR. *Systernaticand Applied Microbiology* 36, 351–358.
Smith, E.F. and Townsend, C.O. (1907) A plant-tumor of bacterial origin. *Science (Washington)* 25, 671–673.
Tolba, I.H. and Zaki, M.F. (2011) Characterization of agrobacterium vitis isolates obtained from galled grapevine plants in Egypt. *Annals of Agricultural Sciences* 56(2), 113–119. DOI: 10.1016/j.aoas.2011.06.001.
USDA (2024) *Fresh Apples, Grapes, and Pears: World Markets and Trade*.

8 The Epidemiology and Management of *Ralstonia solanacearum* and *Ralstonia pseudosolanacearum* in Central and Northern Europe

Karla Cardenas Gomez[1,2*], Daniel Narino Rojas[1,2*], Sophie Alice James[1], Sara Franco Ortega[1], Andrea L. Harper[1,2] and Ville-Petri Friman[3]

[1]*Department of Biology, University of York, York, UK*
[2]*Centre for Novel Agricultural Products, University of York, York, UK*
[3]*Department of Microbiology, University of Helsinki, Helsinki, Finland*

Introduction

Ralstonia solanacearum: Plant hosts, distribution and economic significance

Phytopathogenic *Ralstonia solanacearum* bacterium is considered the second most destructive crop pest in the world (Mansfield *et al.*, 2012). It is responsible for lethal wilts and rots in a wide range of plants globally, including the tropics of the Americas and several countries in Africa, Europe, South East Asia and Oceania. In Europe, the pathogen is present in Belgium, Bulgaria, France, Germany, Hungary, Italy, Poland, Portugal, Romania, Russia, Slovakia, Spain and the United Kingdom, and is transient in the Czech Republic and Ukraine (Fig. 8.1). The unusually broad host range comprises over 250 plant species of different monocot and dicot families of trees, weeds and shrubs, including economically important solanaceous crop plants such as tomatoes and potatoes (Genin and Denny, 2012). Other crop hosts are banana, groundnut plantain, ginger, cotton, eggplant, tobacco, peppers and cucurbits (Osdaghi, 2020). The host range is constantly expanding and new hosts are continuously reported (Genin and Denny, 2012). This bacterium can also survive asymptomatically in secondary plant hosts, and

*These authors contributed equally to the manuscript.

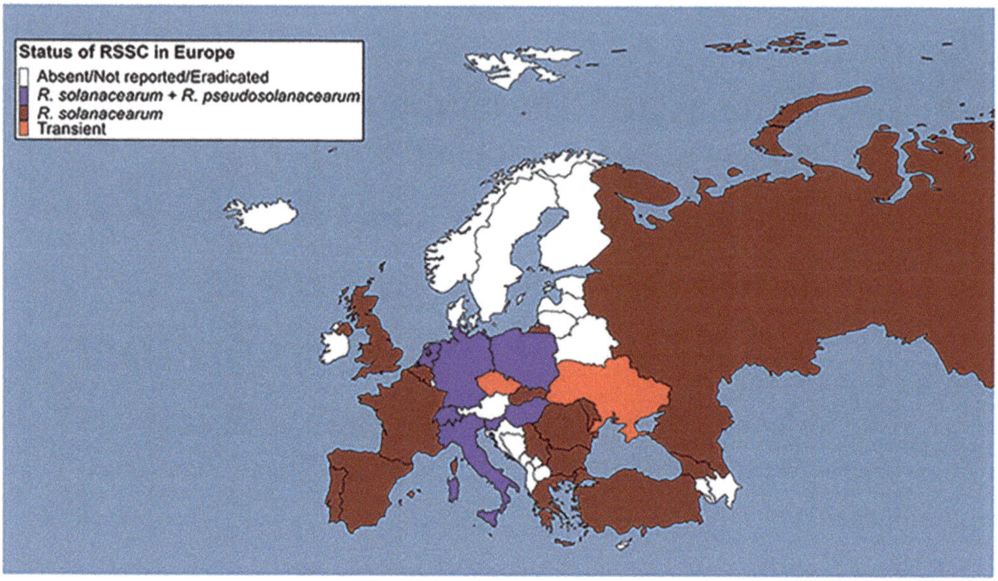

Fig. 8.1. Distribution of the *Ralstonia solanacearum* species complex in Europe (RSSC) (information from EPPO: https://gd.eppo.int/taxon/RALSSO/distribution (accessed 6 December 2023)).

the weed bittersweet (*Solanum dulcamara*), commonly found in riverbanks across North America and Europe, is an important environmental reservoir for this pathogen (Wenneker *et al.*, 1999).

Bacterial wilt is usually a disease of warm temperate regions such as the tropics and subtropics, whereas brown rot commonly occurs in cool temperate regions (Persley, 1985). These diseases rarely occur in regions where the monthly mean temperature is less than 10°C, since the optimal growth temperature range for these bacteria is between 25 to 37°C (Huerta *et al.*, 2015).

Crop losses, as well as disease surveillance and eradication in affected zones, result in considerable costs for growers (Elphinstone, 2001). It is estimated that this pathogen is responsible for an annual global loss of 1 billion USD per year related to potato production alone (Allen *et al.*, 2005), while bacterial wilt accounts for 20 to 100% of yield crop losses over 1.5 million ha in different parts of the world, affecting around 3 million farmers (Elphinstone, 2005; Wang *et al.*, 2023). The economic impact in a particular area depends on many factors, including the local climatic conditions, cropping practices, soil characteristics, crop of choice and the virulence of the specific strains causing the outbreak. Economic losses tend to be higher in zones where the pathogen has quarantine status, as infected potato tubers cannot be sold. For example, after an outbreak in the Netherlands in 1999, it was reported an increase of about £50 per ha for grower levies associated with disease management and eradication (Elphinstone, 2001).

Taxonomic classification

The numerous and heterogeneous group of strains comprising *R. solanacearum* has been grouped into the *Ralstonia solanacearum* species complex (RSSC) (Fegan and Prior, 2005). Historically, the RSSC was divided into races, then biotypes and later into four phylotypes, which roughly correspond to the strain's geographical origin: phylotype I originated from Asia, II from the Americas, III from Africa and IV from Indonesia (Genin and Denny, 2012). Phylotype II has two subclusters: IIA and IIB (Castillo and Greenberg, 2007), the latter can survive at lower temperatures and has been responsible for important outbreaks of potato brown rot in Northern Europe (Sebastià *et al.*, 2021). More recently, comparative genomic analysis conducted on representative strains from each of the four phylotypes divided the RSSC into three species: *R. solanacearum* comprising phylotype II, *R. pseudosolanacearum* grouping phylotypes I and III and *R. syzygii* corresponding to phylotype IV and *R. syzygii* subspecies *syzygii, indonesiensis*, and *celebesensis* (Safni *et al.*, 2014; Prior *et al.*, 2016). Henceforth in this case study, when referring to a particular species, it will be named by their species name (e.g. *Ralstonia syzygii*), while the term RSSC will refer to the three species collectively.

Transmission and life cycle

RSSC spreads mainly through contaminated soil and water, where pathogens can survive for several years (Mansfield *et al.*, 2012). The pathogen can also be transmitted by infected seed potatoes, plant debris and contaminated farming tools. After entering the roots of the susceptible host, the bacterium colonises the root cortex, invades the xylem vessels and, using the vascular system of the plant, spreads to the aerial parts. Consequently, the water flow within the vessels gets constricted and the plant loses its turgidity (Denny, 2007). The symptoms of bacterial wilt resemble water stress (Fig. 8.2), while the symptoms of potato brown rot consist of browning and necrosis of the vascular tissue of the potato tuber (Elphinstone, 2001). The infected plant eventually dies and infectious bacteria remain as saprophytes in the soil, but once in contact with a new host plant, the disease cycle can start again. Irrigation water and farming tools can facilitate further dispersal of the bacterium (Fig. 8.3). River water contamination is facilitated by run-off from contaminated soil or plant materials coming from infested cropping fields or processing plants. Secondary hosts such as the weed bittersweet (*Solanum dulcamara*) (Fig. 8.4), the perennial stinging nettle (*Urtica dioica*) and the annual black nightshade (*Solanum nigrum*) inhabiting river banks across North America, and Central and Northern Europe, can get infected when they come into contact with contaminated river water (Elphinstone *et al.*, 1998). These environmental reservoirs usually remain asymptomatic, only showing wilting symptoms at temperatures exceeding 25°C (van der Gaag *et al.*, 2019). Once infected, these plants can host the bacterium in their roots

Fig. 8.2. Bacterial wilt in tomato caused by *R. solanacearum* strain UW551a model strain of phylotype II in the tomato cv. moneymaker. (a) healthy plant, (b) bacterial wilt symptoms at 21 days post-infection.

for prolonged periods of time and spill-off releases them back into the river, facilitating their transmission and geographical range expansion.

Bacterial Wilt and Brown Rot of Potato in Central and Northern Europe

Ralstonia solanacearum phylotype IIB

In Europe, *R. solanacearum* phylotype IIB sequevar 1 (historically denominated race 3, biovar 2) is the causative agent of potato brown rot (Parkinson *et al.*, 2013). This disease is considered a major threat to potato crop production in temperate areas, due to the pathogen's capacity to survive at low temperatures in primary hosts without visible disease symptoms, and to overwinter in environmental reservoir hosts (Elphinstone *et al.*, 1998). It is highly probable that these cool-adapted strains originated in potato plants in the temperate region of the Andes in South America (Parkinson *et al.*, 2013) and have spread over the past decades by infected seed potatoes around the globe, especially across Europe (Janse, 2012). The prevalence of R. solanacearum phylotype IIB in Central and Northern Europe, particularly in the Netherlands, the UK and Sweden, has been correlated with asymptomatic infections in S. dulcamara (Williamson *et al.*, 2002). In the UK, bacterial population numbers in river water vary depending on the season, with undetectable levels during winter, and higher numbers around summer, when the

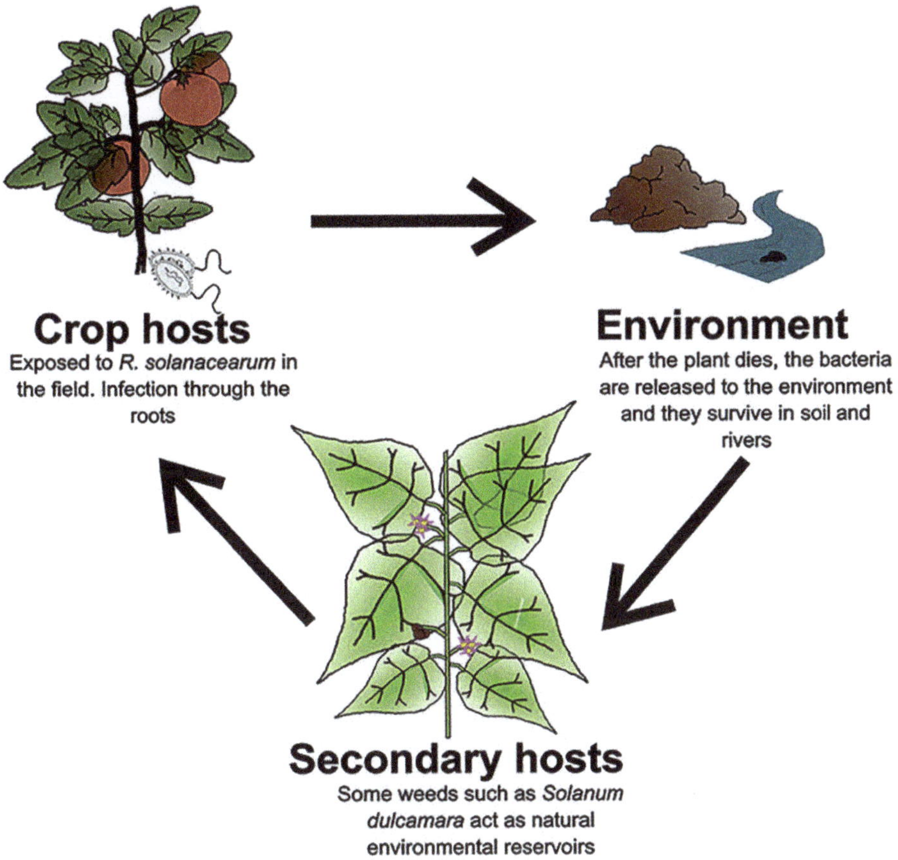

Fig. 8.3. A schematic figure of RSSC life cycle.

water temperature is above 15°C (Elphinstone and Matthews-Berry, 2017). In contrast, the pathogen can be detected all year round in the aquatic roots of *S. dulcamara,* in which the bacterium remains viable even during winter, given the plant's ability to restrict bacterial movement inside the xylem and limit bacterial root colonisation (Sebastià *et al.*, 2021). The mechanisms used by *R. solanacearum* phylotype IIB to overwinter in secondary hosts are not fully understood and require further research.

Currently, *R. solanacearum* phylotype IIB and all the RSSC members (with the exception of *R. syzygii subsp. syzygii*) are classified as quarantine pests in many countries in the EU, and outbreaks caused by this pathogen imply significant political and economic repercussions, including statutory control measures (Parkinson *et al.*, 2013). For instance, the UK and the Netherlands have been conducting annual surveys over the last decades to monitor the bacterium's presence in waterways, in order to restrict the pathogen's spread and prevent new outbreaks (Elphinstone, 2001; Vogelaar *et al.*, 2023). Over the course of those surveys, some rivers have been demarcated as

Fig. 8.4. *Solanum dulcamara* plants growing in a glasshouse.

contaminated in England, and their use for host crop irrigation has been banned (Elphinstone and Matthews-Berry, 2017).

Central and Northern Europe outbreaks and potato infections

Commercial potato cultivars cropped in Northern and Central Europe show low levels of resistance to *R. solanacearum* phylotype IIB, which has resulted in sporadic outbreaks of potato brown rot since the 1970s, including in Sweden in 1972, Belgium in 1989 and the Netherlands in 1995 (Williamson *et al.*, 2002), with the majority of positive testing coming from asymptomatic infected tubers. There have been seven confirmed outbreaks in the UK: two in the Thames Valley in 1992 and 1995, one in Northamptonshire in 1999, one in Kent in 2000, one in Nottinghamshire in 2005, one in Somerset and another in Cornwall in 2010 (Elphinstone and Matthews-Berry, 2017). *R. solanacearum* phylotype IIB also caused bacterial wilt in tomato crops in Bedfordshire in 1997 and 1998 (Elphinstone and Matthews-Berry, 2017). All of these outbreaks have been related to contaminated river water used for crop irrigation, except for the 2010 outbreaks that were linked to infected imported seed potatoes. It is possible that agricultural practices such as monoculture, and the centralization and intensification of potato production have facilitated the spread of *R. solanacearum* across many countries in Central and Northern Europe (Parkinson *et al.*, 2013).

Eradicating *R. solanacearum* phylotype IIB from Central and Northern Europe might be difficult, since it would involve the decontamination of rivers and irrigation sources, as well as the removal of environmental reservoirs (Williamson *et al.*, 2002). In the UK, small-scale removal of *S. dulcamara* from small water courses was proven to be successful in restricting the spread of the pathogen in river water downstream. However, host weed removals at large scale have proven impractical, given the complexity of vast river networks (Elphinstone and Matthews-Berry, 2017).

Potential range expansion and introduction of *R. pseudosolanacearum* in Central and Northern Europe

Human agricultural activity has played a key role in the dissemination of RSSC species, facilitating the presence of multiple strains in a single region (Huerta *et al.*, 2015). During the last decades, warm-adapted strains that usually inhabit the highland tropics of South America, Africa and Asia have been accidentally introduced via infected plant material into Europe and North America (Remenant *et al.*, 2010; Sebastià *et al.*, 2021). In 2015, the warm-adapted *R. pseudosolanacearum* (phylotype I) was detected in greenhouse rose in the Netherlands (EPPO, 2015; Tjou-Tam-sin *et al.*, 2017), an ornamental plant that prior to this outbreak was not a declared host of RSSC. The infected greenhouse roses presented the classic bacterial wilt symptoms, with necrosis on pruned branches, chlorosis, early drop of leaves and, in a few cases, bacteria oozing from the stems (Tjou-Tam-sin *et al.*, 2017). Water surveys conducted in 2020 and 2021 in two different geographical regions in the Netherlands also reported the presence of *R. pseudosolanacearum* (phylotype I) in surface water at a minimum temperature of 15°C. Genetic studies on both incursions suggested that the isolated strains were similar to *R. pseudosolanacearum* (phylotype I) isolates found previously in the Netherlands on greenhouse rose with the ability to overwinter in *S. dulcamara* (Vogelaar *et al.*, 2023). The phylogenetic assignments concluded that the *R. pseudosolanacearum* (phylotype I) isolates from rose clustered with an isolate that originated from eggplant in India (Bergsma-Vlami *et al.*, 2018). In 2021 and 2022, there were additional detections of *R. pseudosolanacearum* in ginger plants in Germany and in watercourses in Hungary, respectively (EPPO, 2021, 2022b). Later in March 2023, this bacterium was also detected in rhizomes of ginger (*Zingiber officinale*) imported for consumption in Germany (EPPO, 2023). These reports suggest that occurrences of *R. pseudosolanacearum* across Central and Northern Europe are increasing, posing a serious threat to crop production in this region. Climate change could also play a significant role in the distribution of RSSC species, since changes in climatic conditions might promote bacterium survival and help it spread to higher latitudes.

Experiments in controlled environment chambers have demonstrated that the severity of wilting in resistant tomato and tobacco lines significantly increases at high temperatures (Krausz and Thurston, 1975; Bittner *et al.*, 2016), which suggests that global warming could make RSSC outbreaks more severe in the future.

Monitoring and Diagnostic Tools

Diagnostic tools with high efficiency, specificity and sensitivity are crucial for detecting RSSC in the environment. In the UK, statutory controls mandate the Scottish Agricultural Science Agency and the Department for Environment, Food and Rural Affairs to annually survey rivers, test plant stocks and provide advice to keep potatoes free of quarantine pathogens (SASA, 2023). Imported material carrying pathogens is prohibited, and potatoes are inspected and tested by Fera Science Ltd in England and similar bodies in Scotland and Northern Ireland. Similarly, local Food and Consumer Product Safety Authorities conduct regular surveys for RSSC across Europe (NVWA, 2023).

Detection methods for RSSC species include selective plating, serological methods and molecular assays (EPPO, 2022c). Plating methods are based on the use of selective media to isolate bacteria from infected plant material or contaminated soil or water. Triphenyl tetrazolium chloride (TTC) and semi-selective SMSA media (Elphinstone *et al.*, 1996) are the gold standard for isolating RSSC species. Bacteria metabolize the TTC in the media producing fluidal creamy colonies with pink or red centres, depending on whether they are virulent or non-virulent, respectively. This technique is widely used in the UK for sampling waterways, but it can be time-consuming and detects any viable cells without suppressing the growth of other Gram-negative bacteria from environmental water samples.

Molecular techniques, such as PCR and qPCR, offer higher specificity by targeting conserved genetic regions of the RSSC, like the flagellum subunit FliC protein (Schönfeld *et al.*, 2003) for RSSC identification and quantification. However, these techniques require dedicated equipment and expertise. Loop-mediated isothermal amplification (LAMP) is gaining popularity due to its simplicity, rapid results and high sensitivity. LAMP can detect the pathogen in various types of samples within an hour and is easier to use than other molecular assays (Notomi *et al.*, 2000). Whole genome sequencing has also become more feasible and is cheaper, allowing analysis of different outbreak strains at the whole genome level (Wang *et al.*, 2024).

Despite advancements in molecular assays, ongoing development is needed for quick detection and differentiation of pathogen species in environmental samples. Once identified, government agencies will report the detection, detailing the pathogen's characteristics, diagnostic tests used and infection extent. Recent findings of *R. pseudosolanacearum* in the Netherlands and elsewhere in Europe highlight the importance of regular surveys and the use of robust diagnostic tests to prevent outbreaks.

Control Methods for RSSC Management

RSSC bacteria have proven to be extremely hard to control. Nonetheless, several approaches aimed at managing bacterial wilt have been employed, such as crop rotation, the use of bioorganic fertilisers and breeding of resistant cultivars (Ahmed *et al.*, 2022). Furthermore, chemical pesticides are frequently used to control pathogens and manage bacterial diseases, but they are often ineffective and their use has serious negative effects on the environment, as well as leading to bacterial resistance and disturbance of the native microbiome (Nicolopoulou-Stamati *et al.*, 2016).

Crop rotation involves growing different types of crops in the same area each year, and this approach is widely used in agriculture to control many diseases caused by soilborne pathogens. However, the ability of RSSC species to survive for long periods of time in soil (Niu *et al.*, 2017) and to infect a wide range of hosts makes this control method unsuitable. Similarly, the breeding of resistant plant cultivars is a highly desirable control approach that has achieved significant progress over the years to generate resistant varieties of some economically important crops, such as tomatoes and potatoes (Huet, 2014). However, based on genetic analyses conducted on some Solanaceous plants, resistance to pathogens seems to involve multiple genes, which might make resistance difficult to breed (Salgon *et al.*, 2017). Some promising results have recently been obtained by genetically engineering resistance genes in potato from *Arabidopsis* and wild potato relatives (Boschi *et al.*, 2017).

Biological Control

Biological control, or biocontrol, is the use of living organisms to control pests and diseases (Pal and Gardener, 2006). Insects and microorganisms, such as bacteria, fungi and viruses, act as natural enemies of pathogens and reduce their densities, minimizing the harmful effects of pathogens in their hosts and managing the incidence of disease. Some microorganisms, such as bacteria, might produce substances or compounds that inhibit the growth of pathogens and outcompete them for shelter and nutrients (Pohjanen *et al.*, 2014).

The use of different bacteria with potential biocontrol effects against RSSC species is currently being explored. Mohammed *et al.* (2020) isolated and identified seven *Pseudomonas* species from the rhizosphere of tomato plants, and they found that six of them inhibited the growth of *R. solanacearum*, including *P. aeruginosa*, *P. syringae* and *P. fluorescens*, and also promoted the growth of tomato plants (Mohammed *et al.*, 2020). Soliman (2020) isolated rhizobacterium *Paenibacillus polymyxa* from the soil of a potato field and reported on its efficacy (80%) in suppressing the incidence of potato brown rot by producing antimicrobial compounds that inhibited *R. solanacearum* growth and competitiveness (Soliman, 2020). Wang *et al.* (2019a) isolated the rhizospheric bacterial strain *Bacillus cereus* AR156 from forest soil, and they investigated the antagonistic effect of this bacterium against *R. solanacearum*. They showed that *B. cereus* AR156 was able to reduce bacterial wilt disease in tomato and showed an efficacy of 51% relative to the control (Wang *et al.*, 2019a).

Interestingly, some species of bacteria display a mutualistic relationship with plants and provide protection against pathogens (Ahmed *et al.*, 2022). Yang *et al.* (2023a) reported on how plants can modify the microbial community composition via root exudation, as particular root exudate profiles led to the recruitment of microbiota able to suppress *R. solanacearum* (Yang *et al.*, 2023a). Overall, these studies provide evidence of the beneficial effects that some bacterial species may offer as biocontrol agents of phytopathogenic bacteria. Growth inhibition of plant pathogenic bacteria due to antagonistic bacteria outcompeting them, plant growth promotion and suppression of disease incidence are some important benefits of microbial biocontrol agents. The challenge of using bacteria as biocontrol agents is that they might face difficulties in colonising and establishing a population in an already crowded microbiome.

Bacteriophages (phages), viruses that infect and kill bacteria, could provide better success by specifically targeting the pathogen (Harper, 2021). During infection, phages can enter in a lytic cycle, where they adsorb to a receptor on the bacterial cell surface, replicate inside the bacterium and finally lyse the cell, resulting in the release of new virions and the death of the bacterium (Mirski *et al.*, 2019). Alternatively, phages can enter a lysogenic cycle, in which the bacterial cell is not destroyed and the phage's genome integrates into the bacterial chromosome (Harper, 2021). Lytic phages can be used as biocontrol agents, by reducing bacterial population densities after each infection cycle. Thus far, the efficacy of a wide variety of lytic phages against RSSC bacteria has been described and reported in the literature (Elhalag *et al.*, 2018; Alvarez *et al.*, 2019; Wang *et al.*, 2019b; Yang *et al.*, 2023b). Elhalag *et al.* (2018) isolated a soilborne lytic phage from Egyptian soil to assess the effect on *R. solanacearum* phylotype IIA *in vitro* and under greenhouse conditions. They found that bacteriolytic activity of the phage was detected at different multiplicity of infection (MOI) and the phage was able to reduce and prevent the formation of exopolysaccharides of *R. solanacearum* during the exponential growth phase. Moreover, the phage proved to be effective in suppressing the bacterium growth under greenhouse conditions, as all phage-treated tomato plants showed no wilting symptoms, whereas all untreated plants wilted after 10 days of infection (Elhalag *et al.*, 2018). Likewise, Alvarez *et al.* (2019) also reported on the use of phages as a biocontrol method against *R. solanacearum* in irrigation water and host plants. Phages proved to be highly specific as to which bacterial species they infect, without damaging non-target bacteria, and were able to reduce bacterial densities when a range of environmental conditions, such as pH, temperature and salinity, were varied (Alvarez *et al.*, 2019).

Phage biocontrol is seldom perfect and one reason behind this could be the rapid evolution of phage resistance, which has been shown to evolve both in the lab and tomato rhizosphere (Wang *et al.*, 2019b). Phage resistance can be highly costly to bacteria, often compromising bacterial fitness and virulence via trade-offs (Burmeister and Turner, 2020). For example, resistance to phages in *R. solanacearum* can result in mutations in the type IV pili or type II secretion systems, which are used as likely receptors by the phage (Wang *et al.*, 2024). However, type IV pili and type II secretion systems are also important virulence factors, and mutations in these genes also result in reduced virulence (Narulita *et al.*, 2016; da Silva Xavier *et al.*, 2022). Spatial heterogeneity of the soil and the presence of competitors might constraint the evolution of phage resistance (Wang *et al.*, 2017), and our understanding of RSSC species phage resistance in natural environments remains limited. Nevertheless, phages remain a promising control strategy against RSSC bacteria but more research is needed, especially at the field scale to assess its efficacy in agricultural systems.

Future Research

1. Obtaining control strategies that stop *R. solanacearum* diseases in the field is a global priority, and future research should focus on finding environmentally friendly and efficient approaches to combat this devastating pathogen. Combined approaches, like cocktails of inhibitory bacteria and phages, should be prioritized, instead of just single biocontrol agents. These combinations would reduce pathogen levels, boost protective bacteria in the soil and potentially help to activate plant defences. This should also be linked to basic research to understand the molecular and genetic mechanisms of phage resistance, phage-bacteria interactions and the processes causing competition between bacteria. This research could help discover new biocontrol agents by allowing studies to focus on microorganisms with specific traits that make them effective against *R. solanacearum*, even under stressful environmental conditions caused by climate change.
2. Developing resistant plant cultivars is crucial for managing bacterial wilt and potato brown rot. While tolerant cultivars are used in some areas, they do not reduce bacterial levels in the soil. Identifying genes linked to resistance is essential; however, no fully resistant cultivars yet exist, due to the complex interaction between plant, bacteria and environment. Future research should focus on finding new resistance genes, engineering them into commercial cultivars and exploring wild plant relatives for potential resistance mechanisms.

Discussion Points

1. Why RSSC is one of the most important bacteria plant pathogens in the world?
2. How human agricultural activities and climate change can facilitate the host and geographic range expansion of RSSC species?
3. Given that it has been difficult to control the spread of RSSC species, which are the best alternatives for disease diagnosis, management and eradication?
4. How effective are the biological control methods to eradicate RSSC species? And, what are their limitations?

Conclusions

The spread of warm-adapted *R. pseudosolanacearum* in riverways and wild and crop plants in temperate countries such as the Netherlands, Hungary and Italy (EPPO, 2022a) poses a new threat to food security, highlighting the ability of this pathogen to potentially adapt to new environmental conditions. Climate change could worsen this situation and new outbreaks with different RSSC species can be expected in new areas in the future. Reliable, quick, cheap and easy-to-use diagnostic tests are required to stop the spread of this phytopathogen. Until now, there is no completely efficient method to eradicate this pathogen, due to its massive host range and its capacity to survive in soil, water, seeds and farming tools. Biocontrol methods have proved environmentally friendly and efficient. These control methods require further efficacy, and their use combined with resistant plant cultivars might attain improved disease control.

Further Reading

EPPO Diagnostic tools. Available at: https://gd.eppo.int/download/standard/145/pm7-021-03-rev-en.pdf (accessed 6 November 2024).

The Netherlands Plant Protection Service. Available at: https://english.nvwa.nl/topics/pest-reporting/pest-reports (accessed 6 November 2024).

Updated distribution of *R. solanacearum* in the world. Available at: https://gd.eppo.int/taxon/RALSSL/distribution (accessed 6 November 2024).

References

Ahmed, W., Yang, J., Tan, Y., Munir, S., Liu, Q. *et al.* (2022) *Ralstonia solanacearum*, a deadly pathogen: Revisiting the bacterial wilt biocontrol practices in tobacco and other Solanaceae. *Rhizosphere* 21, 100479.

Allen, C., Prior, P. and Hayward, A. (2005) *Bacterial Wilt Disease and the Ralstonia solanacearum Species Complex*. American Phytopathological Society (APS Press), St Paul, MN.

Alvarez, B., Lopez, M.M. and Biosca, E.G. (2019) Biocontrol of the major plant pathogen *Ralstonia solanacearum* in irrigation water and host plants by novel waterborne lytic bacteriophages. *Frontiers in Microbiology* 10, 2813.

Bergsma-Vlami, M., van de Bilt, J.L.J., Tjou-Tam-Sin, N.N.A., Westenberg, M., Meekes, E.T.M. *et al.* (2018) Phylogenetic assignment of *Ralstonia pseudosolanacearum(Ralstonia solanacearum* Phylotype I) isolated from *Rosa* spp. *Plant Disease* 102(11), 2258–2267.

Bittner, R.J., Arellano, C. and Mila, A.L. (2016) Effect of temperature and resistance of tobacco cultivars to the progression of bacterial wilt, caused by *Ralstonia solanacearum. Plant and Soil* 408(1/2), 299–310.

Boschi, F., Schvartzman, C., Murchio, S., Ferreira, V., Siri, M.I. *et al.* (2017) Enhanced bacterial wilt resistance in potato through expression of *Arabidopsis* EFR and introgression of quantitative resistance from *Solanum commersonii. Frontiers in Plant Science* 8, 1642.

Burmeister, A.R. and Turner, P.E. (2020) Trading-off and trading-up in the world of bacteria–phage evolution. *Current Biology* 30(19), R1120–R1124.

Castillo, J.A. and Greenberg, J.T. (2007) Evolutionary dynamics of *Ralstonia solanacearum. Applied and Environmental Microbiology* 73(4), 1225–1238.

da Silva Xavier, A., de Melo, A.G., Hendrich, C.G., Tremblay, D.M., Rousseau, G.M. *et al.* (2022) In through the out door: A functional virulence factor secretion system is necessary for phage infection in *Ralstonia solanacearum. mBio* 13(6), e0147522–e0147522.

Denny, T. (2007) *Plant Pathogenic Ralstonia species, Plant-Associated Bacteria*. Springer, Dordrecht, pp. 573–644.

Elhalag, K., Nasr-Eldin, M., Hussien, A. and Ahmad, A. (2018) Potential use of soilborne lytic podoviridae phage as a biocontrol agent against *Ralstonia solanacearum. Journal of Basic Microbiology* 58(8), 658–669.

Elphinstone, J.G. (2001) Monitoring and control of the potato brown rot bacterium (*Ralstonia solanacearum*) in the UK: A case study. In: *2nd European Potato Processing Conference*, Lausanne, Switzerland.

Elphinstone, J. (2005) Bacterial wilt disease and the *Ralstonia solanacearum* species complex. In: Allen, C., Prior, P. and Hayward, A. (eds) *The Current Bacterial Wilt Situation: A Global Overview*. American Phytopathological Society (APS Press), St. Paul, pp. 9–28.

Elphinstone, J. and Matthews-Berry, S. (2017) *Monitoring and Control of the Potato Brown Rot Bacterium in Irrigation Water*. Department for Environment Food and Rural Affairs, London.

Elphinstone, J.G., Hennessy, J., Wilson, J.K. and Stead, D.E. (1996) Sensitivity of different methods for the detection of *Ralstonia solanacearum* in potato tuber extracts. *Bulletin OEPP* 26(3–4), 663–678.

Elphinstone, J.G., Stanford, H.M. and Stead, D.E. (1998) Detection of *Ralstonia solanacearum* in potato tubers, *Solanum dulcamara* and associated irrigation water. In: Prior, P., Allen, C. and Elphinstone, J. (eds) *Bacterial Wilt Disease: Molecular and Ecological Aspects*. Springer, Berlin Heidelberg, Berlin, Heidelberg, pp. 133–139.

EPPO (2015) *Ralstonia Solanacearum (Race 1) Detected in Rosa in the Netherlands*. EPPO Global Database, Paris.

EPPO (2021) *First Report of Ralstonia Pseudosolanacearum in Germany*. EPPO Global Database, Paris.

EPPO (2022a) *First Report of Ralstonia Pseudosolanacearum in Italy*. EPPO Global Database, Paris. Available at: https://gd.eppo.int/reporting/article-7425

EPPO (2022b) *First Report of Ralstonia Pseudosolanacearum in Hungary*. EPPO Global Database, Paris.

EPPO (2022c) PM 7/21 (3) *Ralstonia solanacearum, R. pseudosolanacearum* and *R. syzygii* (*Ralstonia solanacearum* species complex). *Bulletin OEPP* 52(2), 225–261.

EPPO (2023) *New Finding of Ralstonia Pseudosolanacearum in Germany*. EPPO Global Database, Paris.

Fegan, M. and Prior, P. (2005) In: Allen, C., Prior, P. and Hayward, A. (eds) *Bacterial Wilt Disease and the R. Solanacearum Species Complex*. APS Press, St. Paul, MN, p. 449.

Genin, S. and Denny, T.P. (2012) Pathogenomics of the *Ralstonia solanacearum* species complex. *Annual Review of Phytopathology* 50(1), 67–89.

Harper, D.R. (2021) Introduction to bacteriophages. In: Harper, D.R., Abedon, S.T., Burrowes, B.H. and McConville, M.L. (eds) *Bacteriophages*. Springer, Switzerland, pp. 3–16.

Huerta, A.I., Milling, A. and Allen, C. (2015) Tropical strains of Ralstonia solanacearum outcompete race 3 biovar 2 strains at lowland tropical temperatures. *Applied and Environmental Microbiology* 81(10), 3542–3551.

Huet, G. (2014) Breeding for resistances to *Ralstonia solanacearum*. *Frontiers in Plant Science* 5, 715–715.

Janse, J. (2012) Review on brown rot (*Ralstonia solanacearum* race 3, biovar 2, phylotype IIB) epidemiology and control in the Netherlands since 1995: A success story of integrated pest management. *Journal of Plant Pathology* 257–272.

Krausz, J.P. and Thurston, H. (1975) Breakdown of resistance to *Pseudomonas solanacearum* in tomato. *Phytopathology* 65, 1271–1274.

Mansfield, J., Genin, S., Magori, S., Citovsky, V., Sriariyanum, M. *et al.* (2012) Top 10 plant pathogenic bacteria in molecular plant pathology. *Molecular Plant Pathology* 13(6), 614–629.

Mirski, T., Mizak, L., Nakonieczna, A. and Gryko, R. (2019) Bacteriophages, phage endolysins and antimicrobial peptides – The possibilities for their common use to combat infections and in the design of new drugs. *Annals of Agricultural and Environmental Medicine* 26(2), 203–209.

Mohammed, A.F., Oloyede, A.R. and Odeseye, A.O. (2020) Biological control of bacterial wilt of tomato caused by *Ralstonia solanacearum* using *Pseudomonas* species isolated from the rhizosphere of tomato plants. *Archiv Für Phytopathologie Und Pflanzenschut* 53(1–2), 1–16.

Narulita, E., Addy, H.S., Kawasaki, T., Fujie, M. and Yamada, T. (2016) The involvement of the pilq secretin of type IV pili in phage infection in *Ralstonia solanacearum* bacteriophages. *Biochemical and Biophysical Research Communications* 469(4), 868–872.

Nicolopoulou-Stamati, P., Maipas, S., Kotampasi, C., Stamatis, P. and Hens, L. (2016) Chemical pesticides and human health: The urgent need for a new concept in agriculture. *Frontiers in Public Health* 4, 148.

Niu, J., Chao, J., Xiao, Y., Chen, W., Zhang, C. *et al.* (2017) Insight into the effects of different cropping systems on soil bacterial community and tobacco bacterial wilt rate. *Journal of Basic Microbiology* 57(1), 3–11.

Notomi, T., Okayama, H., Masubuchi, H., Yonekawa, T., Watanabe, K. *et al.* (2000) Loop-mediated isothermal amplification of DNA. *Nucleic Acids Research* 28(12), e63–e63.

NVWA (2023) Pest Reports, 2023. Available at: https://english.nvwa.nl/topics/pest-reporting/pest-reports (accessed 24 August 2024).

Osdaghi, E. (2020) *Ralstonia solanacearum (Bacterial Wilt of Potato)*. CABI Digital Library, Wallingford, UK.

Pal, K.K. and Gardener, B.M. (2006) Biological control of plant pathogens. *The Plant Health Instructor* 2, 1117–1142.

Parkinson, N., Bryant, R., Bew, J., Conyers, C., Stones, R. *et al.* (2013) Application of variable-number tandem-repeat typing to discriminate *Ralstonia solanacearum* strains associated with english watercourses and disease outbreaks. *Applied and Environmental Microbiology* 79(19), 6016–6022.

Persley, G.J. (1985) Ecology of pseudomonas solanacearum, the causal agent of bacterial wilt. In: *Proceedings of an International Workshop Held at PCARRD, Los Banos, Philippines, 8–10 October 1985*, ACIAR Proceedings, Los Banos, Philippines.

Pohjanen, J., Koskimäki, J.J., Sutela, S., Ardanov, P., Suorsa, M. *et al.* (2014) Interaction with ectomycorrhizal fungi and endophytic methylobacterium affects nutrient uptake and growth of pine seedlings *in vitro*. *Tree Physiology* 34(9), 993–1005.

Prior, P., Ailloud, F., Dalsing, B.L., Remenant, B., Sanchez, B. *et al.* (2016) Genomic and proteomic evidence supporting the division of the plant pathogen *Ralstonia solanacearum* into three species. *BMC Genomics* 17(1), 90–90.

Remenant, B., Coupat-Goutaland, B., Guidot, A., Cellier, G., Wicker, E. *et al.* (2010) Genomes of three tomato pathogens within the *Ralstonia solanacearum* species complex reveal significant evolutionary divergence. *BMC Genomics* 11(1), 379–379.

Safni, I., Cleenwerck, I., Vos, P., Fegan, M., Sly, L. *et al.* (2014) Polyphasic taxonomic revision of the *Ralstonia solanacearum* species complex: Proposal to emend the descriptions of *Ralstonia solanacearum* and *Ralstonia syzygii* and reclassify current *R. syzygii* strains as *Ralstonia syzygii* subsp. *syzygii* subsp. nov., *R. solanacearum* phylotype IV strains as *Ralstonia syzygii* subsp. *indonesiensis* subsp. nov., banana blood disease bacterium strains as *Ralstonia syzygii* subsp. *celebesensis* subsp. nov. And *R.solanacearum* phylotype I and III strains as *Ralstonia pseudosolanacearum* sp. nov. *International Journal of Systematic and Evolutionary Microbiology* 64(Pt 9), 3087–3103.

Salgon, S., Jourda, C., Sauvage, C., Daunay, M.C., Reynaud, B. *et al.* (2017) Eggplant resistance to the *Ralstonia solanacearum* species complex involves both broad-spectrum and strain-specific quantitative trait loci. *Frontiers in Plant Science* 8, 828–828.

SASA (2023) *Bacteriology, Brown Rot, 2023*. Available at: https://www.sasa.gov.uk/diagnostics/bacteriology (accessed 27 August 2024).

Schönfeld, J., Heuer, H., Van Elsas, J.D. and Smalla, K. (2003) Specific and sensitive detection of *Ralstonia solanacearum* in soil on the basis of PCR amplification of fliC fragments. *Applied and Environmental Microbiology* 69(12), 7248–7256.

Sebastià, P., de Pedro-Jové, R., Daubech, B., Kashyap, A., Coll, N.S. *et al.* (2021) The bacterial wilt reservoir host *Solanum dulcamara* shows resistance to *Ralstonia solanacearum* infection. *Frontiers in Plant Science* 12, 755708.

Soliman, M. (2020) *Paenibacillus polymyxa* and *Bacillus aryabhattai* as biocontrol agents against *Ralstonia solanacearum in vitro* and *in planta*. *Journal of Plant Protection and Pathology* 11(3), 197–203.

Tjou-Tam-sin, N.N.A., van de Bilt, J.L.J., Westenberg, M., Bergsma-Vlami, M., Korpershoek, H.J. *et al.* (2017) First report of bacterial wilt caused by *Ralstonia solanacearum* in ornamental *Rosa* sp. *Plant Disease* 101(2), 378–378.

van der Gaag, D.J., Camilleri, M., Diakaki, M., Schenk, M. and Vos, S. (2019) Pest survey card on potato brown rot, *Ralstonia solanacearum*. *EFSA Supporting Publications* 16(2), n/a.

Vogelaar, M., van de Bilt, J., Blom, N., Pel, C., van Doorn, B. *et al.* (2023) Presence of *Ralstonia pseudosolanacearum* (Phylotype I) in aquatic environments in the Netherlands. *Plant Disease* 107, 2320–2324.

Wang, X., Wei, Z., Li, M., Wang, X., Shan, A. *et al.* (2017) Parasites and competitors suppress bacterial pathogen synergistically due to evolutionary trade-offs. *Evolution* 71(3), 733–746.

Wang, N., Wang, L., Zhu, K., Hou, S., Chen, L. *et al.* (2019a) Plant root exudates are involved in *Bacillus cereus* AR156 mediated biocontrol against *Ralstonia solanacearum*. *Frontiers in Microbiology* 10, 98–98.

Wang, X., Wei, Z., Yang, K., Wang, J., Jousset, A. *et al.* (2019b) Phage combination therapies for bacterial wilt disease in tomato. *Nature Biotechnology* 37(12), 1513–1520.

Wang, Z., Luo, W., Cheng, S., Zhang, H., Zong, J. *et al.* (2023) *Ralstonia solanacearum* – A soil borne hidden enemy of plants: Research development in management strategies, their action mechanism and challenges. *Frontiers in Plant Science* 14, 1141902–1141902.

Wang, J., Wang, X., Yang, K., Lu, C., Fields, B. *et al.* (2024) Phage selection drives resistance-virulence trade-offs in *Ralstonia solanacearum* plant-pathogenic bacterium irrespective of the growth temperature. *Evolution Letters* 8(2), 253–266.

Wenneker, M., Verdel, M.S.W., Groeneveld, R.M.W., Kempenaar, C., van Beuningen, A.R. *et al.* (1999) *Ralstonia (Pseudomonas) solanacearum* race 3 (Biovar 2) in surface water and natural weed hosts: First report on stinging nettle (*Urtica dioica*). *European Journal of Plant Pathology* 105(3), 307–315.

Williamson, L., Nakaho, K., Hudelson, B. and Allen, C. (2002) *Ralstonia solanacearum* race 3, biovar 2 strains isolated from geranium are pathogenic on potato. *Plant Disease* 86(9), 987–991.

Yang, K., Fu, R., Feng, H., Jiang, G., Finkel, O. *et al.* (2023a) RIN enhances plant disease resistance via root exudate-mediated assembly of disease-suppressive rhizosphere microbiota. *Molecular Plant* 16(9), 1379–1395.

Yang, K., Wang, X., Hou, R., Lu, C., Fan, Z. *et al.* (2023b) Rhizosphere phage communities drive soil suppressiveness to bacterial wilt disease. *Microbiome* 11, 16.

9 Global Insights into Bacterial Canker of Stone Fruits Caused by the *Pseudomonas syringae* Species Complex

ANDREA VADILLO-DIEGUEZ[1], ROBERT W. JACKSON[1] AND MICHELLE T. HULIN[2]

[1]*School of Biosciences and the Birmingham Institute of Forest Research, University of Birmingham, Birmingham, UK*
[2]*Department of Plant Soil and Microbial Sciences, Michigan State University, East Lansing, USA*

Introduction

Prunus is a genus from the Rosaceae family which contains more than 400 species. The most important tree species for stone fruit production are organised into three main groups: (A) black cherry (*P. serotina*), sweet cherry (*P. avium*), sour cherry (*P. cerasus*); (B) plums (*P. domestica, P. salicina*), sloes (*P. spinosa*); and (C) apricot (*P. armeniaca*), peach (*P. persica*). Although almond (*P. dulcis*) also belongs to *Prunus* spp., it is considered a nut due to the consumption of the seed rather than the fleshy layers. In the Codex Alimentarius Commission (CAC), managed by the Food and Agriculture Organization of the United Nations (FAO) and World Health Organization (WHO), stone fruits and almonds have different Food Codes: FS0012 and TN0660, respectively.

The consumption of stone fruits is beneficial for human health as they have antioxidant properties and contain high quantities of fibre, vitamins, and minerals, reducing the risk of some oxidation-associated diseases such as cardiovascular diseases, neurodegenerative diseases, and cancer (Tomás-Barberán *et al.*, 2013).

Considering stone fruits as individual commodities, their production has increased over time (Fig. 9.1a), ranking them in the bottom half of the top 30 fruit commodities (FAO, 2024). Since 2020, 14 countries have been responsible for the major production of stone fruits, highlighting the high production

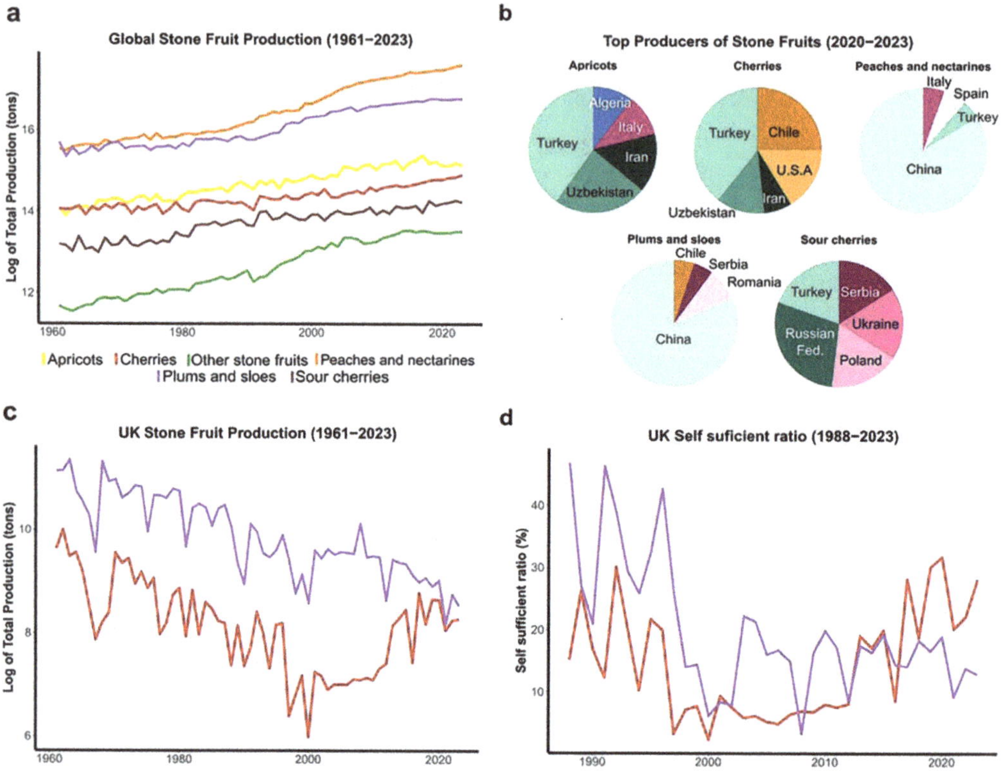

Fig. 9.1. Stone fruit production statistics globally and in the UK. (a) Total production (tons) of stone fruit commodities worldwide: Peaches and nectarines (orange), plums and sloes (purple), apricots (yellow), cherries (red), sour cherries (brown), and other stone fruits (green). Data courtesy of FAOSTATS (2024). (b) Top global producer countries for stone fruit: Asia: Green colours; Europe: pink colours; America: Orange colours; Africa: Blue colours. Data courtesy of FAOSTATS (2024). (c) Total production (tons) of stone fruit commodities in the UK: plums and sloes (purple), cherries (red). Data courtesy of FAOSTATS (2024). (d) Self-sufficient ratio of plums (purple) and cherries (cherries) in the UK (Home production/(Home production + Imports – Exports)*100). Data courtesy of DEFRA Statistics (2024).

rate of peaches in China and the diversity of stone fruits produced in Turkey (Fig. 9.1b). In 2023, Chile was the leader in export quantity and value for cherries, plum, and sloes; and Spain for apricots, peaches and nectarines. Spain was the top exporter by quantity, and Serbia the top exporter by value, for sour cherries. Regarding the importation of stone fruits by quantity and value, Russia and Germany shared the lead for apricots, sour cherries, peaches, and nectarines. Conversely, China was the top importer of cherries, plums and sloes.

In the UK (Fig. 9.1c), the production of stone fruits is limited to plums, sloes and cherries. On one hand, plum and sloe production has decreased since 1960. On the other hand, cherry production has experienced a steep increase since 2000, recovering from the lowest production ever recorded in 2000 due to the loss of orchard land. Production levels are currently approaching those of the 1980s. This recovery has been possible due to the use of dwarf trees, new farming practices, and the control of birds, pests, and diseases (Lang, 2008; Lang *et al.*, 2011). Currently, the UK is increasing the supply of home-produced cherries for the consumer demand during the British production season (mainly July), recently covering up to 30% of the self-supply ratio in a whole calendar year (Fig. 9.1d, DEFRA, 2024). Self-supply ratio is calculated as the percentage of the home-production divided by the total of the sum of home production and importation subtracting the exportation.

However, *Prunus* spp. are not only used for stone fruit production. They are planted as ornamental tree species in urban parks, and they are present in wild woodlands of temperate regions. Stone fruit production and the ornamental tree industry are threatened by several pests and diseases. Here, we will focus on the bacterial canker of stone fruits caused by strains within the *Pseudomonas syringae* (*Ps*) species complex. *P. syringae* is a rod-shaped gram-negative bacterium that was first described as a pathogen of lilac in 1902 and is widespread globally with an extensive host range (Xin *et al.*, 2018).

Economic Impact Due to Bacterial Canker

The *P. syringae* species complex is considered as one of the top 10 most important phytopathogenic bacteria due to significant economic losses and its wide host range (Mansfield *et al.*, 2012). Although economic estimates are sometimes difficult, several countries assessed the economic impact of bacterial canker at the start of the 21st century caused by this bacterium (Lamichhane *et al.*, 2014). The pathogen is particularly devastating in young orchards causing up to 75% loss of trees and economic losses between 10 and 20% among adult cherry trees (Spotts *et al.*, 2010). Outbreaks in Washington in 2023 indicate that the disease is still a threat in some US states (Manna *et al.*, 2024). Rapid death has been observed also in several orchards 1–4 years after planting in the late 2010s in Iran and South Africa (Nosratnezhad *et al.*, 2018; Bophela *et al.*, 2020). In Kyrgyzstan, several disease outbreaks around the country have been reported in 2015–2017 with a disease incidence of 80–90% (Doolotkeldieva and Bobusheva, 2020).

In the UK, a survey on bacterial canker disease in plantations and nurseries of wild cherry (*Prunus avium*) showed that 80% of the plantations and almost half of the nurseries were damaged (Vicente *et al.*, 2004) and new pathovars have been isolated as emerging pathogens in later surveys (Hulin *et al.*, 2023). In 2017, it was reported that bacterial canker could cause UK top-fruit growers to lose 30–50% of their production, having an estimated economic loss of £125–200K per annum in nursery and plantation stages and £5.6 billion per annum overall (Roberts, 2017).

Symptomatology, Pathogen Life Cycle and Causal Agents in *Pseudomonas syringae* Complex Worldwide

Bacterial canker in stone fruits can impact the tree's aerial tissues, or phyllosphere, resulting in reduced fruit yield and plant decline or death. *P. syringae* is pathogen that participates in the water cycle, can be spread through air, and within seeds. Its disease cycle begins in late autumn and winter with the formation of cankers and oozing sap known as gummosis where branches may necrotise, and cause stem die back (Fig. 9.2a). The bacteria reactivate and intensify symptoms in spring and may cause bud and blossom blast

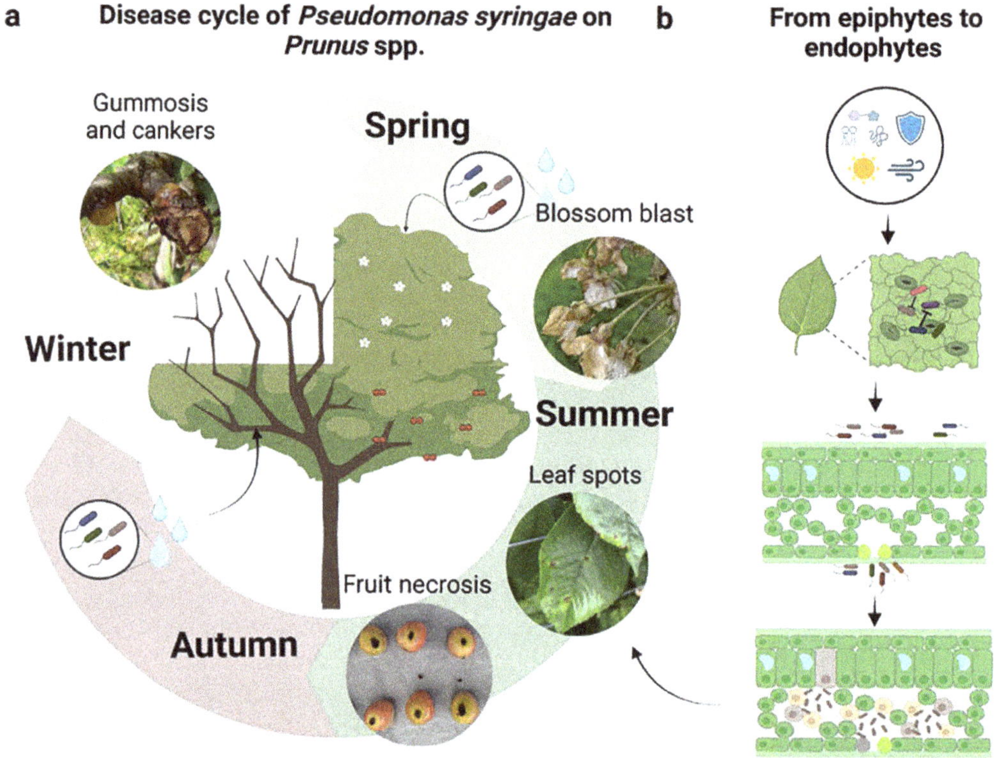

Fig. 9.2. Disease cycle of *Pseudomonas syringae* on *Prunus* spp. (a) Phenology of a cherry tree according to the seasons (dormancy in winter, flowering and leaf emergence in spring, fruiting in summer, leaf decay in autumn). Inoculum of pathogenic strains are represented as rain splashes. Images of the main symptoms are included in the different seasons (gummosis and cankers in winter, blossom blast in spring, leaf spots and fruit necrosis in summer). (b) Epiphytic phase of *Pseudomonas syringae* on the leaf surface. Nutrient availability, plant defences, UV light and weather conditions (represented as the molecules, the shield, the sun and wind respectively) impact epiphytic populations as well as the microbe-microbe interaction (represented as inhibition arrows). *Pseudomonas syringae* enter through the stomata and grows in population prior to disease development. Figure made with Biorender.

affecting fruit yield, followed by a leaf infection phase as leaves emerge. In summer, young leaves are particularly susceptible, developing spots that eventually turn into shotholes, though resistance increases as they mature.

P. syringae strains can live as epiphytes (on the plant surface) and endophytes (inside the host tissues). The epiphytic phase is an important survival stage where the pathogens are exposed to adverse conditions like UV light, nutrient starvation, microbial competition and abiotic stresses (Fig. 9.2b). Epiphytes can be used as a prediction of the presence and diversity of endophytic populations and their potential to cause disease outbreaks (Crosse, 1963; Thomas *et al.*, 2024). They can later enter the host through natural entries like stomata and lenticels or mechanical wounds like leaf scars, pruned twigs or frost damaged tissue. Once *P. syringae* enters the plant and establishes a population within the apoplast (between plant cells), it can disrupt plant immunity and create conditions to trigger disease symptom development (Crosse and Garrett, 1966; Hulin *et al.*, 2020).

The number of causal agents of bacterial canker in stone fruits within the *P. syringae* species complex has increased due to enhanced recent surveillance conducted across five continents on various *Prunus* species (Fig. 9.3a and Table 9.1). These studies identified pathogenic isolates, and their virulence was evaluated through laboratory pathogenicity tests on host tissues. *P. syringae* strains were originally classified as pathogenic varieties or pathovars and named after their host of isolation, leading to ambiguities and sometimes incorrect classifications (Young, 2010). Historical classification of these isolates relied on biochemical techniques (e.g. LOPAT tests: L, Levan production on sucrose medium; O, Oxidase production; P, pectinolitic activity; A, Arginine dihydrolase production; and T, hypersensitivity on Tobacco leaves. GATTa tests: G, Gelatin hydrolysis; A, Aesculin hydrolysis; T, Tyrosinase activity; Ta, Tartaric acid utilisation), microbiological tests (e.g. phenotype of colonies in different media), and molecular biology techniques (e.g.16S rRNA; Rep-PCR) (Young, 2010). DNA-DNA hybridisation was used to classify strains into genomospecies (n=9), where the bacterial population shares more than 70% DNA-DNA hybridisation and has a Tm of less than 5°C (Gardan *et al.*, 1999). Today, genome sequencing of isolates is common, with specific molecular markers used to classify them into phylogroups (n=13), with the aid of bioinformatic tools (Berge *et al.*, 2014; Pritchard *et al.*, 2016). In the *P. syringae* species complex, phylogroups (PG) have been delineated based on the differentiation of bacterial populations that have an average nucleotide identity (ANI) of more than 95% in conserved housekeeping genes (Xin *et al.*, 2018).

It is now understood that at least eight pathovars from five phylogroups can harbour strains with varying host ranges within *Prunus* species (Fig. 9.3b): *P. syringae* pv. *syringae* (*Pss*) from PG2; *P. amygdali* pv. *morsprunorum* (*Pam*), formerly known as *P. syringae* pv. *morsprunorum* Race 1, from PG3; *P. avellanae* (*Pav*), formerly known as *P. syringae* pv. *morsprunorum* Race 2, from PG1; *P. syringae* pv. *avii* (*Pavii*), from PG2; *P. syringae* pv. cerasi (*Pcer*), from PG2; *P. viridiflava* (*Pvir*), from PG7; *P. syringae* pv. *persicae* (*Pper*); *P. syringae* pv. *cerasicola* (*Pscer*); *P. syringae* sp. (*Psp*) from PG2, 4, and 10.

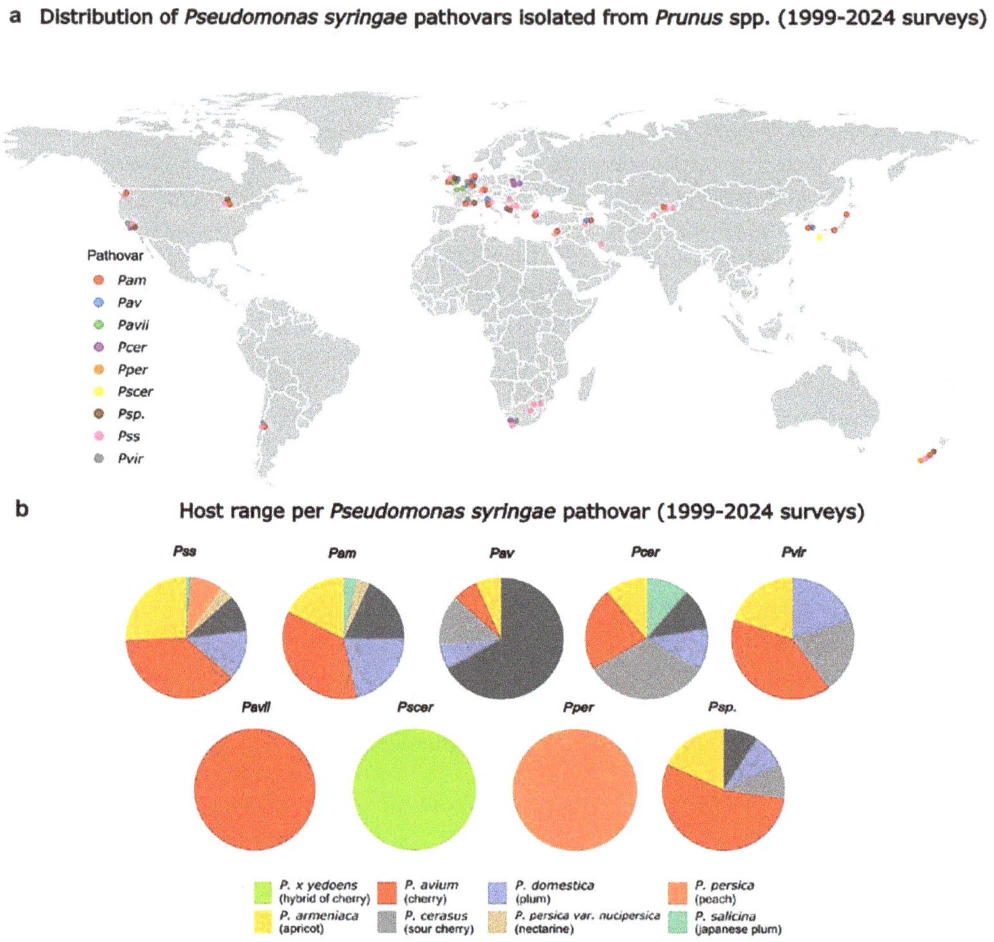

Fig. 9.3. (a) World map showing the different *Pseudomonas syringae* pathovars isolated and characterised from *Prunus* spp. in surveys conducted from 1999 to 2024 (Study References in Table 9.1). (b) *Prunus* spp. host range of different *Pseudomonas syringae* pathovars.

However, not all the phylogroup subgroups have been identified in all continents. For instance, *Pscer* has only been isolated in Japan and *Pper* and is currently considered a GB quarantine pathogen not known to occur in Great Britain. Additionally, not all the pathovars cause the same disease severity. While *Pss* has a wider host range and produces more necrotic symptoms and can produce faster cankers, *Pam* and *Pav* cause water soaking lesions and lesion progression is slower (Hulin *et al.*, 2018b).

In Europe, *Pseudomonas* population surveys from single tree species like sweet cherry, apricot or plum have been commonly performed on symptomatic and asymptomatic tissue. However, there have been multi-host samplings conducted as well. In summary, regional differences were found in bacterial populations from commercial sweet cherry within some

Table 9.1. Recent surveys conducted globally in *Prunus* spp. for the isolation and characterisation of causal agents of bacterial canker of stone fruits in the *Pseudomonas syringae* species complex.

Country	Region	Year	Source	Host	Tissue	Status	Reference
UK	Kent	2000–2001	W, Nu, O	Wild cherry, sweet cherry	Leaves, branches	S	Vicente *et al*. (2004)
UK	Kent, Somerset and Dorset, Staffordshire and Herefordshire	2017	O	Sweet cherry	Leaves, bark strips	A	Hulin *et al*. (2023)
Serbia	Zitorada, Karavukovo	2018–2019	O	Sweet cherry	Leaves, branches	S	Ilicic *et al*. (2021)
Poland	Studzianek, Pamietna, Debowa Góra, Piaski, Strojno	2001, 2007–2009	O	Sour cherry, sweet cherry	Leaves, shoots, flowers and fruits	S, A	Kałuzna *et al*. (2016)
Turkey	Tekirdag	2012–2013	O	Sweet cherry	Branches, shoots	S	Mirik *et al*. (2015)
Turkey	Marmara	2016–2018	O	Sweet cherry	Leaves, branches, flowers, buds	S	Oksel *et al*. (2022)
Turkey	Aras Basin	2021–2022	O	Apricot	Shoots, trunk, leaf, fruit	S	Akbaba *et al*. (2023b)
France	Picardy, Normandy, Ardennes	1991, 1995	NS	Wild cherry	Cankers	S	Ménard *et al*. (2003)
France	Rhône Valley	2011, 2012, and 2015	O	Apricot	Branches	S	Parisi *et al*. (2019)
Germany	South-western Germany	2001, 2003	O	Plum	Shoots, twigs, buds, and leaves	S	Hinrichs-Berger (2004)
Belgium	Not specified	1993–2002	O	Plum, sweet cherry	Fruits, leaves, stems, flowers, and buds	S	Gilbert *et al*. (2009)
Italy	Eastern Emilia-Romagna	2014–2016	O	Apricot	Leaves, fruits, and twigs	S	Giovanardi *et al*. (2018)

Continued

Table 9.1. Continued

Country	Region	Year	Source	Host	Tissue	Status	Reference
Netherlands	Not specified	2009–2010	O	Plum	Stem	S	Wenneker *et al*. (2012)
Montenegro	Not specified	2017–2018	O	Apricot, peach, nectarine, sweet cherry, Japanese plum, almond	Leaves, petioles, fruits, branches, shoots, buds	S	Popović *et al*. (2021)
Lebanon	Mont Lebanon, South and Nabatiyeh, North Lebanon, Bekaa	2013	O	Apricot, peach, nectarine, sweet cherry, Japanese plum, almond	Leaves, fruits, branches	S, A	Moubarak (2016)
Iran	North-western Iran	Not specified	O	Peach, plum, sweet cherry, and sour cherry	Not specified	S	Nosratnezhad *et al*. (2018)
Iran	East Azerbaijan province	2015	O	Apricot	Buds, blossoms, twigs	S	Vasebi *et al*. (2019)
Japan	Miyazaki and Saga prefectures	1994–1996	O	Cherry (P. x yedoensis)	Trunks and twigs	S	Kamiunten *et al*. (2000)
Korea	Japan and Korea	Not specified	O	Japanese apricot, Japanese plum	Not specified	S	Lee *et al*. (2012)
Kyrgyzstan	Chy, Issuk-Kul, and Batken	Not specified	O	Apricot, peach, plum, sweet cherry	Not specified	S	Doolotkeldieva and Bobusheva (2020)

New Zealand	Not specified	Not specified	O	Nectarine	Shoots, trunks, rootstocks, roots, leaves	S	Young (1988)
New Zealand	Central Otago	2015	O	Sweet cherry	Fruit, leaves, buds	S,A	Marroni *et al*. (2024b)
US	California	1995–1996	O	Almond, prune, plum, peach, apricot, and sweet cherry	Flowers, buds, leaves, twigs, and branches	S,A	Little *et al*. (1998)
US	California	2022–2023	O	Sweet cherry	Flowers, buds, leaves, twigs, and branches	S,A	Maguvu *et al*. (2024)
US	Michigan	2003–2004	O	Sweet cherry	Blossom	A	Renick *et al*. (2008)
US	Washington	2023	O	Sweet cherry	Not specified	S	Manna *et al*. (2024)
Chile	Libertador General Bernardo O'Higgins	2017–2018	O	Sweet cherry	Buds and leaves	S	Correa *et al*. (2022); Beltrán *et al*. (2023)
Chile	Los Lagos	2019	O	Sweet cherry	Leaves	S	García *et al*. (2021)
South Africa	Gauteng, Mpumalanga, Free State and Western Cape)	2014–2015	O	Sweet cherry	Leaves, stems, and branches	S	Otto *et al*. (2018)

Source: W: woodland plantation; Nu: nursery; O: orchard; NS: not specified. Status: S: symptomatic; A: asymptomatic.

European countries (Ilicic *et al.*, 2021; Hulin *et al.*, 2023). However, in other countries, regional trends in bacterial diversity were not seen (Ménard *et al.*, 2003). Usually, *Pss* is the most predominant phylogroup (PG2), exhibiting high genetic diversity, with strains clustering into different groups within the phylogroup. *Pam* and *Pav* were found at low frequencies from sweet cherry in the early 2000s but have not been identified in later 2010–2020s studies conducted in the UK and Turkey (Vicente *et al.*, 2004; Mirik *et al.*, 2015; Oksel *et al.*, 2022; Hulin *et al.*, 2023). Novel pathogen isolates are emerging in both sweet cherry and apricot hosts from other phylogroups (*Pavii, Pcer, Pvir*) in several countries (Table 9.1).

Pseudomonas syringae virulence factors

Plants resist infection by most pathogens due to non-host resistance (Panstruga and Moscou, 2020). This phenomenon is defined when "all genotypes of a plant species" suppress the ability of "all genotypes of a pathogen species" to spread and reproduce in the plant. However, pathogens adapted to a host use a wide range of molecules, known as virulence factors, to interact with and overcome plant immunity. The manipulation of host cellular processes is essential to ensure bacterial infection and colonisation of susceptible plants, leading to disease. Recent research has focused on the major virulence factors of *Ps* involved in bacterial canker disease. These include type III effectors, phytotoxins, phytohormones and ice nucleation proteins (Fig. 9.4, Kennelly *et al.*, 2007; Hulin *et al.*, 2020; Marroni *et al.*, 2024a).

Type III effectors are proteins translocated into the plant cell cytoplasm by the bacterium via a Type III secretion system (T3SS), a pilus with a long-needle-like structure. These effector proteins can suppress or induce plant immunity. Signals from the plant environment, often linked to the minimal nutrient environment of the apoplast, induce the expression of regulatory bacterial genes leading to the formation of structural proteins of T3SS and expression of effector genes. Comparative genomics has given insight into the variability of effector genes present in the genomes of the isolates from *Prunus* spp. (Ruinelli *et al.*, 2019; Hulin *et al.*, 2020). Based on their frequency within PGs, convergent patterns of gain and loss of effector genes linked to sweet cherry pathogenicity appear to have occurred e.g. the gain of *hopAR1*, *hopBB1*, *hopBF1* and *hopH1* and the loss of *hopAB1* and *avrPto1* (Hulin *et al.*, 2018a). In addition, effector genes can be encoded on mobile genetic elements and shared among different strains and incorporated into their genomes via a mechanism called horizontal gene transfer. In doing so, this can help pathogens to rapidly adapt to overcome host immunity or access new host plants. Effector proteins have been shown to have functional redundancy (multiple effectors can have the same function or role) and their role in disease severity could be tissue dependent as seen in sweet cherry (Vadillo-Dieguez *et al.*, 2024). Stone fruit pathogens from different PGs differ in the number of effector genes. *Pam* and *Pav* encode a higher number of effector proteins (from 30 to 35). By contrast, *Pss* possesses fewer effectors' genes (15–18) and *Pvir* has

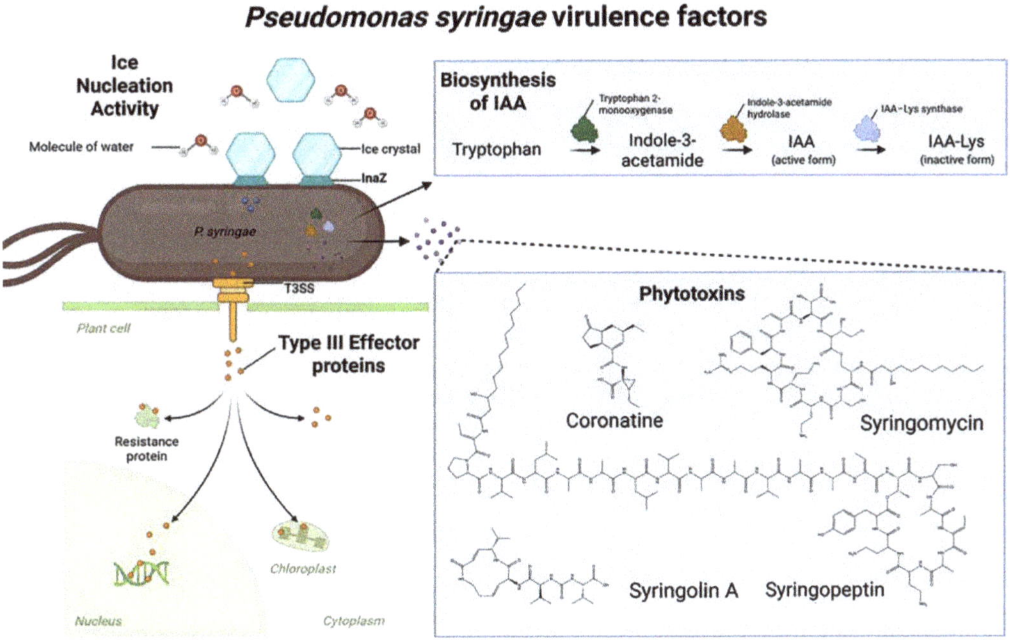

Fig. 9.4. Main virulence factors of *Pseudomonas syringae* species complex in relation to *Prunus* spp. Figure made with Biorender.

even fewer (4) and usually can harbour different types of T3SS (Lipps and Samac, 2022). Although *Pvir* has lower number of effector encoded genes, it has other virulence factor like pectate lyase activity, phenotype variability for the conformation of biofilms and production of ecomycin as an antimycotic toxin (Lipps and Samac, 2022).

Phytotoxins are bacterial molecules commonly produced by strains from PG2, which may compensate for the low number of effectors in *Pss* (Hulin *et al.*, 2018a; Xin *et al.*, 2018). They are structurally diverse, may be constitutively produced or induced by plant signals, and are secreted from the bacterium into the plant apoplast. Gene clusters for biosynthesis of phytotoxins, such as syringomycin, syringopeptin, syringolin A, and mangotoxin, are commonly present in stone fruit pathogens of this phylogroup. Syringomycin and syringopeptin are non-ribosomal lipopeptides that contribute to disease severity by inducing pores in the plant cell membranes leading to ion leakage and necrosis (Bender *et al.*, 1999). The importance of syringomycin and syringopeptin in fruit disease development and necrosis has been previously tested (Scholz-Schroeder *et al.*, 2001; Vadillo-Dieguez *et al.*, 2024). Syringolin A is a peptide derivative that overcomes stomatal immunity and inhibits the plant cell proteasomes (Schellenberg *et al.*, 2010). Mangotoxin is an antimetabolite toxin usually present in *Pss*, which might lead to inhibition of plant amino acids synthesis (Carrion *et al.*, 2012). *Pam* strains can encode genes to produce coronatine, which induces chlorosis and suppresses stomata closure. *Ppe* strains can produce persicomycin that can cause necrosis on peach with unknown function (Barzic and Guittet, 1996; Lee *et al.*, 2013). Some of these toxins are known to be thermoregulated.

Some *P. syringae (Ps)* strains have genes involved in the biosynthesis of phytohormones such as the auxin indoleacetic acid (IAA) (Ruinelli *et al.*, 2019). The production of bacterial phytohormones may disrupt the normal plant growth patterns and be involved in the production of galls and cankers (Rodríguez-Palenzuela *et al.*, 2010).

Cold temperatures in the field can lead to a more severe bacterial canker and apical necrosis. Some *Ps* possess ice nucleation activity (INA), where the protein InaZ can catalyse ice formation, named bionucleation, triggering frost damage just below 0°C. Bionucleation can help *Ps* dissemination through the atmosphere. However, it is unclear if the bacterial-produced frost-injury facilitates entry into the plant and if *Ps* also needs to thaw the frozen tissue before it can enter the damaged openings (Kennelly *et al.*, 2007).

Current Management Practices, Disease Control

Management of bacterial diseases is difficult due to the complex disease life cycle, lack of durable host resistance and ineffective chemical treatments, which might be considered harmful to the environment. For prevention and phytosanitary practices, growers can choose some rootstock and scion combinations that are more tolerant to the disease (Kappel *et al.*, 2012). The variation in cultivar phenology during the season and the age of the orchard might be involved in tolerance to the disease (Hulin *et al.*, 2020). Pruning is recommended during the dry season to cut out diseased material during dormancy and destroy infected material to reduce inoculum levels. Tools should be cleaned before and after use to eliminate the spread of the pathogen to healthy tissues. Clean and efficient irrigation systems also help to reduce the inoculum levels to avoid spreading the pathogen to other areas. In the UK, over-head covers are used to prevent sweet cherry fruit cracking as a result of rain. This also helps trees to be less exposed to rain, which acts as a potential inoculum of pathogens. In summary, the implementation of quarantine and sanitation measures, early detection systems, eradication processes for propagated plant material from nurseries is necessary for the disease control (Puławska *et al.*, 2017).

For chemical treatments, copper compound sprays (copper oxide, copper oxychloride, and copper hydroxide) are commonly used as a bactericide treatment. However, one of the downsides is that it needs to be applied in certain times of the year to align with pathogen epiphyte population and the tree phenology to avoid phytotoxicity. Nevertheless, it is currently limited or banned in European countries due to ecotoxicological and environmental risks as a heavy metal (La Torre *et al.*, 2018) and the potential emergence of copper-resistant strains (Sundin *et al.*, 1989; Beltrán *et al.*, 2021). The use of antibiotics like streptomycin and tetracycline have been applied in the past, but this approach is not widely supported because it drives the emergence of antibiotic-resistant strains in the environment. These chemicals can also have phytotoxicity, which reduces their practicality (Kennelly *et al.*, 2007). Some other chemicals (fosetyl-Al, potassium phosphite, and acibenzolar-S-methyl) have been observed to enhance plant resistance prior to pathogen interaction (Percival and Banks, 2015; Bugiani *et al.*, 2018). However, there are some limitations in fosetyl-Al efficacy as a bactericide (Bultreys *et al.*, 2018).

Durable disease-resistant cultivars adapted to specific climate conditions are the most environmentally friendly approach. However, traditional breeding is challenging due to the time for creating new population generations, the disease screenings trials, the commercialisation and planting acceptance, and the establishment of new orchards.

Few studies have tested a wide selection of tree cultivars for resistance to the disease. This could be due to the challenge of requiring disease resistance in different tissues (in the knowledge that pathogens use different virulence factors in different plant niches) and seasonal environment conditions. The most up-to-date conclusions are that there might be some partial resistance or tolerance in some apricot cultivars: some local apricot cultivars showed resistance towards *Pss* in a study in Turkey and Pakistan (Donmez *et al.*, 2010; Bibi *et al.*, 2022). However, when some cultivars from the Aras Valley were evaluated *in vitro* through imaging technologies, no total tolerance was found (Akbaba *et al.*, 2023a).

The scenario is less promising for cherry cultivars. Sweet and sour cherry cultivars tested in Greece, Iran, and Serbia ranged from high susceptibility to partial tolerance towards different *Ps* strains (Thomidis and Exadaktylou, 2008; Farhadfar *et al.*, 2016; Iličić *et al.*, 2018). In the UK, ornamental and some sweet cherry cultivars were tested with different *Ps* pathovars. Merton Glory variety showed a partial tolerance to the vast majority of isolates and *Prunus incisa*, an ornamental cherry, showed full resistance (Hulin *et al.*, 2022; Lienqueo *et al.*, 2024). Wild species have been used in plant breeding as they act as a source of resistance genes. Screening of wild cherries for resistance to *Ps* strains has been performed displaying only partial resistance (Vicente and Roberts, 2003). Hybrids obtained from *P. incisa* and susceptible cherry cultivars showed that resistance traits could be inherited. Breeding programs in France and Washington (U.S.A) showed reduced susceptibility than parental lines, inferring potential traits inheritance (Santi *et al.*, 2004; Mgbechi-Ezeri *et al.*, 2017). Less is known for other types of *Prunus*. However, some specific cultivars of peach and of plum were found to be resistant in a study in Pakistan (Bibi *et al.*, 2022).

Emerging Research Fields

Pathogen genomic surveillance

Pathogen surveillance is vital to track, diagnose, and gather information on the causal agent of a disease. Historically, biochemical analysis and microbiological techniques were more commonly used. Nowadays, with the decrease in cost of the next-generation sequencing tools, genomic surveillance is more

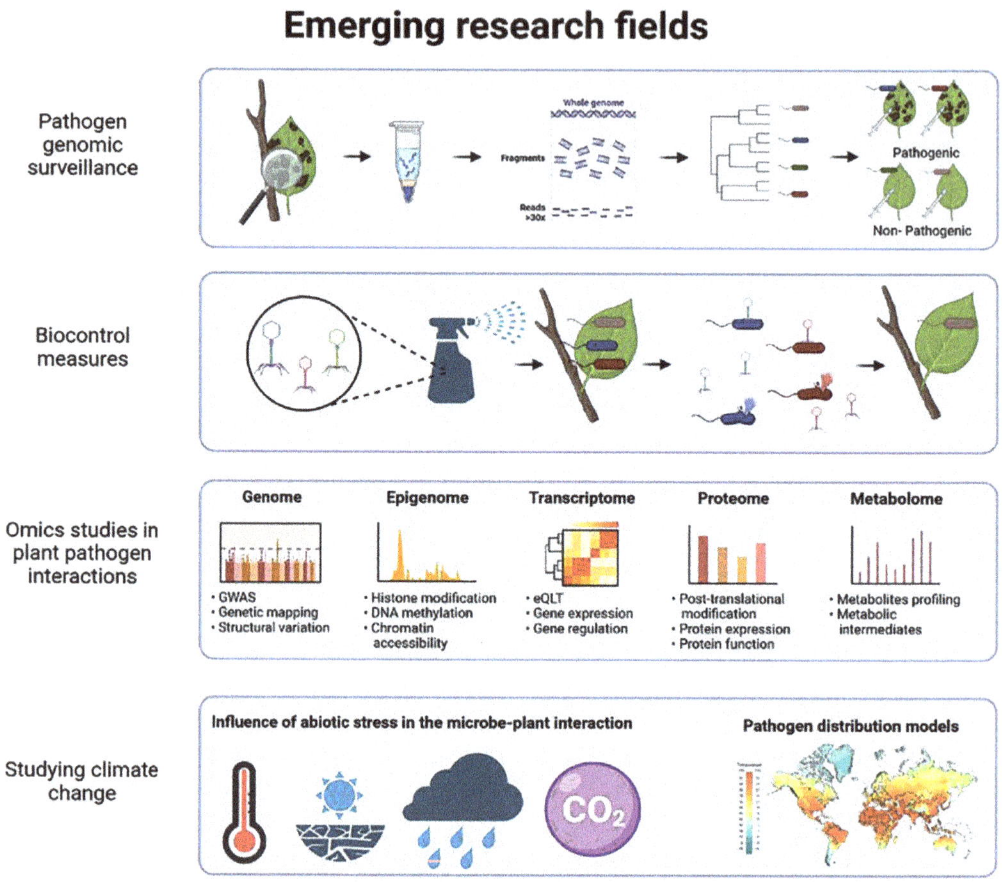

Fig. 9.5. Emerging research fields related to *Pseudomonas-Prunus* pathosystem: Pathogen genomic surveillance, biocontrol measures, omics studies in plant pathogen interactions and studying climate change. GWAS = genome wide association study, eQTL = expression quantitative trait loci. Figure made with Biorender.

informative and powerful for tracking pathogen spread and evolution and may help with targeted application of control measures against particular pathovars. Future research will enhance our understanding of the molecular evolution of plant pathogens and consequently allow improvements in the prediction of pathogen evolution trajectories and the emergence of new disease outbreaks (Fig. 9.5) (Dodds and Rathjen, 2010; Zhou and Zhang, 2020).

Biocontrol measures

The development and implementation of biocontrol measures is an exciting field that holds great promise as an additional way to help control diseases.

Many field sampling campaigns carry out surveys and isolation of epiphytes to study the plant microbiome. This can lead to the discovery of microbes (fungi, bacteria, bacteriophages) that could be utilised for biocontrol. The use of bacteriophages against bacterial cherry canker has been studied (Rabiey *et al.*, 2020; Akbaba and Ozaktan, 2021; Córdova *et al.*, 2023). Phages can target specific strains and kill them by cell lysis. Studies have shown that spraying phage cocktails in glasshouse trials or field grown trees can reduce disease; thus, indicating a promising alternative to current control measures for canker, like copper sprays (Gdanetz *et al.*, 2024). Key challenges for deployment of phage application are to optimise the methodology for application to ensure phages can survive in field conditions. It is also critical to ensure the application of phage (as an antimicrobial) does not lead to the emergence of phage-resistant strains, similar to antibiotic resistance. The use of cocktails of phages (two or more phages that are distinct) may help with this and a recent field trial highlighted that phage-resistant strains could not be detected in cherry orchards after application of phages (Papp-Rupar *et al.*, 2025).

Genomics-driven resistance breeding

Recent advancements in molecular genetic and genomic technologies have significantly transformed breeding programs. They improve traditional breeding programs by identifying genetic variations using molecular markers, speeding up trait identification using genomic selection and enabling targeted modifications using gene editing tools (Kumar *et al.*, 2024). The increasing application of plant genomic-association techniques, such as marker-assisted selection and multi-omics approaches, is facilitating the identification of resistance genes, which can then be introduced into more susceptible varieties (Zambounis *et al.*, 2020). These tools are invaluable for accelerating the selection of superior genotypes, offering high reproducibility and accuracy. For example, a genome-wide association study (GWAS) was performed in several cultivars of apricot correlating canker disease phenotypic data with changes in the cultivar genomes. Candidate resistance genes and areas in their chromosomes, known as quantitative trait loci (QTLs), responsible for resistance were found that might be linked to the plant hormone abscisic acid pathway, important for plant immunity (Omrani *et al.*, 2019).

Omics studies in plant-pathogen interactions

Understanding plant-pathogen interactions is critical for identifying the most effective approaches to uncover resistance genes. Advances in sequencing technologies have greatly expanded genome annotations and enabled transcriptome profiling under pathogen-challenge conditions, which can support research efforts for breeding disease resistance (Cui *et al.*, 2023).

Additionally, whole-genome re-sequencing (WGRS) allows for the characterisation of genetic diversity in cultivars, while genotyping-by-sequencing (GBS) facilitates the identification of single nucleotide polymorphisms (SNPs) and persistent mutations associated with resistance traits across *Prunus* spp. populations (Zambounis *et al.*, 2020).

The tissue-specific roles of effectors and toxins in *Pss* pathogenicity towards sweet cherry have been demonstrated through gene-deletion approaches (Vadillo-Dieguez *et al.*, 2024). The development of a library of virulence factor deletion mutants provides a powerful resource for deciphering molecular interactions and immune responses in *Prunus* species. This approach, known as effectoromics, can accelerate the identification of resistance factors by screening for specific immune responses triggered by individual or grouped effectors. Finally, metabolomics offers another layer of insight into plant-pathogen interactions by identifying key metabolites up- and downregulated in plant-pathogen interactions (Vinchira-Villarraga *et al.*, 2024), thus offering an opportunity to identify those involved in resistance mechanisms (Castro-Moretti *et al.*, 2020). This approach complements genomics and transcriptomics, providing a more comprehensive understanding of the molecular and biochemical pathways underpinning disease resistance.

Studying climate change

Finally, climate change, such as an increase in temperature, CO_2 concentrations in the atmosphere, and more common extreme weather fluctuations, is a topic that needs to be taken into consideration to predict disease dynamics (Hunjan and Lore, 2020). It has been reviewed that temperature impacts Plant-*Pseudomonas* interactions; therefore, it could be affected by climate change (Tribelli and López, 2022). Some virulence factors like the phytotoxin coronatine or phaseolotoxin are thermoregulated with optimum expression at 18–20°C (Smirnova *et al.*, 2001). In addition, the T3SS, T6SS expression and the swimming and swarming motility phenotypes are also influenced by temperature (van Dijk *et al.*, 1999; Arvizu-Gómez *et al.*, 2013; Hockett *et al.*, 2013). Different strains of *P. syringae* optimally express the T3SS at different temperatures (Caullireau *et al.*, 2024). It is known that frost temperatures are important for suitable infection and disease severity in apricot and peach (Weaver, 1978; Klement *et al.*, 1984). However, more sporadic rainfall or adverse weather may lead to more disease. The composition of epiphytic populations, which are the source of pathogens, could change significantly due to differing environment conditions which may favour certain species over others. A study has revealed that sweet cherry plants prioritise their response to cope with summer drought, overshadowing the defence response towards *Pss* (Villalobos-González *et al.*, 2024). It shows that *Pss* reduces sweet cherry growth under well-watered conditions, while drought stress dominates under water deficit, shifting the balance between the hormones abscisic acid (ABA) and salicylic acid (SA), making the trees more susceptible under drought stress. Overall, climate may affect pathogen geographical distribution or tree phenology (Paltineanu and Chitu, 2020). The emergence of mathematical models to provide insights of the potential future distribution of bacterial phytopathogens globally showed that *P. syringae* could expand its geographic range towards the poles (Khalaf *et al.*, 2024).

Discussion Points

1. Discuss the use of field surveys in disease outbreaks prevention and the reason behind the new emergence of pathovars and the countries where they are present.
2. Describe good practices that can be implemented in an orchard to prevent pathogen outbreaks.
3. Explain how new technologies can help in improving breeding lines or generating new biocontrol agents.

Further Readings

Crosse, J.E. (1963) Bacterial canker of stone-fruits. *Annals of Applied Biology* 52, 97–104.

Hulin, M.T., Mansfield, J.W., Brain, P., Xu, X., Jackson, R.W. *et al.* (2018) Characterization of the pathogenicity of strains of *Pseudomonas syringae* towards cherry and plum. *Plant Pathology* 67, 1177–1193.

Hulin, M.T., Jackson, R.W., Harrison, R.J. and Mansfield, J.W. (2020) Cherry picking by pseudomonads: After a century of research on canker, genomics provides insights into the evolution of pathogenicity towards stone fruits. *Plant Pathology* 69, 962–978.

Hulin, M.T., Vadillo Dieguez, A., Cossu, F., Lynn, S., Russell, K. *et al.* (2022) Identifying resistance in wild and ornamental cherry towards bacterial canker caused by *Pseudomonas syringae*. *Plant Pathology* 71, 949–965.

Kennelly, M.M., Cazorla, F.M., De Vicente, A., Ramos, C. and Sundin, G.W. (2007) *Pseudomonas syringae* diseases of fruit trees: Progress toward understanding and control. *Plant Disease* 91, 4–17.

Lamichhane, J.R., Varvaro, L., Parisi, L., Audergon, J.-M. and Morris, C.E. (2014) Chapter four - disease and frost damage of woody plants caused by *Pseudomonas syringae*: Seeing the forest for the trees. In: *Advances in Agronomy*. Academic Press.

Marroni, M.V., Casonato, S., Pitman, A.R., Visnovsky, S.B., Beresford, R.M. *et al.* (2024) Review of pseudomonas species causing bacterial canker of prunus species with emphasis on sweet cherry (*P. avium*) in New Zealand. *European Journal of Plant Pathology* 168, 297–314.

References

Akbaba, M. and Ozaktan, H. (2021) Evaluation of bacteriophages in the biocontrol of *Pseudomonassyringae* pv. *syringae* isolated from cankers on sweet cherry (*Prunus avium* L.) in Turkey. *Egyptian Journal of Biological Pest Control* 31, 35. DOI: 10.1186/s41938-021-00385-7.

Akbaba, M., Hürkan, K. and Karahan, A.E. (2023a) *In vitro* evaluation of apricot cultivars response to *Pseudomonas syringae* pathovars: Image processing as an alternative method. *Journal of Agricultural Sciences* 16–16.

Akbaba, M., Hürkan, K. and Özcan, O. (2023b) Characterization of causal agents of bacterial canker on apricot plantations and risk mapping using GIS in Aras Basin (Türkiye). *Journal of Phytopathology* 171, 517–536.

Arvizu-Gómez, J.L., Hernández-Morales, A., Jofre-Garfias, A.E. and Álvarez-Morales, A. (2013) Temperature, the GacS/GacA system, and host metabolites regulate the type VI secretion system of *Pseudomonas syringae* pv. phaseolicola. *Journal of General Plant Pathology* 79, 194–197.

Barzic, M.-R. and Guittet, E. (1996) Structure and activity of Persicomycins, Toxins produced by a *Pseudomonas syringae* Pv. Persicae/*Prunus* persica isolate. *European Journal of Biochemistry* 239, 702–709.

Beltrán, M.F., Osorio, V., Lemus, G., Millas, P., France, A. *et al.* (2021) Bacterial community associated with canker disease from sweet cherry orchards of central valley of Chile presents high resistance to copper. *Chilean Journal of Agricultural Research* 81, 378–389.

Beltrán, M.F., Correa, F., Moreno, Z., Otárola, J., Sagredo, B. *et al.* (2023) First report of *Pseudomonas* viridiflava causing fruit rot on sweet cherry trees in Chile. *Plant Disease* 107, 1216.

Bender, C.L., Alarcón-Chaidez, F. and Gross, D.C. (1999) *Pseudomonas syringae* phytotoxins: Mode of action, regulation, and biosynthesis by peptide and polyketide synthetases. *Microbiology and Molecular Biology Reviews* 63, 266–292.

Berge, O., Monteil, C.L., Bartoli, C., Chandeysson, C., Guilbaud, C. *et al.* (2014) A user's guide to a data base of the diversity of *Pseudomonas syringae* and its application to classifying strains in this phylogenetic complex. *PLoS ONE* 9, e105547.

Bibi, S., Inam-Ul-Haq, M., Riaz, A. and Malik, S.I. (2022) Identification of host resistance against bacterial canker disease on apricot, plum and peach grown in Punjab and KPK, Pakistan. *International Journal of Phytopathology* 11, 43–48. DOI: 10.33687/phytopath.011.01.4130.

Bophela, K.N., Petersen, Y., Bull, C.T. and Coutinho, T.A. (2020) Identification of *Pseudomonas* isolates associated with bacterial canker of stone fruit trees in the Western Cape, South Africa. *Plant Disease* 104, 882–892.

Bugiani, R., Pradolesi, G. and Donati, G. (2018) *Efficacy Evaluation of Acibenzolar-Methyl in Spring and Autumn Applications Against Pseudomonas syringae pv.* Actinidiae Causal Agent of Kiwifruit Bacterial Canker.

Bultreys, A., Gheysen, I., Rousseau, G., Pitchugina, E., Planchon, V. *et al.* (2018) Antibacterial activity of fosetyl-al, ethyl-phosphite and phosphite against *Pseudomonas syringae* on plant surfaces and *in vitro*. *Plant Pathology* 67, 1955–1966.

Carrion, V.J., Arrebola, E., Cazorla, F.M., Murillo, J. and Vicente, A. (2012) The mbo operon is specific and essential for biosynthesis of mangotoxin in *Pseudomonas syringae*. *PLoS ONE* 7, e36709.

Castro-Moretti, F.R., Gentzel, I.N., Mackey, D. and Alonso, A.P. (2020) Metabolomics as an emerging tool for the study of plant–pathogen interactions. *Metabolites* 10, 52.

Caullireau, E., Danzi, D., Tempo, V.M., Pandolfo, M., Morris, C.E. *et al.* (2024) *Some Like it Hot: Efficiency of the Type III Secretion System and Activation of Plant Defenses have Multiple Thermosensitive Behaviors in the Pseudomonas syringae Complex*. Cold Spring Harbor Laboratory.

Córdova, P., Rivera-González, J.P., Rojas-Martínez, V., Fiore, N., Bastías, R. *et al.* (2023) Phytopathogenic *Pseudomonas syringae* as a threat to agriculture: Perspectives of a promising biological control using bacteriophages and microorganisms. *Horticulturae* 9, 712.

Correa, F., Beltrán, M.F., Millas, P., Moreno, Z., Hinrichsen, P. *et al.* (2022) Genome sequence resources of *Pseudomonas syringae* strains isolated from sweet cherry orchards in Chile. *Molecular Plant-Microbe Interactions* 35, 933–937.

Crosse, J.E. (1963) Bacterial canker of stone-fruits. *Annals of Applied Biology* 52, 97–104. DOI: 10.1111/j.1744-7348.1963.tb03731.x.

Crosse, J. and Garrett, C.M. (1966) Bacterial canker of stone-fruits: Infection experiments with *Pseudomonas* mors-prunorum and *P. syringae*. *Annals of Applied Biology* 58, 31–41.

Cui, W., Fiore, N., Figueroa, F., Rubilar, C., Pizarro, L. *et al.* (2023) Transcriptome analysis of sweet cherry (*Prunus avium* L.) Cultivar 'Lapins' upon infection of *Pseudomonas syringae* pv. *syringae*. *Plants* 12(21), 3718. DOI: 10.3390/plants12213718.

DEFRA (2024) Horticulture statistics 2023. In: Food and Rural Affairs(ed.) *Department for Environment*. Available at: https://www.gov.uk/government/statistics/latest-horticulture-statistics/horticulture-statistics-2023

Dodds, P.N. and Rathjen, J.P. (2010) Plant immunity: Towards an integrated view of plant–pathogen interactions. *Nature Reviews Genetics* 11, 539–548.

Donmez, M.F., Karlidag, H. and Esitken, A. (2010) Identification of resistance to bacterial canker (*Pseudomonas syringae* pv. *syringae*) disease on apricot genotypes grown in Turkey. *European Journal of Plant Pathology* 126, 241–247.

Doolotkeldieva, T. and Bobusheva, S. (2020) Characterization of *Pseudomonas syringae* pv. *syringae* from diseased stone fruits in Kyrgyzstan and testing of biological agents against pathogen. *International Journal of Phytopathology* 9, 21.

FAO (2024) *Agricultural Production Statistics 2010–2023*. FAOSTAT Analytical Briefs, No. 96. Rome. Available at: https://openknowledge.fao.org/handle/20.500.14283/cd3755en

Farhadfar, S., Keshavarzi, M., Bouzari, N., Moghadam, L. and Soleimani, A. (2016) Susceptibility of cherries to bacterial canker (*Pseudomonas syringae* pv. *syringae*) in field and laboratory. *International Journal of Agriculture and Forestry* 6, 20–27.

García, H.M., Miranda, E.M., López, M.A., Parra, S.J., Rubilar, C.F. *et al.* (2021) First report of bacterial canker caused by *Pseudomonas syringae* pv. morsprunorum race 1 on sweet cherry in Chile. *Plant Disease* 105, 3287.

Gardan, L., Shafik, H., Belouin, S., Broch, R., Grimont, F. *et al.* (1999) DNA relatedness among the pathovars of *Pseudomonas syringae* and description of *Pseudomonas tremae* sp. nov. and *Pseudomonas cannabina* sp. nov. (ex Sutic and Dowson 1959). *International Journal of Systematic Bacteriology* 49(Pt 2), 469–478. DOI: 10.1099/00207713-49-2-469.

Gdanetz, K., Dobbins, M.R., Villani, S.M., Outwater, C.A., Slack, S.M. *et al.* (2024) Multisite field evaluation of bacteriophages for fire blight management: Incorporation of ultraviolet radiation protectants and impact on the apple flower microbiome. *Phytopathology* 114, 1028–1038.

Gilbert, V., Legros, F., Maraite, H. and Bultreys, A. (2009) Genetic analyses of *Pseudomonas syringae* isolates from Belgian fruit orchards reveal genetic variability and isolate-host relationships within the pathovar syringae, and help identify both races of the pathovar morsprunorum. *European Journal of Plant Pathology* 124, 199–218.

Giovanardi, D., Ferrante, P., Scortichini, M. and Stefani, E. (2018) Characterisation of *Pseudomonas syringae* isolates from apricot orchards in north-eastern Italy. *European Journal of Plant Pathology* 151, 901–917.

Hinrichs-Berger, J. (2004) Epidemiology of *Pseudomonas syringae* pathovars associated with decline of plum trees in the southwest of Germany. *Journal of Phytopathology* 152, 153–160.

Hockett, K.L., Burch, A.Y. and Lindow, S.E. (2013) Thermo-regulation of genes mediating motility and plant interactions in *Pseudomonas syringae*. *PLoS ONE* 8, e59850.

Hulin, M.T., Armitage, A.D., Vicente, J.G., Holub, E.B., Baxter, L. *et al.* (2018a) Comparative genomics of *Pseudomonas syringae* reveals convergent gene gain and loss associated with specialization onto cherry (*Prunus* avium). *The New Phytologist* 219, 672–696.

Hulin, M.T., Mansfield, J.W., Brain, P., Xu, X., Jackson, R.W. *et al.* (2018b) Characterization of the pathogenicity of strains of *Pseudomonas syringae* towards cherry and plum. *Plant Pathology* 67, 1177–1193.

Hulin, M.T., Jackson, R.W., Harrison, R.J. and Mansfield, J.W. (2020) Cherry picking by pseudomonads: After a century of research on canker, genomics provides insights into the evolution of pathogenicity towards stone fruits. *Plant Pathology* 69, 962–978.

Hulin, M.T., Vadillo Dieguez, A., Cossu, F., Lynn, S., Russell, K. *et al.* (2022) Identifying resistance in wild and ornamental cherry towards bacterial canker caused by *Pseudomonas syringae*. *Plant Pathology* 71, 949–965.

Hulin, M.T., Rabiey, M., Zeng, Z., Vadillo Dieguez, A., Bellamy, S. *et al.* (2023) Genomic and functional analysis of phage-mediated horizontal gene transfer in *Pseudomonas syringae* on the plant surface. *The New Phytologist* 237, 959–973.

Hunjan, M.S. and Lore, J.S. (2020) Climate change: Impact on plant pathogens, diseases, and their management. In: Jabran, K., Florentine, S. and Chauhan, B. (eds) *Crop Protection Under Changing Climate*. Springer, Cham, pp. 85–100. DOI: 10.1007/978-3-030-46111-9_4.

Iličić, R., Balaž, J., Ognjanov, V., Jošić, D., Vlajić, S. *et al.* (2018) Evaluation of cherry cultivar susceptibility to bacterial canker and leaf spot disease. *Journal of Phytopathology* 166, 799–808.

Ilicic, R., Popovic, T., Markovic, S., Jelušić, B., Agi, F. *et al.* (2021) Genetic diversity of *Pseudomonas syringae* pv. *syringae* isolated from sweet cherry in southern and northern regions in Serbia. *Genetika-Belgrade* 53, 247–262.

Kałuzna, M., Willems, A., Pothier, J.F., Ruinelli, M., Sobiczewski, P. *et al.* (2016) Characterization and genetic diversity of causal agent of stone fruit bacterial canker *Pseudomonas cerasi*, a new pathogen of cherry. *Acta Horticulturae* 1149, 9–14. DOI: 10.17660/ActaHortic.2016.1149.2.

Kamiunten, H., Nakao, T. and Oshida, S. (2000) *Pseudomonas syringae* pv. cerasicola, pv. nov., the causal agent of bacterial gall of cherry tree. *Journal of General Plant Pathology* 66, 219–224.

Kappel, F., Granger, A., Hrotkó, K. and Schuster, M. (2012) Cherry. In: Badenes, M. and Byrne, D. (eds) *Fruit Breeding. Handbook of Plant Breeding*. Springer, Boston, MA, pp. 459–504. DOI: 10.1007/978-1-4419-0763-9_13.

Kennelly, M.M., Cazorla, F.M., Vicente, A., Ramos, C. and Sundin, G.W. (2007) *Pseudomonas syringae* diseases of fruit trees: Progress toward understanding and control. *Plant Disease* 91, 4–17.

Khalaf, S.M.H., Alqahtani, M.S.M., Ali, M.R.M., Abdelalim, I.T.I. and Hodhod, M.S. (2024) Using Max Ent modeling to analyze climate change impacts on *Pseudomonas syringae* van Hall, 1904 distribution on the global scale. *Heliyon* 10(24), e41017. DOI: 10.1016/j.heliyon.2024.e41017.

Klement, Z., Rozsnyay, D.S., Báló, E., Pánczél, M. and Prileszky, G. (1984) The effect of cold on development of bacterial canker in apricot trees infected with *pseudomonas syringae* pv. *syringae*. *Physiological Plant Pathology* 24, 237–246.

Kumar, R., Das, S.P., Choudhury, B.U., Kumar, A., Prakash, N.R. *et al.* (2024) Advances in genomic tools for plant breeding: Harnessing DNA molecular markers, genomic selection, and genome editing. *Biological Research* 80, 2024. DOI: 10.1186/s40659-024-00562-6.

Lamichhane, J.R., Varvaro, L., Parisi, L., Audergon, J.-M. and Morris, C.E. (2014) Chapter four - disease and frost damage of woody plants caused by *Pseudomonas syringae*: Seeing the forest for the trees. In: Sparks, D.L. (ed.) *Advances in Agronomy*. Academic Press.

Lang, G.A. (2008) *Sweet Cherry Orchard Management: From Shifting Paradigms To Computer Modeling*. International Society for Horticultural Science (ISHS), Leuven, Belgium, pp. 597–604.

Lang, G., Valentino, T., Demirsoy, H. and Demirsoy, L. (2011) *High Tunnel Sweet Cherry Studies: Innovative Integration of Precision Canopies, Precocious Rootstocks, and Environmental*

Physiology. International Society for Horticultural Science (ISHS), Leuven, Belgium, pp. 717–723.

La Torre, A., Iovino, V. and Caradonia, F. (2018) Copper in plant protection current situation and prospects. *Phytopathologia Mediterranea* 57, 201–236.

Lee, Y.S., Koh, H.S., Sohn, S.H., Koh, Y.J., Jung, J.S. *et al.* (2012) Genetic diversity among Pseudomonas syringae pv. morsprunorum isolates from Prunus mume in Korea and Japan by comparative sequence analysis of 16S rRNA gene. *The Plant Pathology Journal* 28, 295–298.

Lee, S., Ishiga, Y., Clermont, K. and Mysore, K.S. (2013) Coronatine inhibits stomatal closure and delays hypersensitive response cell death induced by nonhost bacterial pathogens. *PeerJ* 1, e34.

Lienqueo, I., Villar, L., Beltrán, F., Correa, F., Sagredo, B. *et al.* (2024) Molecular, phenotypic and histological analysis reveals a multi-tiered immune response and callose deposition in stone fruit rootstocks (*Prunus* spp.) against *Pseudomonas syringae* pv. *syringae* (Pss) infection. *Scientia Horticulturae* 324, 112588.

Lipps, S.M. and Samac, D.A. (2022) *Pseudomonas* viridiflava: An internal outsider of the *Pseudomonas syringae* species complex. *Molecular Plant Pathology* 23, 3–15.

Little, E., Bostock, R. and Kirkpatrick, B. (1998) Genetic characterization of *Pseudomonas syringae* pv. *syringae* strains from stone fruits in California. *Applied and Environmental Microbiology* 64, 3818–3823.

Maguvu, T.E., Frias, R.J., Hernandez-Rosas, A.I., Shipley, E., Dardani, G. *et al.* (2024) Pathogenicity, phylogenomic, and comparative genomic study of *Pseudomonas syringae* sensu lato affecting sweet cherry in California. *Microbiology Spectrum* 12.

Manna, S., Santander, R.D. and Zhao, Y. (2024) First report of *pseudomonas* amygdali pv. morsprunorum causing bacterial canker in sweet cherry Orchards in Washington State. *Plant Disease* 108, 2560.

Mansfield, J., Genin, S., Magori, S., Citovsky, V., Sriariyanum, M. *et al.* (2012) Top 10 plant pathogenic bacteria in molecular plant pathology. *Molecular Plant Pathology* 13, 614–629.

Marroni, M.V., Casonato, S., Pitman, A.R., Visnovsky, S.B., Beresford, R.M. *et al.* (2024a) Review of *Pseudomonas* species causing bacterial canker of *Prunus* species with emphasis on sweet cherry (*Prunus* avium) in new zealand. *European Journal of Plant Pathology* 168, 297–314.

Marroni, M.V., Casonato, S., Visnovsky, S.B., Pitman, A.R., Beresford, R.M. *et al.* (2024b) Pathogenicity and virulence-associated factors of *Pseudomonas syringae* pv. syringae and [P. amygdali pv. morsprunorum] strains from New Zealand sweet cherry (*Prunus* avium) Orchards. *Journal of Phytopathology* e13390.

Ménard, M., Sutra, L., Luisetti, J., Prunier, J. and Gardan, L. (2003) *Pseudomonas syringae* pv. avii (pv. nov.), the causal agent of bacterial canker of wild cherries (*Prunus* avium) in France. *European Journal of Plant Pathology* 109, 565–576.

Mgbechi-Ezeri, J., Porter, L., Johnson, K.B. and Oraguzie, N. (2017) Assessment of sweet cherry (*Prunus* avium L.) genotypes for response to bacterial canker disease. *Euphytica* 213, 145.

Mirik, M., Oksel, C. and Bulbul, M. (2015) Prevalence, isolation and identification of bacterial canker pathogens on sweet cherry trees in Tekirdag. In: *II International Workshop on Bacterial Diseases of Stone Fruits and Nuts 1149*. pp. 43–46.

Moubarak, P. (2016) Survey of bacterial diseases on stone fruits in Lebanon and investigation of phenotypic and genetic diversity of the isolated *Pseudomonas syringae* pathovars.

Nosratnezhad, F., Rouhrazi, K. and Khezrinezhad, N. (2018) Characterization and genetic diversity of *Pseudomonas syringae* isolates from stone fruits in north-western Iran. *Journal of Phytopathology* 166, 516–524.

Oksel, C., Avin, F.A., Mirik, M. and Baysal-Gurel, F. (2022) Identification and genetic characterization of *Pseudomonas syringae* pv. *syringae* from sweet cherry in Turkey. *Plant Disease* 106, 1253–1261.

Omrani, M., Roth, M., Roch, G., Blanc, A., Morris, C.E. *et al.* (2019) Genome-wide association multi-locus and multi-variate linear mixed models reveal two linked loci with major effects on partial resistance of apricot to bacterial canker. *BMC Plant Biology* 19, 31.

Otto, M., Petersen, Y., Roux, J., Wright, J. and Coutinho, T. (2018) Bacterial canker of cherry trees, *Prunus* avium, in South Africa. *European Journal of Plant Pathology* 151, 427–438.

Paltineanu, C. and Chitu, E. (2020) Climate change impact on phenological stages of sweet and sour cherry trees in a continental climate environment. *Scientia Horticulturae* 261, 109011.

Panstruga, R. and Moscou, M.J. (2020) What is the molecular basis of nonhost resistance? *Molecular Plant-Microbe Interactions* 33, 1253–1264.

Papp-Rupar, M., Grace, E.R., Korotania, N., Ciusa, M., Jackson, R.W. *et al.* (2025) Impact of phage therapy on *Pseudomonas syringae* pv. *syringae* and plant microbiome dynamics through coevolution and field experiments. *Environmental Microbiology* 27, e70076. DOI: 10.1111/1462-2920.70076.

Parisi, L., Morgaint, B., Blanco-Garcia, J., Guilbaud, C., Chandeysson, C. *et al.* (2019) Bacteria from four phylogroups of the *Pseudomonas syringae* complex can cause bacterial canker of apricot. *Plant Pathology* 68, 1249–1258.

Percival, G.C. and Banks, J.M. (2015) Phosphite-induced suppression of *Pseudomonas* bleeding canker (*Pseudomonas syringae* pv. *aesculi*) of horse chestnut (*Aesculus hippocastanum* L.). *Arboricultural Journal* 37, 7–20.

Popović, T., Menković, J., Prokić, J., Zlatković, A. and Obradović, A. (2021) Isolation and characterization of *Pseudomonas syringae* isolates affecting stone fruits and almond in Montenegro. *Journal of Plant Diseases and Protection* 128, 391–405.

Pritchard, L., Glover, R.H., Humphris, S., Elphinstone, J.G. and Toth, I.K. (2016) Genomics and taxonomy in diagnostics for food security: Soft-rotting enterobacterial plant pathogens. *Analytical Methods* 8, 12–24.

Puławska, J., Gétaz, M., Kałuzn·a, M., Kuzmanović, N., Obradović,) *et al.* (2017) *Bacterial Diseases. Cherries: Botany, Production and Uses.* CAB International, Wallingford UK.

Rabiey, M., Roy, S.R., Holtappels, D., Franceschetti, L., Quilty, B.J. *et al.* (2020) Phage biocontrol to combat *Pseudomonas syringae* pathogens causing disease in cherry. *Microbial Biotechnology* 13, 1428–1445.

Renick, L.J., Cogal, A.G. and Sundin, G.W. (2008) Phenotypic and genetic analysis of epiphytic *Pseudomonas syringae* populations from sweet cherry in Michigan. *Plant Disease* 92, 372–378.

Roberts, S.J.G.E. (2017) Review of bacterial pathogens of economic importance to UK crops. In: Board, A.A.H.D. (ed.) *Warwick*. Plant Health Solutions Ltd and Fera Science Ltd, York.

Rodríguez-Palenzuela, P., Matas, I.M., Murillo, J., López-Solanilla, E., Bardaji, L. *et al.* (2010) Annotation and overview of the *Pseudomonas* savastanoi pv. savastanoi NCPPB 3335 draft genome reveals the virulence gene complement of a tumour-inducing pathogen of woody hosts. *Environmental Microbiology* 12, 1604–1620.

Ruinelli, M., Blom, J., Smits, T.H.M. and Pothier, J.F. (2019) Comparative genomics and pathogenicity potential of members of the *Pseudomonas syringae* species complex on *Prunus* spp. *BMC Genomics* 20, 172. DOI: 10.1186/s12864-019-5555-y.

Santi, F., Russell, K., Ménard, M. and Dufour, J. (2004) Screening wild cherry (*Prunus* avium) for resistance to bacterial canker by laboratory and field tests. *Forest Pathology* 34, 349–362.

Schellenberg, B., Ramel, C. and Dudler, R. (2010) *Pseudomonas syringae* virulence factor syringolin A counteracts stomatal immunity by proteasome inhibition. *Molecular Plant-Microbe Interactions* 23, 1287–1293.

Scholz-Schroeder, B.K., Hutchison, M.L., Grgurina, I. and Gross, D.C. (2001) The contribution of syringopeptin and syringomycin to virulence of *Pseudomonas syringae* pv. *syringae* strain B301D on the basis of sypA and syrB1 biosynthesis mutant analysis. *Molecular Plant-Microbe Interactions* 14, 336–348.

Smirnova, A., Li, H., Weingart, H., Aufhammer, S., Burse, A. *et al.* (2001) Thermoregulated expression of virulence factors in plant-associated bacteria. *Archives of Microbiology* 176, 393–399.

Spotts, R.A., Wallis, K.M., Serdani, M. and Azarenko, A.N. (2010) Bacterial canker of sweet cherry in Oregon—Infection of horticultural and natural wounds, and resistance of cultivar and rootstock combinations. *Plant Disease* 94, 345–350.

Sundin, G., Jones, A. and Fulbright, D.W. (1989) Copper resistance in *Pseudomonas syringae* pv. *syringae* from cherry orchards and its associated transfer *in vitro* and in planta with a plasmid. *Phytopathology* 79, 861–865.

Thomas, G., Kay, W.T. and Fones, H.N. (2024) Life on a leaf: The epiphyte to pathogen continuum and interplay in the phyllosphere. *BMC Biology* 22, 168.

Thomidis, T. and Exadaktylou, E. (2008) Susceptibility of 30 cherry (*Prunus* avium) genotypes to the bacterium *Pseudomonas syringae* pv. *syringae*. *New Zealand Journal of Crop and Horticultural Science* 36, 215–220.

Tomás-Barberán, F.A., Ruiz, D., Valero, D., Rivera, D., Obón, C. *et al.* (2013) Health benefits from pomegranates and stone fruit, including plums, peaches, apricots and cherries. In: Skinner, M. and Hunter, D. (eds) *Bioactives in Fruit*. DOI: 10.1002/9781118635551.ch7.

Tribelli, P.M. and López, N.I. (2022) Insights into the temperature responses of *Pseudomonas* species in beneficial and pathogenic host interactions. *Applied Microbiology and Biotechnology* 106, 7699–7709.

Vadillo-Dieguez, A., Zeng, Z., Mansfield, J.W., Grinberg, N.F., Lynn, S.C. *et al.* (2024) Genetic dissection of the tissue-specific roles of type III effectors and phytotoxins in the pathogenicity of *Pseudomonas syringae* pv. *Syringae* to cherry. *Molecular Plant Pathology* 25, e13451. DOI: 10.1111/mpp.13451.

Van Dijk, K., Fouts, D.E., Rehm, A.H., Hill, A.R., Collmer, A. *et al.* (1999) The Avr (effector) proteins HrmA (HopPsyA) and AvrPto are secreted in culture from *Pseudomonas syringae* pathovars via the Hrp (type III) protein secretion system in a temperature- and pH-sensitive manner. *Journal of Bacteriology* 181, 4790–4797.

Vasebi, Y., Khakvar, R., Faghihi, M.M. and Vinatzer, B.A. (2019) Genomic and pathogenic properties of *Pseudomonas syringae* pv. *syringae* strains isolated from apricot in East Azerbaijan province, Iran. *Biocatalysis and Agricultural Biotechnology* 19, 101167.

Vicente, J.G. and Roberts, S.J. (2003) *Screening Wild Cherry Micropropagated Plantlets for Resistance to Bacterial Canker*. Springer, Netherlands.

Vicente, J.G., Alves, J.P., Russell, K. and Roberts, S.J. (2004) Identification and discrimination of *Pseudomonas syringae* isolates from wild cherry in England. *European Journal of Plant Pathology* 110, 337–351.

Villalobos-González, L., Carreras, C., Beltrán, M.F., Figueroa, F., Rubilar-Hernández, C. *et al.* (2024) Sweet cherry plants prioritize their response to cope with summer drought, overshadowing the defense response to *Pseudomonas syringae* pv. *Syringae*. *Plants* 13, 1737.

Vinchira-Villarraga, D., Dhaouadi, S., Milenkovic, V., Wei, J., Grace, E.R. *et al.* (2024) Metabolic profiling and antibacterial activity of tree wood extracts obtained under variable extraction conditions. *Metabolomics* 21.

Weaver, D. (1978) Interaction of *Pseudomonas syringae* and freezing in bacterial canker on excised peach twigs. *Phytopathology* 68, 1460–1463.

Wenneker, M., Janse, J., Bruine, A.D., Vink, P. and Pham, K. (2012) Bacterial canker of plum caused by *Pseudomonas syringae* pathovars, as a serious threat for plum production in the Netherlands.

Xin, X.F., Kvitko, B. and He, S.Y. (2018) *Pseudomonas syringae*: What it takes to be a pathogen. *Nature Reviews. Microbiology* 16, 316–328.

Young, J. (1988) *Pseudomonas syringae* pv. persicae from nectarine, peach, and Japanese plum in New Zealand 1. *EPPO Bulletin* 18, 141–151.

Young, J.M. (2010) Taxonomy of *Pseudomonas syringae*. *Journal of Plant Pathology* 92, S5–S14.

Zambounis, A., Ganopoulos, I., Aravanopoulos, F., Hilioti, Z., Madesis, P. *et al.* (2020) Genomics opportunities and breeding strategies towards improvement of climate-smart traits and disease resistance against pathogens in sweet cherry. In: *Genomic Designing of Climate-Smart Fruit Crops*. pp. 385–404.

Zhou, J.M. and Zhang, Y. (2020) Plant immunity: Danger perception and signaling. *Cell* 181, 978–989.

10 Kiwifruit Bacterial Canker: A Recent Disease for Modern Plant Pathologists

Davide Danzi

Università degli studi di Verona, Italy

Introduction

The evolution of kiwifruit (*Actinidia chinensis* Planch 1847) into a commercial crop started less than 100 years ago; just after some kiwifruit seeds were brought from China to New Zealand for horticultural purposes in 1904. Those plants produced the first fruits in 1910, and then the first commercial orchard was planted in 1930 (Huang, 2016). There, it found the approval as a new crop due to its delicious fruits, which occur in large amounts on plants. Kiwifruit got its name in New Zealand since it was previously known as monkey peach in China or as Chinese gooseberry in other countries. The details about the origin name are mysterious, but starting from the 1950s, after its importation in the USA, it was mutually agreed as kiwifruit. The term kiwifruit includes at least three commercial varieties, which can be easily identified by the colour of the flesh; indeed, we distinguish kiwifruit carrying green, yellow or red pulp. The worldwide kiwifruit market size accounts for more than 9 billion USD in 2024, and it is expected to increase in the coming years.

The leading countries in kiwifruit production are China, New Zealand, and Italy, with green-fleshed varieties being the most cultivated (Vanneste, 2017). Kiwifruit cultivation was rapidly adopted all over the world, especially in Italy, due to its higher profitability compared to other crops and to the absence of relevant diseases (Testolin, 2015). For these reasons, the growth of the kiwifruit market did not face any significant arrest till 2008 when, in Italy, some plants showing spots on leaves and bud rot were sampled, and a *Pseudomonas syringae* strain was isolated (Testolin, 2015). Until then, kiwifruit has never been severely affected by diseases, even though *P. syringae* strains have been already isolated from symptomatic plants in the early 1980s. Indeed, already in 1984, in the Shizuoka prefecture in Japan, some pathogenic strains were isolated. From the first characterization, the isolates were attributed to the LOPAT group I and subsequently was proposed to classify it as a new pathovar that is pathovar *actinidiae*. The isolate ICMP 9617 was thus considered the pathotype for this new pathovar. To control the disease, authors reported that the application of kasugamycin, streptomycin and copper reduced disease severity. Similar symptoms on kiwifruit were reported before 1984 in China and Japan, but the problem was investigated later due to the increased number of reports of severe damage to orchards. The disease spread rapidly all over the world and was reported in South Korea and Italy in 1994 (Scortichini *et al.*, 2012). The bacterial canker severely compromised kiwifruit production in Korea and Japan, while in Italy, only negligible effects were reported at the time (Scortichini *et al.*, 2012). *Pseudomonas syringae* pv *actinidiae* (Psa) was isolated from *A. chinensis* var. *deliciosa*, the green-fleshed kiwifruit, till the beginning of 2000. This fruit variety dominated the market, but after the introduction of the yellow kiwifruit, Psa started to attack mainly this variety, which was later found to be more susceptible (Scortichini *et al.*, 2012). Due to its high susceptibility, the orchards started to be replaced with the green variety Hayward, which is still the most common and more resistant to Psa. However, severe epidemics also occurred on this variety, which worried Italian growers, for a good reason, since in 2008, it was isolated for the first time in central Italy, a very aggressive strain of Psa (Vanneste, 2017). Later, this was recognized as the most aggressive biovar of *P. syringae* pv. *actinidiae*, and the reasons will be discussed in the dedicated section. Since then, strains belonging to this aggressive biovar (Psa 3) have been feared and highly monitored. Two years later, in 2010, this biovar was isolated in New Zealand, changing the kiwifruit production.

Psa is distributed all over the world where kiwifruit is cultivated, and the disease was reported in China, Chile, Italy, France, New Zealand, Japan, South Korea and also in other European countries (Fig. 10.1).

In most of the countries where Psa3 is found, the production of kiwifruit was subject to an unprecedented massive loss of fruits. It has been estimated that in New Zealand alone, the outbreaks of bacterial canker disease caused 310–410 million NZ$ (180–240 million $) in the first 5 years from the first outbreak (Vanneste, 2017). The impact on kiwifruit production was huge all over the world and pushed the government of New Zealand to establish the Kiwifruit Vine Health Inc. (KVH) to monitor the disease and help growers. The KVH is still a resource that helps New Zealand growers to keep healthy kiwifruit orchards. Today, the impact of the disease is less important, and the replacement of different varieties of yellow-fleshed kiwifruit, luckily more tolerant to Psa with respect to "Hort16A", restricts the spread of the disease (Vanneste, 2017). Moreover, in the last 14 years, a lot of knowledge has been acquired and turned into practical information, which is really helpful in coexisting with Psa (Scortichini *et al.*, 2012; Vanneste, 2017).

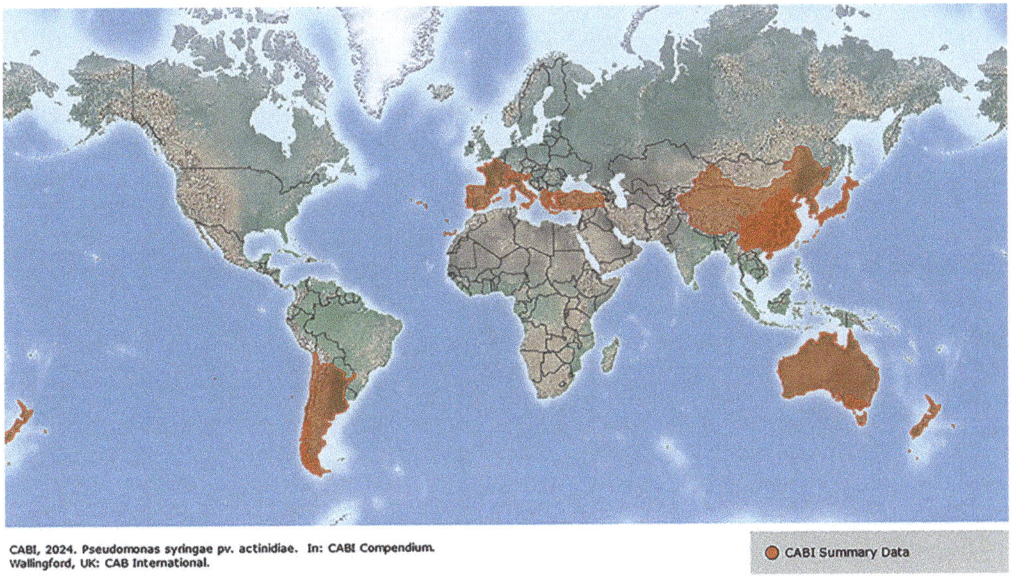

Fig. 10.1. Global distribution of Psa. Map obtained from CABI website (Available at: https://www.cabidigitallibrary.org/doi/full/10.1079/cabicompendium.45002), accessed 8 August 2024.

The Pathogen

Psa is the causal agent of the bacterial canker of kiwifruit. It belongs to the phylogroup 1b of the *P. syringae* complex, one of the largest bacterial plant-pathogen species. Since 2012, Psa has been listed in the A2 EPPO list, which includes pathogens that are already present in the EPPO territory and are considered a threat to agricultural production and for which quarantine measures are suggested. Psa is further divided into five biovars (Psa 1–6) that are distinguished for their genetic elements and are named in chronological order according to the year of their isolation (Chapman *et al.*, 2012; McCann *et al.*, 2013; Vanneste, 2017). Despite the existence of different nomenclatures, this one concerning the chronological order is now unanimously accepted, but it is frequent to find the suffixes V for virulent and LV for low virulent to distinguish the strains. The biovar 4 is absent from the list since it has been redefined as pathovar *actinidifoliorum*. It was isolated from kiwifruit but did not cause severe symptoms and was generally indicated as low virulent. It has been frequently found in kiwifruit orchards but does not cause as much damage as the other biovars, and symptoms remain limited to the leaves. Psa 1 includes strains isolated in Japan and Italy during the first epidemics of the 1880s and 1990s, while Psa 2 seems restricted to South Korea. Psa 3 is the most widespread and has a worldwide distribution. It originated in China, and apparently, individual transmission events occurred in New Zealand, Italy and Japan (McCann *et al.*, 2017). Biovars 5 and 6 are endemic to Japan, and there is no detailed information about their virulence. However, they

Table 10.1. Year of first Psa strains isolation in different countries. It must be noted that all Psa biovar except Psa 3 have a restricted geographical distribution.

Country/Region	Strain (year)
Japan	1984 (Psa1), 2012 (Psa 5), 2014 (Psa 3), 2015 (Psa 6)
South Korea	1980s (Psa 2)
China	1984 (Psa 1)
Italy	1992 (Psa 1), 2008 (Psa 3)
New Zealand	2010 (Psa 3)
France	2011 (Psa 3)
Spain	2011 (Psa 3)
Portugal	2010 (Psa 3)
Chile	2010 (Psa 3)
Argentina	2015 (Psa 3)

do not seem as aggressive as biovars 1 and 3 (Sawada and Fujikawa, 2019). A unique feature among Psa biovar is that Psa 6 can produce both phaseolotoxin and coronatine and display full T3SS (type three secretion system) activity, despite its effector repertoire differs from other Psa biovars (Fujikawa and Sawada, 2019; Sawada and Fujikawa, 2019). Psa 1 was first isolated in Japan, then in South Korea and Italy. It acquired the ability to synthesize the phaseolotoxin, most likely by horizontal gene transfer from other bacteria (Sawada and Fujikawa, 2019). It is considered a virulent strain of Psa, but its distribution seems now restricted to Japan (Sawada and Fujikawa, 2019). Psa 2 is a poor virulent lineage of Psa and was only found in South Korea, it can produce coronatine, but it does not produce phaseolotoxin and shows no activation of the T3SS when put in minimal medium (Sawada and Fujikawa, 2019; Vandelle *et al.*, 2021). However, it produces high levels of biofilm, a characteristic that seems inversely correlated with the synthesis of T3SS components (Vandelle *et al.*, 2021). Psa 3 is the most aggressive biovar of Psa and has a worldwide distribution. It was first isolated in Italy in 2008 and was later found in 12 other countries (Fig. 10.1, Table 10.1). Interestingly, all epidemic lineages seem to have a monophyletic origin and have a restricted distribution, while in China the population of Psa 3 is polyphyletic, which further suggests that Psa 3 spread from this country and established in other countries forming closely related populations (McCann *et al.*, 2017). At the time of this report, only Psa3, among all Psa biovars, is considered a threat to kiwifruit production, being the most widespread and the most aggressive and considering that Psa 1, the other virulent Psa biovar, has not been isolated since 1998.

The disease cycle lifecycle

Psa can colonize kiwifruit plants all year long, reaching the highest spread during the cold season (Fig. 10.2). During the winter, Psa can survive within

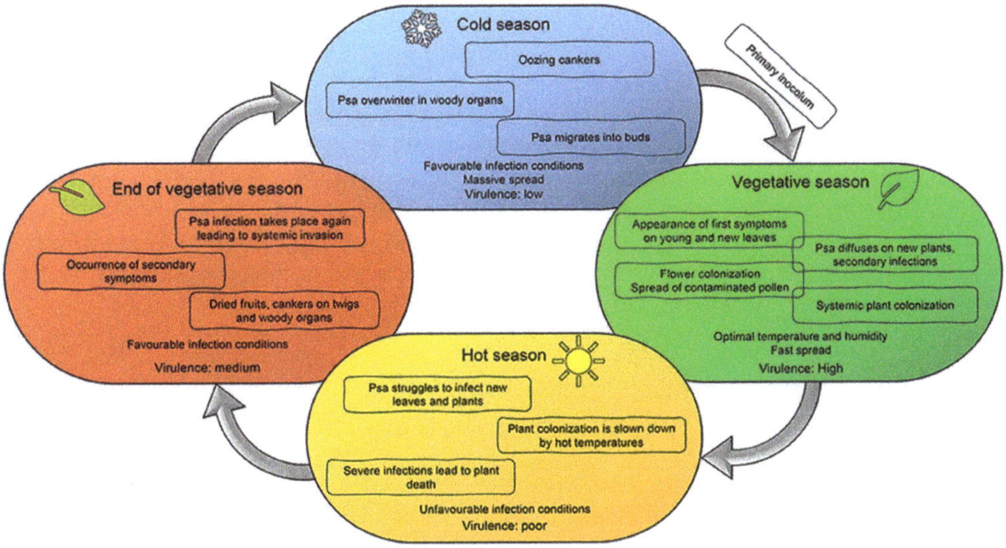

Fig. 10.2. Life cycle of *Pseudomonas syringae* pv. *actinidiae*. The schematic representation of the Psa life cycle starts with the primary source of the inoculum in the cold season, which then spreads heavily during the vegetative season. The first symptoms appear on young plant organs, such as leaves and buds. The spread of the pathogen is then hampered according to temperature increase. As the temperature decreases, Psa starts to spread again and finally overwinters within woody organs during the cold season.

plant roots or other overwintering organs, then the first infections occur during the early spring. The systemic movement of Psa within the host is a peculiarity of this pathovar. Indeed, such a way of spreading and infecting is not common for *P. syringae* (Scortichini *et al.*, 2012). In this season, Psa can move systemically from roots to buds due to the high root pressure (Donati *et al.*, 2020). Moreover, during plant dormancy, the water pressure inside the plant, together with the cold temperatures, favours the development of wounds, ultimately facilitating the spread of Psa (Donati *et al.*, 2020). Canker exudates are indeed an important source of inoculum for primary infections since they are easily spread with rain and wind. Wounds caused by frost or winter pruning constitute another way of penetration of Psa into host tissues where Psa can survive until late winter. At the end of the winter, the first symptoms appear as shoot and bud dieback and as necrotic spots on new leaves (Vanneste, 2017; Donati *et al.*, 2020). At the beginning of the spring, Psa can spread rapidly in the orchard due to favourable environmental conditions like high humidity, low temperature and wind. The presence of a water layer on the leaf surface is fundamental for Psa to penetrate through stomata; thus, rainfalls and wind are the primary causes of Psa spreading in the field (Scortichini *et al.*, 2012). Psa is able to survive and colonize pollen, which further increases its diffusion in the field. The involvement of contaminated pollen in Psa spread has recently been investigated, highlighting another aspect that makes Psa 3a very able colonizer (Donati *et al.*, 2018). Kiwifruit is a dioecious plant, and to ensure proper fertilisation of flowers, a ratio of 6–10 female/male plants is maintained. However, this is not sufficient, and farmers employ artificial pollination. Symptoms in male plants are generally more pronounced, but because there is no difference in susceptibility between male and female plants, the number of males is reduced. The spread of the disease is slowed down when temperatures rise over 22°C, during the hot season symptoms development occur less frequently and Psa enters in a latency state (Donati *et al.*, 2020). Psa spread takes place rapidly in autumn due to the high level of inoculum in the orchard and the increasing number of wounds originating from detached leaves and fruits.

Symptoms

Symptoms on leaves occur as irregular, often angular, brown and water-soaked necrotic spots often surrounded by a chlorotic halo (Fig. 10.3G–L (Son *et al.*, 2016)). On canes and leaders, Psa induces the formation of soaked

Fig. 10.3. Symptoms of Psa on kiwifruit plants. (a-b) Bacterial white-milky ooze on leaders, (c, d, e) reddish exudates oozing from woody organs, (f) difference between a light brown healthy and a dark-brown death portion of a leader, (g-l) symptoms on kiwifruit leaves, showing different severity degrees, occurring as irregular necrotic spots surrounded by a chlorotic halo, (m, n) white exudates oozing from blossoms and necrotic blossom on canes. Credit: Son *et al.* (2016).

lesions with cracks from which oozes a pale or reddish gummy exudate (Fig. 10.3C–E, Son *et al.*, 2016). Cankers also appear on trunks at the end of the growing season and can extend to a large part of the plant. Sometimes, lesions and cankers blacken and produce large amounts of exudates (Fig. 10.3C–F, Son *et al.*, 2016). Psa also causes flowers to wither and finally fall from the plants, white exudates oozing from flowers can be observed (Fig. 10.3M–N, Son *et al.*, 2016). Although fruits are not directly attacked by Psa, when an infection is ongoing, they appear dehydrated and shrivelled up. When plants are severely affected, Psa causes the wilting of canes, leaders and also of entire branches. In the most severe cases, the entire plant can die.

Virulence of Psa

Among Psa biovars, the most virulent is Psa 3, followed by Psa 1, which, however, has never been isolated since 1998. Psa 1 produces phaseolotoxin and shows a weak activation of the T3SS with respect to Psa 3, which likely explains its reduced virulence compared with Psa 3. Therefore, the T3SS is considered the main virulence factor for Psa and its rapid activation is specific to Psa 3 (Vandelle *et al.*, 2021). Thus, this paragraph explores the principal mechanisms of virulence found in Psa 3.

Psa, similarly to other *P. syringae* pathovars, shows a dual lifestyle phase, starting as an epiphyte onto host leaves; it then turns into a pathogenic endophyte when it reaches the apoplast. During the epiphytic phase, its survival depends on the ability to withstand the scarcity of nutrients and to rapidly reach the apoplast, overcoming the stomatal defence (McAtee *et al.*, 2018). Psa 3 possesses high flagellar motility, which can be observed by *in-vitro* experiments (Vandelle *et al.*, 2021). Moreover, it was shown to upregulate genes related to nutrient acquisition very soon after being placed over kiwifruit leaves (McAtee *et al.*, 2018). This should be considered as a rapid adaptation to a new environment and suggests the fast response of Psa 1 to the surroundings. When put in a minimal medium, motility-related genes are also upregulated in Psa 1 but not in Psa2 (Vandelle *et al.*, 2021). A similar trend was observed for T3SS, which is fundamental to achieving full virulence (Vanneste, 2017; McAtee *et al.*, 2018; Vandelle *et al.*, 2021). Indeed, Psa 3 rapidly activates the entire *hrp-hrc* clusters, which are poorly induced in Psa 1 and Psa 2. This aspect could reflect the higher virulence of Psa 3 among all Psa biovars (Vandelle *et al.*, 2021; Puttilli *et al.*, 2022). Consistently, the same behaviour of Psa 3 was observed when the pathogen was inoculated directly on kiwifruit plants (McAtee *et al.*, 2018). Researchers observed that Psa3 activates genes related to the T3SS within 4 hour, and their expression peaks around 12 hour (McAtee *et al.*, 2018). The *hrpA1* gene, coding for the main protein that constitutes the pilus of the T3SS, is among the earliest induced genes both *in-vitro* and *in-planta* (McAtee *et al.*, 2018; Vandelle *et al.*, 2021). Thus, its promoter was used to guide the expression of the green fluorescent protein (GFP) and to investigate Psa 3 responsiveness to different environmental conditions. With this system, it was shown that

Psa responds to plant signals that anticipate induction of the T3SS, although the mechanism of this response remains to be elucidated (Puttilli *et al.*, 2022; Caullireau *et al.*, 2024). Moreover, a temperature dependence on virulence induction was observed, highlighting that Psa 3 prefers low temperatures and specifically that temperature influences effector secretion more than their transcription (Puttilli *et al.*, 2022). Despite the importance of effectors in *P. syringae* pathogenicity being of uttermost importance, their function is, in large part, still unknown. Moreover, their relative contribution to virulence is still far from being elucidated. However, two effectors, namely AvrE1 and HopR1, seem pivotal to Psa 3 virulence (Jayaraman *et al.*, 2020; Ishiga *et al.*, 2022). Indeed, mutants defective for these effectors can poorly colonize kiwifruit tissues and cause symptoms (Jayaraman *et al.*, 2020; Ishiga *et al.*, 2022). AvrE1 is generally associated with increased water availability in the apoplast, which is essential for the colonization of host tissues. The role of HopR1 in Psa has also been recently investigated, and it was shown to be related to stomatal aperture despite its role seeming different according to the Psa strain (Ishiga *et al.*, 2022). Large transposon mutagenesis revealed multiple important players for Psa virulence, and most of them were related to effectors of T3SS. Thus, this is another piece of evidence that the T3SS is the most important virulence factor in Psa (Ishiga *et al.*, 2022). Consistently, the loss of a 51 kb region within the EEL (Exchangeable effector locus) caused the adaptation of Psa3 to *A. arguta*, which is known for its resistance to the bacterial canker (Hemara *et al.*, 2022). This loss of effector within the EEL did not affect the virulence of *A. chinensis*, highlighting that *A. arguta* can recognize some Psa effectors and that these effectors have a negligible effect on the Psa virulence of *A. chinensis*.

Another aspect that could contribute to Psa virulence is the quorum sensing (QS) which, in this case, is not mediated by a canonical system but instead by three LuxR solos, namely PsaR1, PsaR2 and PsaR3 (Patel *et al.*, 2014). The LuxR solo PsaR3 is carried by a large plasmid specific to Psa biovar 3, which was supposed to be involved in Psa virulence. Single deletion mutants of the three LuxR solos showed impaired *in-planta* growth; thus, all these genes are considered important for Psa fitness (Patel *et al.*, 2014). Although the role of these LuxR solos is still unknown, it has been suggested that PsaR1 and PsaR3 probably respond to AHL produced by other bacteria while PsaR2 is most likely involved in interkingdom signalling (Patel *et al.*, 2014; Cellini *et al.*, 2020).

Toxin production in Psa does not seem closely related to its virulence since Psa 3 does not produce any phytotoxins, but it is the most virulent biovar (Fig. 10.4). Consistently, despite Psa 6 being the only biovar possessing functional AvrE1 and HopR1 effectors together with the ability to produce both coronatine and phaseolotoxin, it is not as virulent as Psa3 (Fujikawa and Sawada, 2019). This further suggests that phytotoxins are not essential for Psa virulence. Another specificity of Psa 3 is the presence of HopZ5, which has been found in bacterial pathogens of woody plants, but it is not found in the other Psa biovars (Zhao *et al.*, 2019). The central role of the T3SS for Psa virulence is further witnessed by recent pieces of evidence about genetic

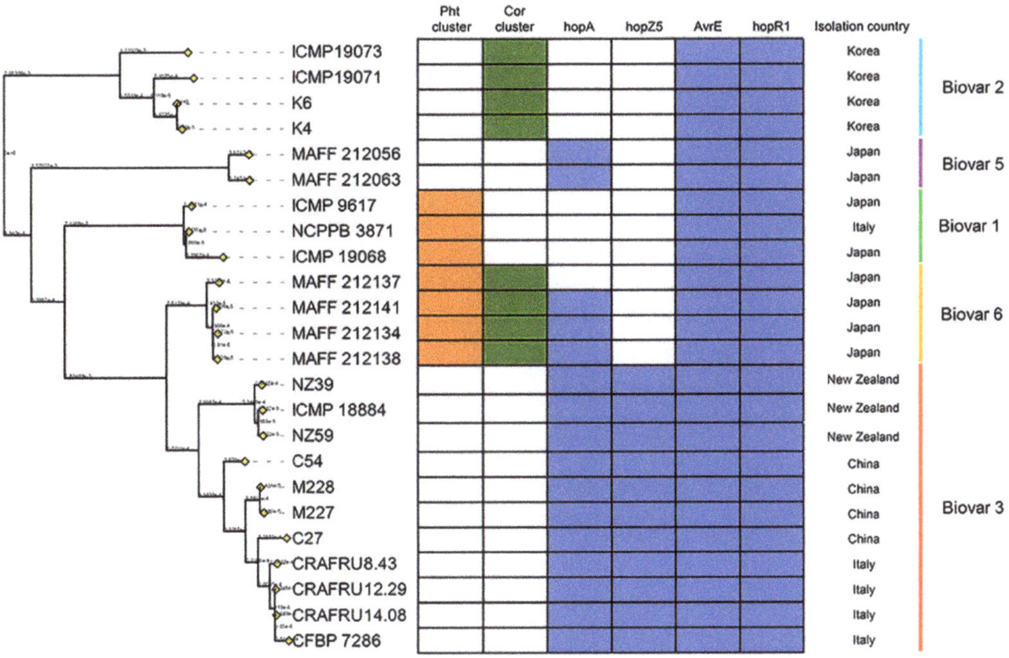

Fig. 10.4. Virulence factors of Psa biovars. The ability to produce phytotoxins is considered important for some pathovars of the *P. syringae* complex, while their role in shaping Psa virulence appears negligible. The effectors AvrE1 and HopR1 are of fundamental importance for *P. syringae* virulence and are conserved in the primary phylogroups of *P. syringae*. Psa belongs to phylogroups 1b and shows effector variability among different biovars. Indeed, approximately 50% of effectors are shared among biovars represented by the showed strains. However, Psa virulence is not only associated with the number of effectors as briefly detailed in the section "Virulence of Psa".

variants within Psa 3 populations that abolish or severely impair its virulence (Zhao *et al.*, 2019; Li *et al.*, 2021; Xie *et al.*, 2023). Finally, not only the presence of virulence factors is important for Psa virulence but also their interaction and regulation. More information about these aspects can be found in the cited literature.

Kiwifruit Bacterial Canker Disease Management

Integrated pest management must be carried out to protect the orchards from bacterial canker disease. Since Psa infection is favoured by relatively cold temperatures (10–18°C) and high humidity, cultural practises should be directed to avoid such conditions to maintain healthy orchards.

Cultural practises

- *Water irrigation*: Kiwifruit is very sensitive to water stress and especially to waterlogging. Thus, irrigation should be performed when necessary to avoid the formation of high humidity and, when possible, to use uncontaminated water since irrigation could be an important source of inoculum.
- *Plant fertilization*. It was observed that nitrogen fertilisation could influence the severity of the disease. However, no difference was observed according to the form of nitrogen supply.
- *Canopy structure*. These are generally selected according to the environmental conditions of the implant site, which can influence the orchard's productivity. However, well-aerated orchards should be maintained independently from the training system of choice, allowing light penetration and humidity reduction. Both these choices also influence the temperature of leaves and fruits. On the one hand, Psa is favoured by relatively low temperatures (15–20°C), while, on the other hand, high temperatures could negatively impact the yield. Pruning should be avoided during the periods of plant high susceptibility (spring and autumn), while it should be performed in winter to increase fruit production and in summer to guarantee a healthy orchard
- *Pollination*. Pollination should be managed and performed using healthy material since Psa can be transmitted through pollen dissemination (Donati *et al.*, 2018). Artificial pollination is common for kiwifruit, so healthy pollen should be distributed over the orchards to avoid Psa infections.

Chemical control

Chemical control of plant bacterial diseases is largely performed with copper-based compounds, especially in the European Union, where the use of antibiotics in agriculture is forbidden. Copper mixtures effectively control Psa spread in the field, although the dose and frequency of treatments can have adverse effects (Cameron *et al.*, 2014). To control the disease, New Zealand approved the use of kasugamycin in 2013, while streptomycin has been used for a long time in China, Japan and South Korea (Cameron *et al.*, 2014). Despite their initial efficacy, chemical treatments become less effective due to the emergence of resistant strains. Resistance against streptomycin and copper was already observed in the late 1980s in Asian isolates and recently also in Europe (Cameron *et al.*, 2014; Vanneste, 2017). Another strategy that proved its efficacy in controlling the spread of Psa is the induction of plant defence mechanisms through the foliar application of ASM (acid-benzol-S methyl), which is commercially available under the name of Bion® and Actigard® (Reglinski *et al.*, 2013). This strategy strongly reduces the probability of the development of resistance mechanisms since its activity is not directly against Psa. Finally, the control of Psa is also performed with biocontrol agents (BCA) and specifically with strains of *Bacillus subtilis* and *B. amyloliquefaciens* (Pereira *et al.*, 2021).

Genetic resources

Actinidia breeding has been pushing towards the development of Psa-resistant cultivars since its emergence as the major threat to kiwifruit production. The occurrence of resistant species and the availability of different genotypes are of uttermost importance for finding resistance traits. Rapid screening methodologies have been proposed to investigate this trait (Wang *et al.*, 2020). Resistance towards Psa seems positively correlated with the degree of ploidy of the kiwifruit cultivar. Recent investigations into the mechanisms associated with Psa resistance will help breeders to develop resistant cultivars possibly associated with better agronomic traits. The knowledge about the genetic determinants shaping Psa resistance will help also in gene editing programs.

What's coming to control Psa?

A short summary of the most recent approaches to control the kiwifruit bacterial canker disease follows. Since the occurrence of antimicrobial resistance, finding alternatives to control plant bacterial diseases has become urgent. Indeed, pathogens evolve rapidly in the field and can exchange genetic elements with disease-related and unrelated bacteria (Colombi *et al.*, 2024). Thus, agriculture is facing a very challenging task in advancing research towards long-lasting, evolution-proof (if possible) and cost-effective solutions. Nanotechnology is one of the most promising alternatives to common plant protection products since aims to combine old solutions with the most recent new technologies. Fabrication of nano-sized metal particles enhances their antimicrobial activity due to a larger surface-to-size ratio.

Antimicrobial peptides

Antimicrobial peptides constitute a natural source of organisms' self-defence, devoid of killing or inhibiting microbes' growth. They attracted more attention, especially in the last 20 years since the issue of antimicrobial resistance became urgent. There are many examples showing AMP potential to protect plants, from crop genetic engineering to their direct application on plant tissues, despite their development as valid alternatives, is slowed down by their poor environmental stability. However, they attract a lot of attention, and in the context of kiwifruit protection, a few examples exist in the literature showing the potency of the AMP tested to inhibit Psa growth (Cameron *et al.*, 2014).

Essential oils

Essential oils (EOs) are a complex mixture of several compounds with multiple biological purposes. Some of these are recognized antimicrobial compounds that could have bactericidal activity or can interfere with virulence and quorum-sensing mechanisms. Essential oils are generally not considered hazardous as they are used daily for multiple purposes, from cosmetics to the food industry. Moreover, some of them are recognized as GRAS (Generally recognized as safe) in the USA by the Food and Drug Administration (FDA) and in small amounts, are harmless to humans (Zhang *et al.*, 2020). The most common antimicrobial activity of EOs is to cause membrane permeabilization, which leads to the leakage of cytosolic compounds and, ultimately, to cell death. Such activity was also assessed against bacterial plant pathogens as well as against Psa (Song *et al.*, 2016). However, their poor water solubility and labile nature prevent their application in open fields and slow down their development as crop protection products. However, recent advances propose to encapsulate EOs to produce an emulsion in which the active compounds are protected by an organic polymer layer, making them stable over time and biodegradable (Zhang *et al.*, 2020). Due to their multiple modes of action and activity against bacterial membranes, they are not expected to be overcome by antimicrobial resistance in the short term.

Biotechnological approaches

PHAGE THERAPY. Phage therapy consists of exploiting the bacteriophages as natural bacterial killers to control plant pathogens in the field. This approach is very promising due to its proven efficacy against a variety of bacterial plant pathogens, both in laboratory conditions and in the field, as recently reviewed elsewhere (Holtappels *et al.*, 2021). Against Psa, the application of phages in planta led to a reduction of symptoms but not to a lower endophytic pathogenic population and was likely due to a reduced phage survival rate (reviewed in Santos *et al.*, 2024). Phage application has many advantages compared to other new strategies, namely the recognition of GRAS status from the FDA, the high specificity towards the target pathogen and the very low environmental footprint (Holtappels *et al.*, 2021). Their application in open fields is hampered by a few limitations, which include their stability in the environment, especially temperature and UV irradiation, persistence in planta and emergence of resistant strains. Indeed, it was often observed that bacteria can evolve resistance against bacteriophages and the underlying genetic variants responsible for the resistance are investigated to understand how bacteria and phages coevolve and thus to develop evolution-proof resistance strategies (Holtappels *et al.*, 2021). Thus, lytic phages are often preferred due to their higher antimicrobial activity and mode of action. Currently, mixing different phages is considered the most promising approach to avoid the development of antimicrobial resistance and to ensure long-lasting bacterial control products (Rabiey *et al.*, 2024). It must be noted that, often, such experiments are performed in *in-vitro* conditions using nutrient-rich media, which are very different from real scenarios. Indeed, the occurrence of resistance against phages is less frequently observed *in planta*. As with other new strategies to control bacterial pathogens, the administration of the active ingredients is pivotal to ensure their application. In this context, phages have sometimes been administrated within nanocarriers or nanocapsules and even within non-pathogenic strain carriers (Pereira *et al.*, 2021).

Inhibition of virulence factors

The increased understanding of virulence factors regulation in plant pathogenic bacteria allows the development of pathogen-specific crop protection products. In the context of Psa, the principal virulence factor most likely is the T3SS, which can be induced by several plant signals as outlined in much research concerning *P. syringae* virulence (O'Malley and Anderson, 2021). At the same time, secondary metabolites inhibiting the T3SS functionality have also been observed, and it is widely recognized that hampering pathogen virulence can deeply reduce the number of symptoms in host tissues (Anderson, 2023). The approach of inhibiting pathogens' virulence factors aims to reduce the occurrence of antimicrobial resistance insurgence. Turning off positive virulence regulators should not severely affect pathogen survival within plant tissues, even though in some plant-pathogen interactions, secreted effectors stimulate water release in the apoplast, thus promoting bacterial growth. However, this strategy does not limit bacterial survival, and thus, it should not push selective pressure on pathogen populations. Moreover, targeting specific pathogen features like T3SS has the advantage of being harmless to non-pathogenic strains. Indeed, the targets of virulence inhibition are often players in the perception of the surroundings, like extra-cytoplasmic sigma factors, two-component systems (TCS), quorum sensing regulators and biosynthetic genes for the production of phytotoxins (O'Malley and Anderson, 2021). Moreover, pathogens belonging to the same species are likely to respond similarly to the same plant metabolite(s) to regulate virulence. For example, sugars and organic acids are known to trigger virulence in *P. syringae* as well as some polyphenols act as inhibitors (Anderson, 2023). Considering the conserved structure of the T3SS among pathogenic bacteria, finding molecules targeting directly the T3SS components could also be effective on other pathogenic bacterial species (Hotinger *et al.*, 2021). However, developing strategies to control bacterial plant pathogens based on virulence inhibition is not trivial and requires an accurate understanding of the mechanisms responsible for such inhibition and a biotechnological approach to turn this knowledge into a real product.

Discussion Points

Considering that kiwifruit is one of the latest domesticated crops, there is a huge amount of information about it since it became a commercial crop. At the same time, the evolution of Psa as a pathogen is quite recent and covers the last 40 years. These two factors are not so common and thus constitute an almost unique example to reflect on modern agricultural practices. Thus, we leave the following discussion points to open a common discussion.

1. What are the lessons coming from kiwifruit bacterial canker disease management?
2. How do agricultural practices shape microorganism populations?
3. The natural variation in the genetic diversity of Psa 3, which makes it, at the same time, the most aggressive biovar and, in some isolates, a non-virulent one, is it still possible to consider it a pathovar? Can these strains be pathogenic to other hosts?
4. Is it possible to produce evolution-proof control measures?

Further Reading

De Assis, J.C.S., Gonçalves, O.S., Fernandes, A.S., de Queiroz, M.V., Bazzolli, D.M.S. *et al.* (2022) Genomic analysis reveals the role of integrative and conjugative elements in plant pathogenic bacteria. *Mobile DNA* 13, 19.

EPPO Overview of Psa (n.d.). Available at: https://gd.eppo.int/taxon/PSDMAK (accessed 1 August 2024).

Kah, M., Tufenkji, N. and White, J.C. (2019) Nano-enabled strategies to enhance crop nutrition and protection. *Nature Nanotechnology* 14, 532–540.

Kiwifruit Vine Health (n.d.). Available at: https://kvh.org.nz/

Santos, M.G., Nunes Da Silva, M., Vasconcelos, M.W. and Carvalho, S.M. (2024) Scientific and technological advances in the development of sustainable disease management tools: A case study on kiwifruit bacterial canker. *Frontiers in Plant Science* 14, 1306420.

Takikawa, Y., Serizawa, S., Ichikawa, T., TSUYUMU, S. and Goto, M. (1989) *Pseudomonas syringae* pv. *actinidiae* pv. nov.: The causal bacterium of canker of kiwifruit in Japan. *Japanese Journal of Phytopathology* 55, 437–444.

References

Anderson, J.C. (2023) I11 communication: Host metabolites as virulence-regulating signals for plant-pathogenic bacteria. *Annual Review of Phytopathology* 61, 49–71.

Balestra, G.M., Mazzaglia, A., Quattrucci, A., Renzi, M. and Rossetti, A. (2009) Current status of bacterial canker spread on kiwifruit in Italy. *Australasian Plant Disease Notes* 4, 34–36.

Cameron, A., Zoysa, G.H.D. and Sarojini, V. (2014) Antimicrobial peptides against *Pseudomonas syringae* pv. *actinidiae* and *Erwinia amylovora*: Chemical synthesis, secondary structure, efficacy, and mechanistic investigations. *Biopolymers* 102, 88–96.

Caullireau, E., Danzi, D., Tempo, V.M., andolfo, M., Morris, C.E. *et al.* (2024) Some like it hot: Efficiency of the type III secretion system and activation of plant defenses have multiple thermosensitive behaviors in the *Pseudomonas syringae* complex. DOI: 10.1101/2024.05.31.596889.

Cellini, A., Donati, I., Fiorentini, L., Vandelle, M., Polverari, A. *et al.* (2020) N-Acyl Homoserine Lactones and Lux Solos Regulate Social Behaviour and Virulence of *Pseudomonas syringae* pv. actinidiae. Microbial Ecology 79, 383–396.

Chapman, J.R., Taylor, R.K., Weir, B.S., Romberg, M.K., Vanneste, J.L. *et al.* (2012) Phylogenetic relationships among global populations of *Pseudomonas syringae* pv. *actinidiae*. *Phytopathology®* 102, 1034–1044.

Colombi, E., Bertels, F., Doulcier, G., McConnell, E., Pichugina, T. *et al.* (2024) Rapid dissemination of host metabolism–manipulating genes via integrative and conjugative elements. *Proceedings of the National Academy of Sciences of the United States of America* 121, e2309263121.

Donati, I., Cellini, A., Buriani, G., Mauri, S., Kay, C. *et al.* (2018) Pathways of flower infection and pollen-mediated dispersion of *Pseudomonas syringae* pv. *actinidiae*, the causal agent of kiwifruit bacterial canker. *Horticulture Research* 5, 56.

Donati, I., Cellini, A., Buriani, G., Mauri, S., Kay, C. *et al.* (2020) *Pseudomonas syringae* pv. *actinidiae*: Ecology, infection dynamics and disease epidemiology. *Microbial Ecology* 80, 81–102.

Fujikawa, T. and Sawada, H. (2019) Genome analysis of *Pseudomonas syringae* pv. *actinidiae* biovar 6, which produces the phytotoxins, phaseolotoxin and coronatine. *Scientific Reports* 9, 3836.

Hemara, L.M., Jayaraman, J., Sutherland, P.W., Montefiori, M., Arshed, S. *et al.* (2022) Effector loss drives adaptation of *Pseudomonas syringae* pv. *actinidiae* biovar 3 to *Actinidia arguta*. Coaker G (ed.). *PLoS Pathogens* 18, e1010542.

Holtappels, D., Fortuna, K., Lavigne, R. and Wagemans, J. (2021) The future of phage biocontrol in integrated plant protection for sustainable crop production. *Current Opinion in Biotechnology* 68, 60–71.

Hotinger, J.A., Pendergrass, H.A. and May, A.E. (2021) Molecular targets and strategies for inhibition of the bacterial type III secretion system (T3SS); inhibitors directly binding to T3SS components. *Biomolecules* 11, 316.

Huang, H. (2016) *Kiwifruit: The Genus Actinidia*. Elsevier/AP, Academic Press is an imprint of Elsevier; Science Press, Amsterdam; Boston: Beijing.

Ishiga, T., Sakata, N., Usuki, G., Nguyen, V.T., Gomi, K. *et al.* (2022) Large-scale transposon mutagenesis reveals type III secretion effector HopR1 is a major virulence factor in *Pseudomonas syringae* pv. *actinidiae*. *Plants* 12, 141.

Jayaraman, J., Jones, W.T., Harvey, D., Hemara, L.M., McCann, H.C. *et al.* (2020) Variation at the common polysaccharide antigen locus drives lipopolysaccharide diversity within the *Pseudomonas syringae* species complex. *Environmental Microbiology* 22, 5356–5372.

Li, Y., Zhu, Q., Zhi, T., Fan, R., Xie, T. *et al.* (2021) Genetic causes of non-pathogenic *Pseudomonas syringae* pv. *actinidiae* isolates in kiwifruit orchards. *Frontiers in Microbiology* 12, 650099.

McAtee, P.A., Brian, L., Curran, B., Linden, O., Nieuwenhuizen, N.J. *et al.* (2018) Reprogramming of *Pseudomonas syringae* pv. *actinidiae* gene expression during early stages of infection of kiwifruit. *BMC Genomics* 19, 822.

McCann, H.C., Rikkerink, E.H., Bertels, F., Fiers, M., Lu, A. *et al.* (2013) Genomic analysis of the kiwifruit pathogen *Pseudomonas syringae* pv. *actinidiae* provides insight into the origins of an emergent plant disease. *PLoS Pathogens* 9, e1003503.

McCann, H.C., Li, L., Liu, Y., Li, D., Pan, H. *et al.* (2017) Origin and evolution of the kiwifruit canker pandemic. *Genome Biology and Evolution* 9, 932–944.

O'Malley, M.R. and Anderson, J.C. (2021) Regulation of the *Pseudomonas syringae* type III secretion system by host environment signals. *Microorganisms* 9, 1227.

Patel, H.K., Ferrante, P., Covaceuszach, S., Lamba, D., Scortichini, M. *et al.* (2014) The kiwifruit emerging pathogen *Pseudomonas syringae* pv. *actinidiae* does not produce ahls but possesses three luxR solos. *PLoS ONE* 9, e87862.

Pereira, C., Costa, P., Pinheiro, L., Balcão, V.M. and Almeida, A. (2021) Kiwifruit bacterial canker: An integrative view focused on biocontrol strategies. *Planta* 253, 49.

Puttilli, M.R., Danzi, D., Correia, C., Brandi, J., Cecconi, D. *et al.* (2022) Plant signals anticipate the induction of the type III secretion system in *Pseudomonas syringae* pv. *actinidiae*, facilitating efficient temperature-dependent effector translocation. *Microbiology Spectrum* 10, e02073–22.

Rabiey, M., Grace, E.R., Pawlos, P., Bihi, M., Ahmed, H. *et al.* (2024) Coevolutionary analysis of *Pseudomonas syringae* – phage interactions to help with rational design of phage treatments. *Microbial Biotechnology* 17, e14489.

Reglinski, T., Vanneste, J.L., Wurms, K., Gould, E., Spinelli, F. *et al.* (2013) Using fundamental knowledge of induced resistance to develop control strategies for bacterial canker of kiwifruit caused by *Pseudomonas syringae* pv. *actinidiae*. *Frontiers in Plant Science* 4. DOI: 10.3389/fpls.2013.00024.

Santos, M.G., Silva, M., Vasconcelos, M.W. and Carvalho, S.M. (2024) Scientific and technological advances in the development of sustainable disease management tools: A case study on kiwifruit bacterial canker. *Frontiers in Plant Science* 14, 1306420.

Sawada, H. and Fujikawa, T. (2019) Genetic diversity of *Pseudomonas syringae* pv. *actinidiae*, pathogen of kiwifruit bacterial canker. *Plant Pathology* 68, 1235–1248.

Scortichini, M., Marcelletti, S., Ferrante, P., Petriccione, M. and Firrao, G. (2012) *Pseudomonas syringae* pv. *actinidiae*: A re-emerging, multi-faceted, pandemic pathogen. *Molecular Plant Pathology* 13, 631–640.

Serizawa, S., Ichikawa, T., Takikawa, Y., Tsuyumu, S. and Goto, M. (1989) Occurrence of bacterial canker of kiwifruit in Japan. *Japanese Journal of Phytopathology* 55, 427–436.

Son, K.I., Kim, G.H., Choi, E.D., Kim, K.H., Lee, Y.S. *et al.* (2016) Seasonal prevalence of bacterial canker of kiwifruit in Boseong in 2014 and 2015. *Research in Plant Disease* 22(4), 217–226.

Song, Y.R., Choi, M.S., Choi, G.W., Park, I.K. and Oh, C.S. (2016) Antibacterial activity of cinnamaldehyde and estragole extracted from plant essential oils against *Pseudomonas syringae* pv. *actinidiae* causing bacterial canker disease in kiwifruit. *The Plant Pathology Journal* 32, 363–370.

Testolin, R. (2015) Kiwifruit (*actinidia* spp.) in Italy: The history of the industry, international scientific cooperation and recent advances in genetics and breeding. *Acta Horticulturae* 1096, 47–61. DOI: 10.17660/ActaHortic.2015.1096.2.

Vandelle, E., Colombo, T., Regaiolo, A., Maurizio, V., Libardi, T. *et al.* (2021) Transcriptional profiling of three *Pseudomonas syringae* pv. *actinidiae* biovars reveals different responses to apoplast-like conditions related to strain virulence on the host. *MPMI* 34, 376–396.

Vanneste, J.L. (2017) The scientific, economic, and social impacts of the New Zealand outbreak of bacterial canker of kiwifruit (*Pseudomonas syringae* pv. *actinidiae*). *Annual Review of Phytopathology* 55, 377–399.

Wang, F.M., Mo, Q.H., Ye, K.Y., Gong, H.J., Qi, B.B. *et al.* (2020) Evaluation of the wild *Actinidia* germplasm for resistance to *Pseudomonas syringae* pv. *actinidiae*. *Plant Pathology* 69, 979–989.

Xie, T., Wu, X., Luo, L., Qu, Y., Fan, R. *et al.* (2023) Natural variation in the *hrpL* promoter renders the phytopathogen *Pseudomonas syringae* pv. *actinidiae* nonpathogenic. *Molecular Plant Pathology* 24, 262–271.

Zhang, D., Gan, R.Y., Zhang, J.R., Farha, A.K., Li, H.B. *et al.* (2020) Antivirulence properties and related mechanisms of spice essential oils: A comprehensive review. *Comprehensive Reviews in Food Science and Food Safety* 19, 1018–1055.

Zhao, Z., Chen, J., Gao, X., Zhang, D., Zhang, J. *et al.* (2019) Comparative genomics reveal pathogenicity-related loci in *Pseudomonas syringae* pv. *actinidiae* biovar 3. *Molecular Plant Pathology* 20, 923–942.

11 The Benefits and Risks of Antibiotics Use in Plant Agriculture

Toi Ketehouli[1], Erica M. Goss[1,2], Fernanda Batistel[3] and Samuel J. Martins[1]

[1]Department of Plant Pathology, University of Florida, Gainesville, FL, 32611, USA
[2]Emerging Pathogens Institute, University of Florida, Gainesville, FL, 32610, USA
[3]Department of Animal Sciences, University of Florida, Gainesville, FL, 32611, USA

Introduction

The use of antibiotics in agriculture is emerging as a significant subject of debate, particularly concerning problems of human, animal, and plant health, environmental sustainability, and agricultural productivity (McManus *et al.*, 2002). Historically, antibiotics have been examined thoroughly and regulated in medicine and animal husbandry, where their application has raised significant worries about the emergence of antibiotic resistance. Conversely, their utilization in plant agriculture, especially in managing bacterial diseases in crops, is a comparatively under-researched domain. Indeed, in the 1950s, antibiotics such as streptomycin were applied to manage fire blight disease caused by *Erwinia amylovora* in apple trees (McManus *et al.*, 2002). Since then, antibiotics have been widely used to control various plant bacterial diseases, such as bacterial spot disease caused by *Xanthomonas* spp. on tomato and pepper plants. In areas such as Florida, where citrus cultivation encounters significant obstacles from bacterial infections such as *Candidatus* Liberibacter asiaticus (Clas), the responsible bacterium for citrus greening or Huanglongbing (HLB), antibiotics such as oxytetracycline have been utilized within integrated disease management approaches since 2016 (Batuman *et al.*, 2024).

Proponents of antibiotic use in agriculture contend that years of usage have not directly caused notable human health problems or ecological disturbances; however, recent studies highlight the potential long-term risks (Wu *et al.*, 2022). Because the antibiotics used in agriculture are the same as in medical settings, a major concern is that agricultural antibiotics may foster the emergence of antibiotic-resistant bacteria, which can transfer resistance genes to human pathogens (Yin *et al.*, 2023). Moreover, there are increasing apprehensions regarding the unintentional consequences of antibiotics on the rhizosphere, the essential region of soil encircling plant roots where intricate microbial interactions facilitate plant health and nutrient cycling. This case study examines the benefits of antibiotics in plant agriculture and their associated impacts and risks to plant health and the environment. We examine how antibiotics have been used in plant agriculture in the last decade, explain the risks and concerns about plant and environmental health, and provide insight into sustainable farming methods.

Benefits of Antibiotic Use in Plants

Antibiotics have been used in plant farming since the middle of the 20th century to protect valuable crops from bacterial diseases (Fair and Tor, 2014). The use of antibiotics has been considered essential for keeping plant diseases under control and ensuring healthy crops. Streptomycin and oxytetracycline are among the two most commonly and widely used antibiotics for managing plant bacterial diseases (Table 11.1).

In the USA, antibiotics applied to plants account for less than 0.5% of total antibiotic use. Studies have shown that these antibiotics can be translocated into plant tissues, influencing the plant microbiome and potentially posing health risks to humans (McManus *et al.*, 2002; Yin *et al.*, 2023). Antibiotic residues in commercially grown vegetables have also been documented, with manure application and irrigation practices being significant factors (Yin *et al.*, 2023). In Florida, since 2016, citrus growers have been allowed to use oxytetracycline to manage citrus greening or HLB disease. This disease is caused by a bacterial pathogen called *Candidatus* Liberibacter asiaticus, transmitted by *Diaphorina citri* (Asian citrus psyllid) (Hu *et al.*, 2018). Inadequate production of fruit, premature fruit drop, and reduced fruit size and quality are all consequences of citrus greening, which leads to significant losses in crop yield. Although some antibiotics applied through the foliage have improved fruit quality and decreased fruit drop, findings have been mixed, and farmers have not seen a return on investment. The effects of reducing bacterial load on other measures of tree health, such as fruit drop, yield, and fruit quality, may vary between young and older trees (Archer *et al.*, 2023). While the use of antibiotics has been essential in managing bacterial diseases in crops, their application raises concerns about antibiotic resistance and potential health risks.

Table 11.1. Examples of commonly used antibiotics in plant production and distribution of antibiotic resistance bacteria (ARB).

Antibiotics	Crops on which antibiotics are used	Countries where antibiotics use is permitted	Countries with ARB detection	References
Streptomycin	Solanaceous crops (tomato, pepper, potato...), apples, pears, Iran, and oranges	Canada, Chile, China, Argentina, Mexico, New Zealand, Israel, South Korea, and USA	USA, Canada, Chile, China, South Korea	Xu *et al.*, 2010; Hyun *et al.* (2012); Förster *et al.* (2015); Tancos and Cox (2016); Tancos *et al.* (2016); Sundin and Wang (2018); Li *et al.* (2021); Miller *et al.* (2022); Sundin *et al.* (2023); Ham *et al.* (2024)
Oxytetracycline	Citrus, apples, pears, peaches, brassica crops, and palms	Costa Rica, Guatemala, El Salvador, Mexico, Honduras, China, and USA	China and USA	Li *et al.* (2011) ; Förster *et al.* (2020); Zhang *et al.* (2021); Herbert *et al.* (2022); Sundin *et al.* (2023)
Gentamicin	Solanaceous crops (tomato, eggplants...), apples, pears, and brassica crops	Chile, Costa Rica, Honduras, Guatemala, El Salvador, and Thailand	Thailand	Heuer *et al.* (2002); Yadav *et al.* (2017); Srichamnong *et al.* (2021)
Oxolinic acid	Apples and rice	Israel, Japan, and South Korea	Israel, Japan, and South Korea	Kleitman *et al.* (2005); Kwon *et al.* (2010); Ham *et al.* (2022)
Kasugamycin	Apples, pears, rice, cherries, and walnuts	Japan, New Zealand, and USA	Japan	Yoshii *et al.*, 2012; Slack *et al.*, 2020
Penicillin	Brassica crops, grasses (Poaceae)	USA	–	Minden *et al.* (2017a)

Adapted from Batuman *et al.* (2024).

Risks Related to Antibiotic Use in Plant Agriculture

Antibiotics resistant bacteria

Antibiotic resistance in pathogenic or non-pathogenic bacteria is a significant concern for both plant health and environmental safety. The use of antibiotics in agricultural plants has led to the emergence and spread of antibiotic-resistant bacteria, posing challenges to managing plant diseases and maintaining ecological balance (Ifedinezi *et al.*, 2024). Antibiotics used in agriculture create selective pressure that fosters the development of antibiotic-resistant bacteria (ARB) and the spread of antibiotic-resistant genes (ARGs) (Skandalis *et al.*, 2021). Urban wastewater treatment plants are also significant sources of ARB and ARGs, as they release these elements into the environment, contributing to the spread of resistance (Yuan *et al.*, 2021). The rhizosphere (narrow region of soil that surrounds and is influenced by plant roots) and the rhizoplane (surface of the plant roots, including the root epidermis and adhering soil particles) are essential environments where ARB can flourish. The selective forces that these two environments apply to bacterial colonization are distinct, resulting in differing distributions of ARB. As an illustration, the prevalence of ARB in the rhizoplane is typically higher than in the rhizosphere. It is possible to trace this discrepancy to the unique properties of these ecosystems, which influences the quantity and resistance profiles of the bacteria that colonize them (Yang *et al.*, 2022). The rhizoplane exhibits a higher prevalence of polymyxin-resistant isolates, but the rhizosphere contains a more significant number of amphenicol-resistant isolates (Yang *et al.*, 2022). This vital distinction underscores the complexity of plant-microbe-soil interactions and the need for targeted studies to understand better the mechanisms driving antibiotic resistance in these environments.

There are multiple processes by which antibiotic resistance in plant-pathogenic bacteria develops. These processes involve the acquisition of resistance determinants through horizontal gene transfer and mutations (Sundin and Wang, 2018). Under the pressure of antibiotic use, resistance genes can be transferred from non-pathogenic epiphytic bacteria to pathogenic ones. For instance, the *strAB* streptomycin-resistance genes have been discovered in various plant diseases, including *Erwinia amylovora*, *Pseudomonas syringae*, and *Xanthomonas campestris* (Sundin and Wang, 2018). This set of genes is thought to have been acquired from bacteria that are not harmful and are found close to plant hosts (McManus *et al.*, 2002). Another example is the transfer of extended-spectrum beta-lactamase (ESBL) genes. ESBL-producing *Escherichia coli*, a human pathogenic bacterium, can transfer these resistance genes to non-pathogenic epiphytic bacteria (like *Pseudomonas fluorescens*) found on plant surfaces. This transfer can occur through plasmids that are mobile genetic elements that facilitate the exchange of genetic material between bacteria (Zalewska *et al.*, 2021).

Another concern is the potential for ARB to enter the human food chain by consuming antibiotic treated crops (Fig. 11.1). While some studies have shown negligible health risks from vegetable or fruit consumption (Tadić *et al.*, 2021), there are other studies that highlight potential concerns (Allydice-Francis and Brown, 2012; Serwecińska, 2020). In Florida, even though antibiotic resistance in plant-pathogenic bacteria, such as *Xanthomonas citri*, has been reported in the management of diseases like citrus canker, no studies so far have shown the bacteria resistance to oxytetracycline injection into the citrus trunk in HLB management. However, a high prevalence of antibiotic-resistant bacteria has been observed in the rhizosphere and rhizoplane of *Citrus medica*, with different resistance profiles depending on the environment, suggesting that citrus plants can be reservoirs for antibiotic-resistant

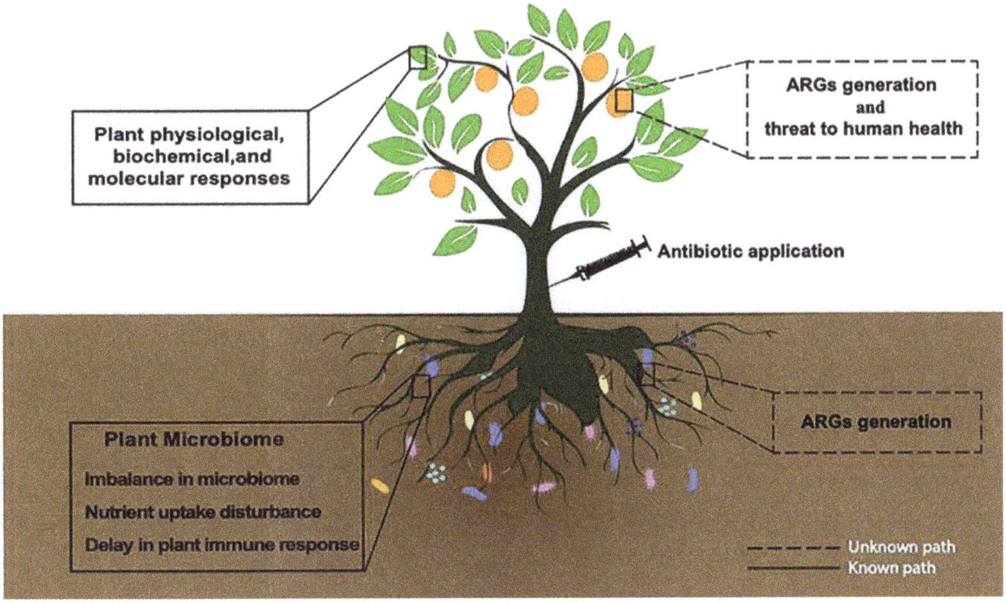

Fig. 11.1. Impact of antibiotic application on plant health and its implications. The diagram depicts the impact of antibiotic use on plant-microbe interactions. Antibiotics can disturb the plant microbiome, resulting in imbalances that influence plant physiological, biochemical, and molecular responses. This disruption may lead to nutrient absorption complications, delayed plant immune response, and the development of antibiotic resistance microorganisms (ARM) and antibiotic resistance genes (ARGs). The dissemination of ARM/ARGs presents a potential risk to human health *via* food contamination.

bacteria (Fang *et al.*, 2023). Thus, researchers on citrus greening management are beginning to consider the possible emergence of oxytetracycline-resistant bacteria in the coming years.

Impact of Antibiotics on Rhizosphere Microbial Communities

In the same way that different bacteria live in our intestines as the gut microbiome, the plant microbiome is linked to microorganisms in the rhizosphere. Dysbiosis is known as an imbalance of microbes that occurs in both animals and plants. In humans, dysbiosis, an imbalance of microbes, can lead to other diseases, such as inflammatory bowel disease and obesity (Amabebe *et al.*, 2020). In plants, dysbiosis can result from abiotic stresses, such as antibiotic use or nutrient deficiency, negatively impacting plant health. Studies have shown that when tomato plants had rhizospheric dysbiosis, there was a decline in the *Firmicutes* and *Actinobacteria* populations and a rise in bacterial wilt caused by *Ralstonia solanacearum* and a significant increase in the *Fusarium oxysporum* wilt (Zhang *et al.*, 2021; Lee *et al.*, 2022). Additionally, Ketehouli *et al.* (2024b) found that dysbiosis caused by antibiotic application to the rhizosphere had a systemic negative effect on plants, which became more susceptible to bacterial spot, a foliar disease. Genes involved in primary and secondary metabolism in plants were down-regulated, while genes involved in plant defense associated with systemic resistance were up-regulated (Ketehouli *et al.*, 2024a, b).

Current developments in culture-independent approaches and metagenomics have facilitated the characterization of microbiomes across various plant tissues. Recent research has demonstrated the existence of beneficial, neutral, and harmful prokaryotic and eukaryotic microbiomes in plants' rhizosphere and bark microbiome. The microbial populations of the rhizosphere have been studied in many crops, including citrus (Xu *et al.*, 2018). This region has elevated microbial activity, which is significant for crop yield. Rhizosphere microbiota can stimulate plant systemic resistance, improve nutrient absorption, safeguard against infections, and augment tolerance to various abiotic stresses. A study has demonstrated that the application of oxytetracycline impeded plant growth by diminishing nutrient absorption, decreased the total bacterial population in the rhizoplane, and increased both fungal populations and the prevalence of oxytetracycline-resistant bacteria in the rhizoplane (Yin *et al.*, 2023). Ketehouli *et al.* (2024b), *via* high-throughput sequencing of bacterial 16S rRNA genes, demonstrated significant shifts in bacterial community structures post-antibiotic treatment of the citrus rhizosphere. Notably, the decrease in the *Pseudomonas* population after streptomycin and oxytetracycline applications indicated susceptibility to both antibiotics. Conversely, the increase in the number of *Streptomyces* under these antibiotics suggests a potential resistance or competitive advantage over *Pseudomonas*. This increase may be linked to *Streptomyces'* ability to produce antibiotics, which could enhance their tolerance to antibiotic exposure (Ketehouli *et al.*, 2024b).

Impact of Antibiotics on Plant Physiology

The use of antibiotics in plant agriculture has significant environmental implications, raising concerns about ecological balance, plant health, and agricultural sustainability. Antibiotics released into the environment through manure, direct application, and irrigation can affect non-target species and alter plant species composition in natural field. Antibiotics can affect plant traits even at small concentrations, leading to delayed germination and lower biomass allocation (Minden *et al.*, 2017). Plants exposed to antibiotics showed negative effects on their initial establishment and development. In *Arabidopsis thaliana*, for example, exposure to antibiotics resulted in radial expansion of the root cells and the production of unusually long root hairs (Rocha *et al.*, 2021). The number of adventitious roots increased, and reductions in the dimensions of vascular bundles were observed in corn plants exposed to antibiotics. These effects are linked to interactions with reactive oxygen species (ROS) formation and their signaling to phytohormones such as auxins (Rocha *et al.*, 2021). Another phenomenon observed is antibiotic-induced hormesis, where low doses of antibiotics enhance plant health while high doses suppress it. Although administering oxytetracycline or streptomycin may effectively target Clas in HLB-diseased citrus (Li *et al.*, 2019), it can also adversely impact plant physiology and growth, as antibiotics influence beneficial rhizobacteria that enhance plant development and physiology (Hu and Wang, 2016; Ketehouli *et al.*, 2024a). Antibiotic treatment significantly reduces photosynthesis and transpiration in plants compared to controls. These processes are essential for plant health but can produce harmful ROS (Kapoor *et al.*, 2019). Chemical stress from antibiotics disrupts metabolic activities, prompting plants to downregulate energy-intensive photosynthesis to conserve resources (Ketehouli *et al.*, 2024a, b). Photosynthesis and transpiration are interlinked, with transpiration decreasing as photosynthesis slows, particularly under water stress or chemical damage. This reduction is a stress response to preserve vital physiological functions and survival (Ketehouli *et al.*, 2024a).

Eco-Friendly Farming Approaches

Eco-friendly farming methods emphasize sustainable practices to protect the environment, boost biodiversity, and ensure long-term agricultural productivity. These methods include biological control agents to manage pests and diseases, reducing the need for chemical compounds, such as antibiotics. One promising strategy against antibiotic-resistant bacteria is the use of bioprotectants derived from natural substances, such as specialized metabolites, essential oils, and essential oils, which enhance plants' natural defenses against diseases and pests. Biofertilization involves using living microorganisms like plant growth-promoting rhizobacteria (PGPR), mycorrhizal fungi, to enhance plant nutrient update and growth, and mitigate stress factors including under field conditions (Martins *et al.*, 2015, 2018a, 2018b, 2018c). A diverse array of prevalent PGPR species that inhabit plant roots and augment the plants' capacity to withstand most abiotic and chemical stresses encompasses *Pseudomonas, Bacillus, Rhizobium, Enterobacter, Azospirillum, Azotobacter, Arthrobacter, Mesorhizobium, Xanthomonas, Acinetobacter, Alcaligenes, Burkholderia, Erwinia,* and *Flavobacterium* (Glick, 1995; Chieb and Gachomo, 2023). For example, *Bacillus velezensis*, an endophytic bacterium extracted from citrus plants, exhibits significant antagonistic properties against streptomycin-resistant *Xanthomonas citri* subsp. *citri*, the pathogen responsible for citrus bacterial canker (Rabbee *et al.*, 2022). This bacterium was able to diminish the incidence of canker, indicating its potential as a biocontrol agent (Rabbee *et al.*, 2022). Naringenin, a flavonoid in citrus fruits, demonstrates antibacterial efficacy against resistant bacteria and fungi (Duda-Madej *et al.*, 2022). Natural chemicals like terpenes and flavonoids have shown promise in fighting germs and reducing their resistance to antibiotics (Khameneh *et al.*, 2021). Recently, the attention of biological approaches has turned into combining different species or strains of microbes with complementary and/or synergistic effects into consortia or synthetic microbial communities (SynComs) (Martins *et al.*, 2023).

Additionally, bacteriophages (viruses that specifically target and lyse bacterial cells) are emerging as a promising tool for controlling bacterial pathogens in agriculture (Collinge *et al.*, 2022). Their use as biocontrol agents offers several advantages over traditional chemical pesticides and antibiotics. Phages are usually highly specific to their bacterial hosts, which means they can eliminate harmful bacteria without affecting beneficial microorganisms or the environment. This specificity reduces the risk of collateral damage to non-target species, a common issue with broad-spectrum chemical pesticides. Besides, phages have shown effectiveness against multidrug-resistant bacterial strains, which are increasingly problematic in agriculture. Unlike antibiotics, which bacteria can develop resistance to, phages can evolve alongside their bacterial hosts, potentially overcoming resistance mechanisms (Azam and Tanji, 2019). Phage therapy is considered environmentally friendly because phages are naturally occurring and biodegradable, leaving no harmful residues in the soil or water (Azam and Tanji, 2019; Collinge *et al.*, 2022). Despite their potential, challenges such as the narrow host range of phages and regulatory hurdles must be addressed. Using phage cocktails and continued research on phage stability and delivery methods can help overcome these challenges. The effectiveness of conventional antibiotics could be improved, and resistance to these antibiotics could be decreased by combining them with biological approach.

Discussion Points

1. Antibiotic use in plants: A friend or a foe? This question has been the subject of controversial discussions in recent years. Although antibiotics offer distinct advantages in plant cultivation, their application must be meticulously regulated to prevent adverse effects.
2. How can ARB contamination occur at various stages, from cultivation to consumption, and what impact does this have on plant health and food safety?
3. Understanding the genetic basis of antibiotic resistance and the role of the plant microbiome can inform the development of management strategies.
4. Research into plant-based phytochemicals or biological agents and their mechanisms of action can provide sustainable alternatives to traditional antibiotics.

Conclusion

The use of antibiotics in plant agriculture presents short-term benefits and both short and long-term risks. While they effectively control plant bacterial diseases, the emergence of antibiotic-resistant strains and environmental impact are significant concerns. Sustainable agricultural approaches offer promising alternatives, including plant-based phytochemicals or biological agents. A balanced approach that considers both the benefits and risks of antibiotic use is essential for ensuring plant health and environmental sustainabilityMartins *et al.*, 2018c.

References

Allydice-Francis, K. and Brown, P.D. (2012) Diversity of antimicrobial resistance and virulence determinants in *Pseudomonas aeruginosa* associated with fresh vegetables. *International Journal of Microbiology* 2012(1), 426241.

Amabebe, E., Robert, F.O., Agbalalah, T. and Orubu, E.S.F. (2020) Microbial dysbiosis-induced obesity: Role of gut microbiota in homoeostasis of energy metabolism. *BritishJournalofNutrition* 123(10), 1127–1137.

Archer, L., Kunwar, S., Alferez, F., Batuman, O. and Albrecht, U. (2023) Trunk injection of oxytetracycline for Huanglongbing management in mature grapefruit and sweet orange trees. *Phytopathology* 113(6), 1010–1021.

Azam, A.H. and Tanji, Y. (2019) Bacteriophage-host arm race: An update on the mechanism of phage resistance in bacteria and revenge of the phage with the perspective for phage therapy. *AppliedMicrobiology and Biotechnology* 103, 2121–2131.

Batuman, O., Britt-Ugartemendia, K., Kunwar, S., Yilmaz, S., Fessler, L. *et al.* (2024) The use and impact of antibiotics in plant agriculture: A review. *Phytopathology* 114(5), 885–909. DOI: 10.1094/PHYTO-10-23-0357-IA.

Chieb, M. and Gachomo, E.W. (2023) The role of plant growth promoting rhizobacteria in plant drought stress responses. *BMC Plant Biology* 23(1), 407. DOI: 10.1186/s12870-023-04403-8.

Collinge, D.B., Jensen, D.F., Rabiey, M., Sarrocco, S., Shaw, M.W. *et al.* (2022) Biological control of plant diseases–what has been achieved and what is the direction? *Plant Pathology* 71(5), 1024–1047.

Duda-Madej, A., Stecko, J., Sobieraj, J., Szymańska, N. and Kozłowska, J. (2022) Naringenin and its derivatives—Health-promoting phytobiotic against resistant bacteria and fungi in humans. *Antibiotics* 11(11), 1628.

Fair, R.J. and Tor, Y. (2014) Antibiotics and bacterial resistance in the 21st century. *Perspectives in Medicinal Chemistry* 6, MC–S14459.

Fang, Y., Chen, C., Cui, B. and Zhou, D. (2023) Self-rescue of nitrogen-cycling bacteria under β-lactam antibiotics stress during managed aquifer recharge (MAR): Microbial collaboration and anti-resistance. *Water Research* 231, 119623.

Förster, H., McGhee, G.C., Sundin, G.W. and Adaskaveg, J.E. (2015) Characterization of streptomycin resistance in isolates of *Erwinia amylovora* in California. *Phytopathology* 105(10), 1302–1310.

Förster, H.C., Belisle, R.J., Sundin, G.W. and Adaskaveg, J.E. (2020) Oxytetracycline resistance in *Erwinia amylovora* detected in commercial pear orchards in California. *Plant Plant Health 2020 Online.*

Glick, B.R. (1995) The enhancement of plant growth by free-living bacteria. *Canadian Journal of Microbiology* 41(2), 109–117.

Ham, H., Oh, G.-R., Park, D.S. and Lee, Y.H. (2022) Survey of oxolinic acid-resistant *Erwinia amylovora* in Korean apple and pear orchards, and the fitness impact of constructed mutants. *The Plant Pathology Journal* 38(5), 482.

Ham, H., Oh, G.-R., Lee, B.W., Lee, Y.H. and Lee, Y.H. (2024) Changes of sensitivity to streptomycin in *Erwinia amylovora* isolated from 2019 to 2023 in Korea. *Research in Plant Disease* 30(2), 199–205.

Herbert, A., Hancock, C.N., Cox, B., Schnabel, G., Moreno, D. *et al.* (2022) Oxytetracycline and streptomycin resistance genes in *Xanthomonas arboricola* pv. Pruni, the causal agent of bacterial spot in peach. *Frontiers in Microbiology* 13, 821808.

Heuer, H., Krögerrecklenfort, E., Wellington, E.M.H., Egan, S., Van Elsas, J.D. *et al.* (2002) Gentamicin resistance genes in environmental bacteria: Prevalence and transfer. *FEMS Microbiology Ecology* 42(2), 289–302.

Hu, J. and Wang, N. (2016) Evaluation of the spatiotemporal dynamics of oxytetracycline and its control effect against citrus Huanglongbing via trunk injection. *Phytopathology* 106(12), 1495–1503. DOI: 10.1094/PHYTO-02-16-0114-R.

Hu, J., Jiang, J. and Wang, N. (2018) Control of citrus Huanglongbing via trunk injection of plant defense activators and antibiotics. *Phytopathology* 108(2), 186–195.

Hyun, J.W., Kim, H.J., Yi, P.H., Hwang, R.Y. and Park, E.W. (2012) Mode of action of streptomycin resistance in the citrus canker pathogen (*Xanthomonas smithii* subsp. citri) in Jeju Island. *The Plant Pathology Journal* 28(2), 207–211.

Ifedinezi, O.V., Nnaji, N.D., Anumudu, C.K., Ekwueme, C.T., Uhegwu, C.C. *et al.* (2024) Environmental antimicrobial resistance: Implications for food safety and public health. *Antibiotics* 13(11), 1087.

Kapoor, D., Singh, S., Kumar, V., Romero, R., Prasad, R. *et al.* (2019) Antioxidant enzymes regulation in plants in reference to reactive oxygen species (ROS) and reactive nitrogen species (RNS). *Plant Gene* 19, 100182.

Ketehouli, T., Goss, E.M., Ascunce, M.S. and Martins, S.J. (2024a) Metabolic and physiological effects of antibiotic-induced dysbiosis in citrus. *Ecotoxicology and Environmental Safety* 287, 117325. DOI: 10.1016/j.ecoenv.2024.117325.

Ketehouli, T., Pasche, J., Buttrós, V.H., Goss, E.M. and Martins, S.J. (2024b) The underground world of plant disease: Rhizosphere dysbiosis reduces above-ground plant resistance to bacterial leaf spot and alters plant transcriptome. *Environmental Microbiology* 26(7), e16676. DOI: 10.1111/1462-2920.16676.

Khameneh, B., Eskin, N.A.M., Iranshahy, M. and Fazly Bazzaz, B.S. (2021) Phytochemicals: A promising weapon in the arsenal against antibiotic-resistant bacteria. *Antibiotics* 10(9), 1044.

Kleitman, F., Shtienberg, D., Blachinsky, D., Oppenheim, D., Zilberstaine, M. *et al.* (2005) *Erwinia amylovora* populations resistant to oxolinic acid in Israel: Prevalence, persistence and fitness. *Plant Pathology* 54(2), 108–115.

Kwon, H.R., Choi, G.J., Choi, Y.H., Jang, K.S., Sung, N. *et al.* (2010) Suppression of pine wilt disease by an antibacterial agent, oxolinic acid. *Pest Management Science: Formerly Pesticide Science* 66(6), 634–639.

Lee, S.-K., Chiang, M.-S., Hseu, Z.-Y., Kuo, C.-H. and Liu, C.-T. (2022) A photosynthetic bacterial inoculant exerts beneficial effects on the yield and quality of tomato and affects bacterial community structure in an organic field. *Frontiers in Microbiology* 13, 959080.

Li, Z., Xie, X., Zhang, S. and Ling, Y. (2011) Negative effects of oxytetracycline on wheat (*Triticum aestivum* L.) growth, root activity, photosynthesis, and chlorophyll contents. *Agricultural Sciences in China* 10(10), 1545–1553.

Li, J., Pang, Z., Duan, S., Lee, D., Kolbasov, V.G. *et al.* (2019) The in planta effective concentration of oxytetracycline against '*Candidatus liberibacter asiaticus*' for suppression of citrus Huanglongbing. *Phytopathology* 109(12), 2046–2054.

Li, J., Kolbasov, V.G., Lee, D., Pang, Z., Huang, Y. *et al.* (2021) Residue dynamics of streptomycin in citrus delivered by foliar spray and trunk injection and effect on '*Candidatus liberibacter asiaticus*' titer. *Phytopathology* 111(7), 1095–1103.

Martins, S.J., Medeiros, F.H.V., Souza, R.M., Faria, A.F., Cancellier, E.L. *et al.* (2015) Common bean growth and health promoted by rhizobacteria and the contribution of magnesium to the observed responses. *Applied Soil Ecology* 87, 49–55.

Martins, S.A., Schurt, D.A., Seabra, S.S., Martins, S.J., Ramalho, M.A.P. *et al.* (2018a) Common bean (*Phaseolus vulgaris* L.) growth promotion and biocontrol by rhizobacteria under rhizoctonia solani suppressive and conducive soils. *Applied Soil Ecology* 127, 129–135.

Martins, S.J., Rocha, G.A., de Melo, H.C., de Castro Georg, R., Ulhôa, C.J. *et al.* (2018b) Plant-associated bacteria mitigate drought stress in soybean. *Environmental Science and Pollution Research* 25, 13676–13686.

Martins, S.J., Rocha, G.A., de Melo, H.C., de Castro Georg, R., Ulhôa, C.J. *et al.* (2018c) Plant-associated bacteria mitigate drought stress in soybean. *Environmental Science and Pollution Research* 25, 13676–13686.

Martins, S.J., Pasche, J., Silva, H.A.O., Selten, G., Savastano, N. *et al.* (2023) The use of synthetic microbial communities to improve plant health. *Phytopathology* 113(8), 1369–1379.

McManus, P.S., Stockwell, V.O., Sundin, G.W. and Jones, A.L. (2002) Antibiotic use in plant agriculture. *Annual Review of Phytopathology* 40, 443–465.

Miller, S.A., Ferreira, J.P. and LeJeune, J.T. (2022) Antimicrobial use and resistance in plant agriculture: A one health perspective. *Agriculture* 12(2), 289.

Minden, V., Deloy, A., Volkert, A.M., Leonhardt, S.D. and Pufal, G. (2017) Antibiotics impact plant traits, even at small concentrations. *AoB Plants* 9(2), lx010.

Rabbee, M.F., Islam, N. and Baek, K.-H. (2022) Biocontrol of citrus bacterial canker caused by *Xanthomonas citri* subsp. citri by *Bacillus velezensis*. *Saudi Journal of Biological Sciences* 29(4), 2363–2371.

Rocha, D.C., da Silva Rocha, C., Tavares, D.S., de Morais Calado, S.L. and Gomes, M.P. (2021) Veterinary antibiotics and plant physiology: An overview. *Science of the Total Environment* 767, 144902.

Serwecińska, L. (2020) Antimicrobials and antibiotic-resistant bacteria: A risk to the environment and to public health. *Water* 12(12), 3313.

Skandalis, N., Maeusli, M., Papafotis, D., Miller, S., Lee, B. *et al.* (2021) Environmental spread of antibiotic resistance. *Antibiotics* 10, 1–14. Available at: https://api.semanticscholar.org/CorpusID:235298956 (accessed 5 May 2025).

Slack, S.M., Walters, K.J., Outwater, C.A. and Sundin, G.W. (2020) Effect of Kasugamycin, Oxytetracycline, and streptomycin on in-orchard population dynamics of *Erwinia amylovora* on apple flower stigmas. *Plant Disease* 105(6), 1843–1850. DOI: 10.1094/PDIS-07-20-1469-RE.

Srichamnong, W., Kalambaheti, N., Woskie, S., Kongtip, P., Sirivarasai, J. *et al.* (2021) Occurrence of antibiotic-resistant bacteria on hydroponically grown butterhead lettuce (*Lactuca sativa* var. *capitata*). *Food Science and Nutrition* 9(3), 1460–1470.

Sundin, G.W. and Wang, N. (2018) Antibiotic resistance in plant-pathogenic bacteria. *Annual Review of Phytopathology* 56, 161–180. Available at: https://api.semanticscholar.org/CorpusID:46918802 (accessed 5 May 2025).

Sundin, G.W., Peng, J., Brown, L.E., Zeng, Q., Förster, H. *et al.* (2023) A novel IncX plasmid mediates high-level oxytetracycline and streptomycin resistance in *Erwinia amylovora* from commercial pear orchards in California. *Phytopathology* 113(12), 2165–2173.

Tadić, Đ., Hernandez, M.J.B., Cerqueira, F., Matamoros, V., Piña, B. *et al.* (2021) Occurrence and human health risk assessment of antibiotics and their metabolites in vegetables grown in field-scale agricultural systems. *Journal of Hazardous Materials* 401, 123424.

Tancos, K.A. and Cox, K.D. (2016) Exploring diversity and origins of streptomycin-resistant *Erwinia amylovora* isolates in New York through CRISPR spacer arrays. *Plant Disease* 100(7), 1307–1313.

Tancos, K.A., Villani, S., Kuehne, S., Borejsza-Wysocka, E., Breth, D. *et al.* (2016) Prevalence of streptomycin-resistant *Erwinia amylovora* in New York apple orchards. *Plant Disease* 100(4), 802–809.

Wu, J., Wang, J., Li, Z., Guo, S., Li, K. *et al.* (2022) Antibiotics and antibiotic resistance genes in agricultural soils: A systematic analysis. *Critical Reviews in Environmental Science and Technology* 53, 847–864. Available at: https://api.semanticscholar.org/CorpusID:250295751 (accessed 5 May 2025).

Xu, Y., Zhu, X., Zhou, M., Kuang, J., Zhang, Y. *et al.* (2010) Status of streptomycin resistance development in Xanthomonas oryzae pv. oryzae and Xanthomonas oryzae pv *oryzicola* in China and their resistance characters. *Journal of Phytopathology* 158(9), 601–608.

Xu, J., Zhang, Y., Zhang, P., Trivedi, P., Riera, N. *et al.* (2018) The structure and function of the global citrus rhizosphere microbiome. *Nature Communications* 9(1), 4894.

Yadav, N., Sharma, S., Sharma, S. and Sharma, K. (2017) Critical analysis of protective role of plants against gentamicin induced nephrotoxicity. *Indian Journal of Environmental Science* 21(1), 1–34.

Yang, F., Wang, Y., Liu, Q., Xu, B., Chen, H. *et al.* (2022) High prevalence and varied distribution of antibiotic-resistant bacteria in the rhizosphere and rhizoplane of citrus medica. *Microorganisms* 10, 1–9. Available at: https://api.semanticscholar.org/CorpusID:251876228

Yin, L., Wang, X., Li, Y., Liu, Z., Mei, Q. *et al.* (2023) Uptake of the plant agriculture-used antibiotics oxytetracycline and streptomycin by cherry radish-effect on plant microbiome and the potential health risk. *Journal of Agricultural and Food Chemistry* 71(11), 4561–4570.

Yoshii, A., Moriyama, H. and Fukuhara, T. (2012) The novel kasugamycin 2′-N-acetyltransferase gene aac (2′)-IIa, carried by the IncP island, confers kasugamycin resistance to rice-pathogenic bacteria. *Applied and Environmental Microbiology* 78(16), 5555–5564.

Yuan, L., Wang, Y., Zhang, L., Palomo, A., Zhou, J. *et al.* (2021) Pathogenic and indigenous denitrifying bacteria are transcriptionally active and key multi-antibiotic-resistant players in wastewater treatment plants. *Environmental Science and Technology* 55, 10862–10874. Available at: https://api.semanticscholar.org/CorpusID:229293847 (accessed 5 May 2025).

Zalewska, M., Błaze·jewska, A., Czapko, A. and Popowska, M. (2021) Antibiotics and antibiotic resistance genes in animal manure–consequences of its application in agriculture. *Frontiers in Microbiology* 12, 610656.

Zhang, Z., Su, R., Chang, C., Cheng, X., Peng, Q. *et al.* (2021) Effects of oxytetracycline on plant growth, phosphorus uptake, and carboxylates in the Rhizosheath of alfalfa. *Plant and Soil* 461, 501–515.

Part III: Fungal Diseases of Cereals and Oilseed Crops

Fungal pathogens of cereals and oilseed crops pose a persistent threat to global food security, with wheat, maize, and oilseed rape serving as essential commodities that sustain billions and support agricultural economies worldwide. This section explores both historical pathogens and emerging threats through six case studies covering wheat, maize, and oilseed rape.

Ergot disease caused by *Claviceps purpurea* offers historical context, as ergot epidemics devastated medieval populations through mycotoxin-contaminated grain. Although modern practices have reduced its occurrence, ergot remains a quality concern with strict regulatory thresholds. Most unexpectedly, the emergence of *Trichoderma* species as maize pathogens in European fields has been observed. Long celebrated as beneficial biocontrol agents, some isolates have shown pathogenic capacity, causing ear rot and yield losses. This shift in understanding challenges assumptions about fungal-plant interactions and raises questions regarding the safety assessment of biological control agents, illustrating how agricultural intensification can reveal unexpected pathogenic behaviours.

The biotrophic rust fungi *Puccinia striiformis* and *P. triticina*, which cause yellow rust and leaf rust respectively, demonstrate an ongoing evolutionary arms race between host defenses and pathogen virulence. Yellow rust has spread into warmer regions once deemed unsuitable, indicating its adaptive ability and raising climate change concerns. Both fungi show high genetic diversity and constantly evolve new races that bypass resistance genes, leading to boom-and-bust cycles where previously resistant cultivars become vulnerable within a few years.

Necrotrophic pathogens like *Pyrenophora tritici-repentis*, which causes wheat tan spot, and *Plenodomus lingam*, responsible for phoma stem canker in oilseed rape, pose unique management challenges. They can survive saprophytically on crop debris and produce necrotrophic effectors that induce host cell death. These pathogens flourish when residue management is poor and crop rotation is inadequate. Control strategies must be integrated, including crop rotation, resistant varieties when available, and strategic fungicide applications that use multiple modes of action to prevent leaf infection.

The case studies in this section provide essential insights into fungal pathogen biology, disease epidemiology, and evidence-based management strategies, enabling readers to tackle these economically important threats through sustainable, integrated approache

12 Ergot (*Claviceps* Spp.) of Cereals in Western Canada

SAMIA BERRAIES[1], SEAN WALKOWIAK[2,3], LONE BUCHWALDT[4] AND JAMES G. MENZIES[5]

[1]*Agriculture and Agri-Food Canada, Swift Current Research and Development Centre Swift Current, Saskatchewan, Canada*
[2]*Grain Research Laboratory, Canadian Grain Commission, Winnipeg, Canada*
[3]*Department of Plant Sciences, University of Manitoba, Winnipeg, Canada*
[4]*Agriculture and Agri-Food Canada, Saskatoon Research and Development Centre, Saskatoon, Saskatchewan, Canada*
[5]*Agriculture and Agri-Food Canada, Morden Research and Development Centre, Morden, Manitoba, Canada*

Causal Agent

Claviceps purpurea (FR.) Tul. is an obligate biotrophic fungal plant pathogen. *Claviceps* is Latin for "club head" and *purpurea* means "purple", both words refer to the purple-black sclerotia (resting bodies) that form in place of cereal grains, also known as ergot (Fig. 12.1). There are more than 50 species of *Claviceps* that have been organized into four main sections: *Claviceps*, *Pusillae*, *Citrinae*, and *Paspalorum* (Wyka *et al*., 2021) based on the size of their host range, ecological, morphological and metabolic traits. In Canada, *C. purpurea* is the species most frequently associated with ergot in cereal grains. The pathogen belongs to the *Claviceps* section, which has a broad host range (Menzies, 2004; Wyka *et al*., 2021). For references on the distribution of ergot in other countries please see Walkowiak *et al*. (2022).

The History of 'Ergotism'

Ergot is a fungal pathogen of both historical and modern importance worldwide, particularly in cereals such as rye and wheat (Menzies and Turkington, 2015). Ergot sclerotia can contain secondary metabolites, some of which are mycotoxins, primarily alkaloids such as ergotamine, ergovaline,

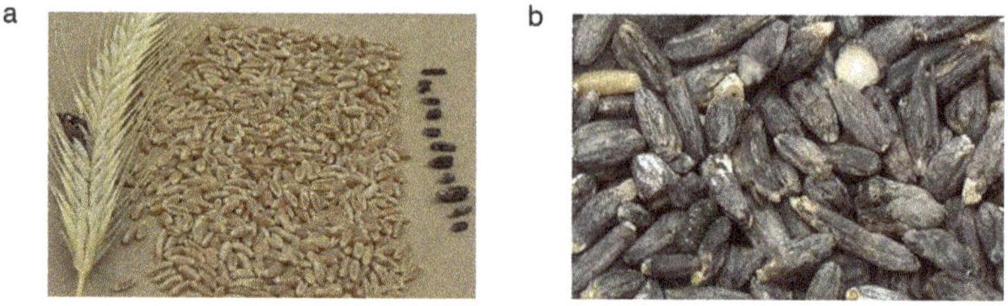

Fig. 12.1. Ergot (sclerotia) produced by *Claviceps purpurea* is black and therefore easily recognizable among grains of wheat and other cereals. (a) Comparison of colour and size of wheat grain and ergot; note the single infected floret in the wheat head. (b) Close-up image of ergot sclerotia.

ergocryptine, ergocornine and ergocristine. Disease symptoms in humans and animals caused by the consumption of ergot-contaminated grain are collectively called 'ergotism'. The toxicity of ergot cannot be reduced by the processing of the grain, baking or cooking, and there is no cure or antidote once it is ingested. Symptoms in humans include burning sensation in the arms and legs, convulsions, hallucinations, and gangrene; in addition, women may suffer reduced fertility and possibly miscarriage. Symptoms in livestock include decreased growth rate, followed by constriction of blood flow to extremities causing redness and swelling at the coronary bands above the hooves leading to hind leg lameness. Ergotism is the earliest recorded example of mycotoxin affecting humans and animals and likely dates as far back as 600 BC. Epidemics are known from the middle ages when rye was common in the diet of the poorer part of the population. In France, a hospital was built in 1039, where monks from the order of St Anthony took care of people suffering from ergotism; the human disease became known as St Anthony's fire.

Life Cycle and Epidemiology

An obvious sign of ergot infection is the pathogen's large black/purple sclerotia or ergot bodies that replace normal grain kernels of cereal and grass species. Ergot sclerotia are resting bodies that consist of condensed mycelium surrounded by a protective rind similar to other sclerotia-forming fungi (Fig. 12.1). They are easily dislodged from mature heads during harvest, and those that are not removed with the grain, are later incorporated into the soil by tillage or ploughing. In addition, ergot on infected grass species in adjacent areas dislodges and fall to the ground. Most ergot survives in the soil for about 1 year, but some may survive longer (Menzies and Turkington, 2015). Ergot requires 4 to 8 weeks of vernalization at around 0°C to induce the formation of stroma, which are stipes with knob-shaped heads (capitula). Since *C. purpurea* is homothallic, the fungi form the sexual stage without mating, forming perithecia that contain club-shaped asci and ascospores. During periods with wet weather in the spring or early summer, and temperature

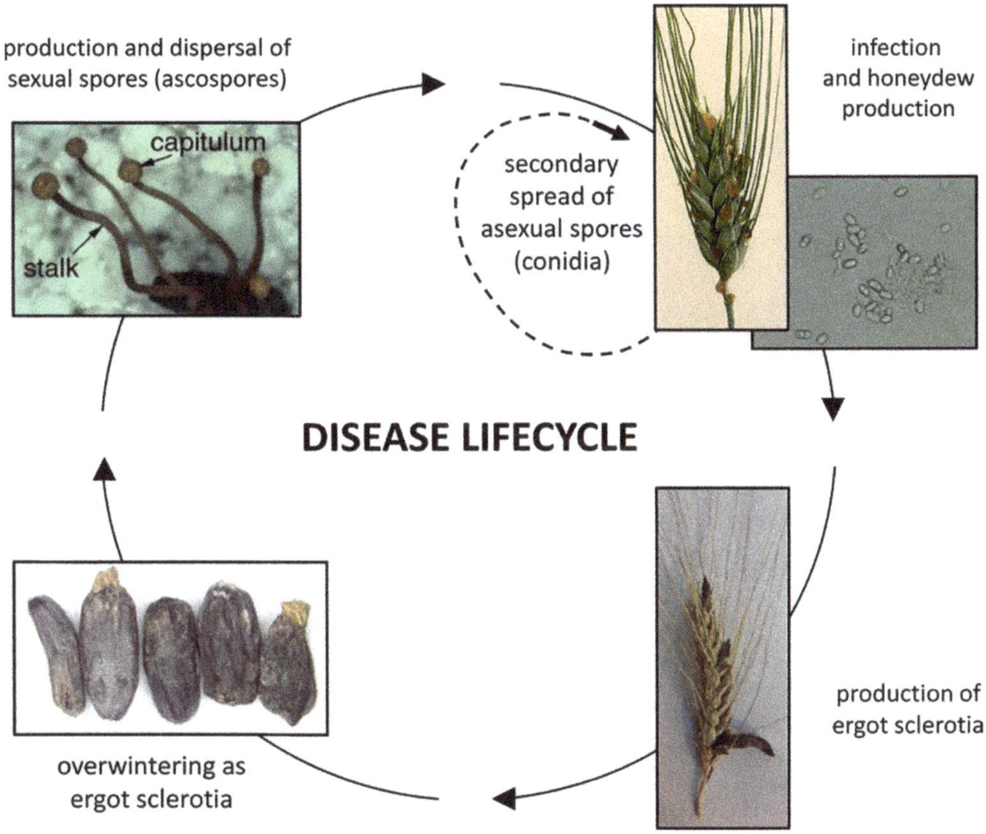

Fig. 12.2. Lifecycle of *Claviceps purpurea* on wheat. The fungus overwinters for about 1 year in the soil as sclerotia. In the spring, they form stroma with perithecia and asci (sexual stage). Airborne ascospores infect the heads of wheat through open florets and trigger formation of a sugary mass (honeydew) that contains conidiospores (asexual stage). This honeydew attracts insects which spread the conidia to other plants (secondary spread). The fungus grows and forms new sclerotia seen as black, hard ergots in place of grain. Sclerotia are either harvested with the grain or fall to the ground. The photo of a stroma is courtesy of the USDA (Alderman, 2006).

around 18°C, each ergot sclerotia produce one or more stroma (Fig. 12.2, top left). The ascospores are forcibly ejected and dispersed by wind currents and rain splash to neighbouring plants. Ascospores land on the florets of a susceptible grass host, resulting in infection of the ovary within 24 hours. The cool wet weather conditions that promote the ascospore stage also extend the flowering period of host plants. The plants are susceptible as long as florets stay open and ovaries are unfertilized. The infected ovary surface turns into a mycelial mass called a sphacelium, upon which millions of asexual conidiospores are produced. The conidia become submerged in a sugary sap called "honeydew", which is exuded from the floret and is often the first sign of infection (Fig. 12.2, top right). The honeydew is responsible for the secondary spread of the pathogen to other susceptible florets through rain splash, wind and insects which are attracted to the honeydew, and subsequently, carry conidia to other plants. The mycelium in the floret continues to grow and eventually forms an elongated mass that turns into hard purplish-black ergot sclerotium (Fig. 12.2, bottom right). The size of the ergot sclerotia vary depending on the host species and environmental conditions, and range from smaller than the host kernel to many times larger (Fig. 12.2, bottom left). Ergot maturation and vernalization complete the pathogen's lifecycle.

Disease Management

Several cultural practices and recommendations exist to manage ergot both pre- and post-harvest. Pre-harvest management in Canada focuses on reducing the number of sclerotia in the soil, which will produce the primary inoculum. This can be achieved by sowing ergot-free seed and adhering to a crop rotation with non-cereal crops, since most sclerotia only survive for about one year. Deep ploughing after the harvest of an infested crop may also help reduce primary inoculum since stipes on ergots buried deeper than 5 cm have difficulty reaching the soil surface (Menzies and Turkington, 2015). Mowing of headlands and ditches around cereal fields can reduce honeydew in flowering grass species, thereby attracting less insects to the crop, which may limit some of the secondary spread. Already pollinated florets are not accessible to *C. purpurea*, in contrast to sterile flowers, which are susceptible to ascospore infection. In addition, crops with an excessive amount of tillers may extend the flowering period and give the pathogen a longer window of infection. Thus, management measures that create a healthy and uniform crop stand to contribute to reducing floret sterility and tillering. Optimal fertilizer rates, particularly boron and copper, and proper timing and rates of herbicide applications may also reduce ergot. Fungicide applications for the control of *C. purpurea* have not been validated, and no fungicides are currently registered for this purpose in any crop type in Canada.

Growers can choose to harvest infected areas separately from ergot-free areas, making sure the two seed lots are kept separate. Another option is to delay the harvest of parts of the field with a high frequency of ergot-infected plants, which on one hand will result in many ergot bodies falling to the ground, but on the other hand, will reduce infestation of the harvested grain. Most infested seed lots can subsequently be cleaned based on the different sizes and weights of ergots (gravity-type cleaners) or colour difference (colour-sorter equipment); however, seed cleaning comes at a cost. Severely ergot-infested seed lots may have to be disposed of, so there is no risk for human and animal ingestion.

Host Resistance and Pathogen Genetic Diversity

Cereal species that are open-pollinated are particularly affected by *Claviceps*, including rye (*Secale cereale*), ryegrasses (genus *Lolium*) and *Triticale*, which is a hybrid of wheat (*Triticum*) and rye (*Secale*). Varieties of these species that have a shorter flowering period may have less disease, mainly because they escape infection; not due to a genetic or molecular defense resistance mechanism. Genetic resistance has been identified by screening a large number of wheat and durum lines with *C. purpurea* isolates collected from different host species and countries. Although *C. purpurea* is an obligate pathogen in nature, mycelium can be grown on potato dextrose agar-plates, and conidia can be increased in potato dextrose broth. A screening method was developed whereby a conidial suspension was injected into florets with a syringe. Differences in resistance became evident by the amount of honeydew which was assessed after 14 days using a 1–4 scale, and by the average weight of sclerotia per head measured at maturity (Menzies, 2004). This method identified lines with a higher level of genetic resistance in three lines of hexaploid spring wheat (*T. aestivum*) Kenya Farmer, Lee and HY630, and two lines of durum wheat (*T. aestivum* var. *durum*) Kyle and 9260B-173A. At the same time, these lines were part of a differential set that showed different virulence phenotypes among *C. purpurea* isolates (Menzies, 2004). While it is known that diversity exists within *C. purpurea* both for aggressiveness and toxin potential, there is limited knowledge of the population diversity in Canada and elsewhere. More information is also needed regarding the rate of genetic change in virulence phenotypes, to understand the future threats this pathogen poses. Future breeding of cereal varieties with improved genetic resistance to *C. purpurea* would be of economic benefit, would support growers, and would complement cultural practices in reducing ergot on the farm. Reducing ergot at the farm level also limits the need for post-harvest seed cleaning.

Regulations Around Ergot in Grain

The Canadian Grain Commission (CGC) has established science-based quality standards that maintain Canada's reputation as a consistent and reliable source of high-quality cereal grain, oilseeds and pulses (CGC, 2023a). A detailed Grain Grading Guide ensures that farmers receive fair compensation for high-quality seed lots during transactions, while downgraded seed lots result in a financial penalty (CGC, 2023b). Evaluation of grain samples for the presence of ergot sclerotia is one of several grading factors in the grain handling system. Samples of minimum of 500g are examined for both whole ergots and fragments, and the result is given as percent sclerotia weight compared to grain weight. Table 12.1 shows the allowable amount of ergot in grain from Canada (CGC, 2023c), the USA, the European Union, and internationally as determined by the Codex Alimentarius Commission.

Table 12.1. Tolerance levels for ergot (% weight) in cereal grain samples in different parts of the world compared to the international standard (Codex Alimentarius Commission).

Grain	Canada	USA	European Union	Codex
Wheat	0.04	0.05	0.02	0.05
Durum	0.02	0.05	0.02	0.05
Rye	0.05	0.30	0.05	n/a
Barley	0.02	0.10	0.02	0.02
Oats	0.00	0.10	0.02	0.05

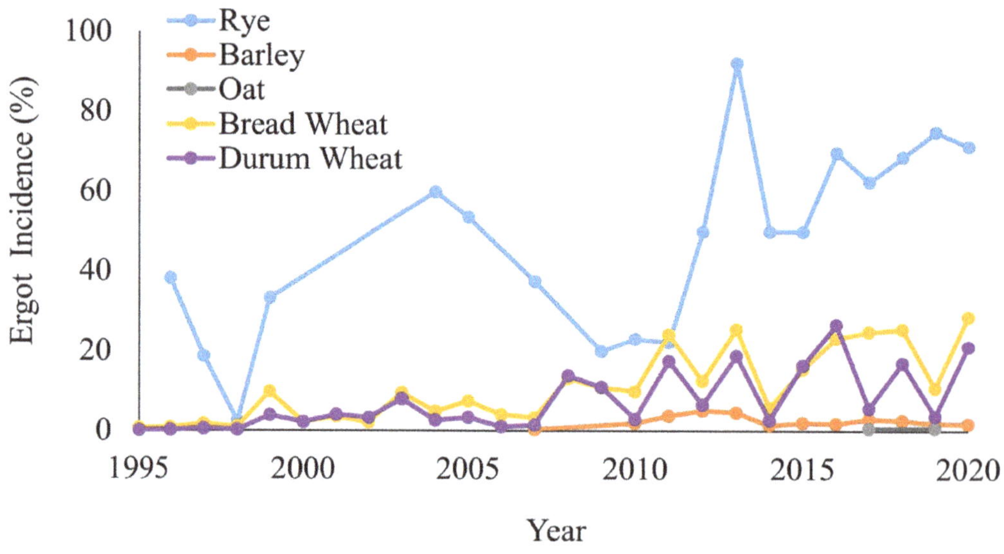

Fig. 12.3. Annual ergot incidence in five cereal grain types based on inspection of samples submitted to the Canadian Grain Commission from 1995 to 2020 (Walkowiak *et al.*, 2022).

These levels correspond to approximately 10–15 ergot bodies per kilogram seed. The Codex Alimentarius Commission was established in 1963 jointly by FAO, WHO and the UN. Counties that have signed on to the Codex follow the agreed quality standards for food and feed which help to ensure that traded grain is suitable for consumption.

Scientists within the Canadian Grain Commission conduct research to understand all aspects of grain quality and grain safety. A study comprised of more than 230,000 grain samples collected in Western Canada between 1995 and 2020 showed that ergot incidence in harvested grain samples was greatest in rye, followed by bread wheat, durum wheat, barley and oat (Fig. 12.3). The data also pointed to an increase in the occurrence of ergot in the last 15 years (Walkowiak *et al.*, 2022). Between 1990 and 2009 ergot incidence was 27.6% in rye, 4.2% in bread wheat, and 2.9% in durum wheat; in contrast to 2010–2020, when ergot incidence was 65.6% in rye, 19.5% in bread wheat, and 13.1% in durum wheat. Differences in ergot incidence may be due to a combination of several factors, including yearly and regional variation in weather, especially rainfall from mid-June to mid-July which affects sclerotial germination, ascospore production and the amount of sclerotia that will be produced for germination in the following year (Walkowiak *et al.*, 2022). Changes in crop rotation and on-farm management practices may also have caused an increase in ergot disease. Fortunately, the average amount of ergot bodies in grain samples was generally below the allowable threshold in Table 12.1, meaning that most seed lots would not have been severely downgraded, and through grain handing, there was not a negative impact on trade and export.

Future Perspectives

Automated ergot assessment in grain samples: Accurate and efficient assessment of ergot and alkaloids is necessary to ensure consistent delivery of high-quality grain for food and feed. Currently, ergot assessment is done manually by trained experts. New automated imaging technologies may increase the throughput of assessing ergot and may allow more accurately measured than other grading factors given the colour difference between ergot sclerotia and healthy grain.

Ergot-resistant or tolerant varieties: In the past, there has been less emphasis on breeding for ergot resistance relative to other cereal diseases, and the level of resistance or tolerance in the current Canadian cereal varieties is largely unknown. Studies on the nature of different sources of resistance, and development of molecular markers to aid plant breeders in incorporating resistance in breeding programs are underway, but warrant further research and years of field testing.

Ergot modelling and forecasting: This plant health case shows that the incidence of ergot has increased in Western Canada since the 1990s; however, the reasons for this are unclear. The collection of new data from a large number of specific fields over time, followed by the analysis of ergot incidence and severity could help model the complex relationships to different cereal species, crop variety, weather parameters, management practices and geography. These relationships are becoming even more important in light of climate change. In the future, it would benefit both growers and the industry to forecast the ergot-risk the following year or even longer.

Discussion Points

1. List the different negative effects ergot has on grain producers, livestock, grain trade and export, producers of baked goods and other grain products, and consumers in general.
2. Research historical examples of ergot epidemics. Discuss if similar epidemics and their negative effects on human health could happen today, or may already be happening.

3. Create information materials for farmers on how to reduce the amount of ergot in the field and harvested grain.
4. What do you think are the most important research areas that will help to control ergot in the future?

Further Reading

Haarmann, T., Rolke, Y. and Giesbert, S. (2009) Ergot: From witchcraft to biotechnology. *Molecular Plant Pathology* 10(4), 563–577. DOI: 10.1111/j.1364-3703.2009.00548.x.

References

Alderman, S. (2006) *Ergot: Biology and Contro*. USDA-ARS National Forage Seed Production Research Center, Corvallis, OR. Available at: https://www.ars.usda.gov/ARSUserFiles/81/ErgotDVDtranscript.pdf

Canadian Grain Commission (CGC) (2023a) *Harvest Sample Program*. Canadian Grain Commission, Winnipeg, MB. Available at: https://www.grainscanada.gc.ca/en/grain-quality/harvest-sample

Canadian Grain Commission (CGC) (2023b) *Official Grain Grading Guide*. Canadian Grain Commission, Winnipeg, MB. Available at: https://www.grainscanada.gc.ca/en/grain-quality/official-grain-grading-guide

Canadian Grain Commission (CGC) (2023c) *Ergot's Effect on Wheat Grades*. Canadian Grain Commission, Winnipeg, MB. Available at: https://grainscanada.gc.ca/en/grain-quality/grain-grading/grading-factors/ ergot.html

Menzies, J.G. (2004) The reactions of Canadian spring wheat genotypes to inoculation with *Claviceps purpurea*, the causal agent of ergot. *Canadian Journal of Plant Science* 84, 625–629. DOI: 10.4141/P03-086.

Menzies, J.G. and Turkington, T.K. (2015) An overview of the ergot (*Claviceps purpurea*) issue in Western Canada: Challenges and solutions. *Canadian Journal of Plant Pathology* 37, 40–51. DOI: 10.1080/07060661.2014.986527.

Menzies, J.G., Klein-Gebbinck, H.W., Gordon, A. and O'Sullivan, D.M. (2017) Evaluation of *Claviceps purpurea* isolates on wheat reveals complex virulence and host susceptibility relationships. *Canadian Journal of Plant Pathology* 39, 307–317. DOI: 10.1080/07060661.2017.1355334.

Walkowiak, S., Taylor, D., Fu, B.X., Drul, D., Pleskach, K. *et al.* (2022) Ergot in Canadian cereals – Relevance, occurrence, and current status. *Canadian Journal of Plant Pathology* 44, 793–805. DOI: 10.1080/07060661.2022.2077451.

Wyka, S.A., Mondo, S.J., Liu, M., Dettman, J., Nalam, V. *et al.* (2021) Whole-genome comparisons of ergot fungi reveals the divergence and evolution of species within the genus *Claviceps* are the result of varying mechanisms driving genome evolution and host range expansion. *Genome Biology and Evolution* 13(2). DOI: 10.1093/gbe/evaa267.

13 Tan Spot (*Pyrenophora tritici-repentis*) of Wheat: A Minireview

Marwa Laribi[1], Reem Aboukhaddour[2] and Stephen E. Strelkov[1]

[1]*Department of Agricultural, Food and Nutritional Science, University of Alberta, Edmonton, Alberta, Canada*
[2]*Agriculture and Agri-Food Canada Lethbridge Research and Development Centre, Lethbridge, Alberta, Canada*

Introduction

Tan spot, caused by the fungus *Pyrenophora tritici-repentis*, is a destructive foliar disease of wheat. Its global emergence was first reported in the 1970s, coinciding with the adoption of no-till or reduced tillage practices and the widespread cultivation of modern cultivars. The rise of tan spot as a major disease has spurred research into this pathosystem, including the biology and life cycle of *P. tritici-repentis*, the identification of its races and virulence factors, plant-host interactions, and resistance genetics. The tan spot pathogen exhibits physiological specialization, with isolates of *P. tritici-repentis* classified into eight races based on their virulence on four common wheat differentials (Lamari and Strelkov, 2010). Disease development is characterized by the formation of necrotic and/or chlorotic foliar lesions on susceptible hosts, and under conducive conditions, yield losses can reach up to 50% (Rees and Platz, 1983). *P. tritici-repentis* can also infect kernels and may be seed transmitted, causing a condition known as red smudge (Fernandez *et al.*, 1994). Despite efforts to develop modern tan spot-resistant cultivars, success has been limited due to the incomplete understanding of the *P. tritici-repentis*-wheat interaction, which is now recognized as more complex than a simple inverse gene-for-gene relationship. In addition, there is often an over-reliance on susceptible cultivars, which are heavily bred for higher yield or resistance to diseases such as yellow rust or fusarium head blight. As a result, less attention and fewer resources are devoted to tan spot, which is ranked lower in priority by breeders and the cereal industry, partly due to its perceived lesser impact compared to other diseases.

Causal Agent

P. tritici-repentis was first identified as a saprophyte in 1823 and later recognized as a pathogen in 1923 when it was found to cause yellow spots on wheat (De Wolf *et al.*, 1998). Over the years, the fungus was referred to by several different names, with its nomenclature remaining conflicted until 1962 when Shoemaker assigned it its current name (De Wolf *et al.*, 1998). *P. tritici-repentis* belongs to the genus *Pyrenophora,* within the family Pleosporaceae, order Pleosporales, class Dothideomycetes, and phylum Ascomycota (Ciuffetti *et al.*, 2014). Dothideomycetes is the largest class of ascomycete fungi, encompassing a highly diverse group of more than 19,000 species, both pathogenic and non-pathogenic, that infect many hosts. Notable examples are cereal-infecting pathogens including *P. tritici-repentis, Parastagonospora nodorum, Pyrenophora teres, Zymoseptoria tritici,* and *Bipolaris sorokiniana.*

P. tritici-repentis has a wide host range and can overwinter as sexual fruiting bodies called pseudothecia. These structures can persist on cereal crop debris, including wheat, barley, rye, and oats (Fig. 13.1a), as well as on secondary hosts comprising over 26 grass species (Ciuffetti *et al.*, 2014). The pseudothecia are black, erumpent, 200–700 µm in diameter, and covered with dark spines. They contain asci bearing eight sexual spores known as ascospores (Fig. 13.1b). Each ascus ranges in size from 180 to 250 µm in length and 20 to 38.5 µm in width (Orabi *et al.*, 2002). Ascospores are yellow-brown and oval-shaped, measuring 40–70 µm in length and 15–28 µm in width. Each ascospore has three transverse septa and a longitudinal septum in the median cells (Fig. 13.1b). The asexual spores, or conidia, of *P. tritici-repentis* are hyaline to light brown and measure 45–245 µm by 12–24 µm (Fig. 13.1c and d) (Orabi *et al.*, 2002). They are distinguished from those of closely related fungi by the conical tapering of the basal cell of the conidium, which resembles a snake's head (Wegulo, 2011; Ciuffetti *et al.*, 2014) (Fig. 13.1c and d).

Tan Spot Occurrence and Disease Cycle

Tan spot emerged as a major wheat disease in the 1970s, coinciding with the shift to conservation tillage practices. Since then, the disease has been reported in various regions around the world, including North and South America, Australia, Asia, Europe, and Africa (Lamari and Strelkov, 2010). In addition to the shift in agronomic practices towards reduced or zero-tillage, several other factors have contributed to the increased occurrence of tan spot worldwide. These include the cultivation of susceptible wheat cultivars and the horizontal transfer of the necrotrophic effector-encoding gene *ToxA* from *P. nodorum* to *P. tritici-repentis,* which likely occurred sometime before 1940 (Friesen *et al.*, 2006).

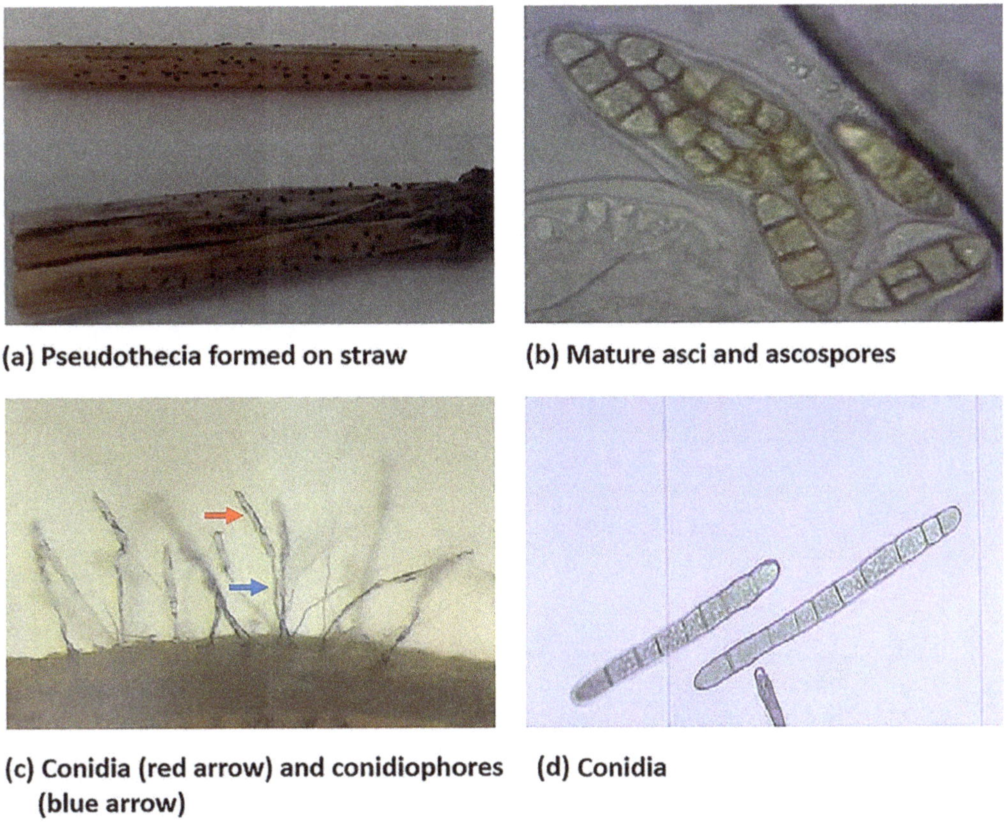

Fig. 13.1. Morphological characteristics of *Pyrenophora tritici-repentis*, including pseudothecia produced on stubble (a), asci and ascospores (b), conidia borne on conidiophores (c), and conidia (d). Panels (a) and (b) are from Laribi *et al.* (2022b).

P. tritici-repentis is a stubble-borne pathogen that overwinters on wheat straw and debris as pseudothecia containing ascospores (Fig. 13.2). Under conducive conditions, the pseudothecia mature with alternating light cycles of day and night, eventually erupting and releasing ascospores early in the growing season. These ascospores are then spread by wind and rain-splash. When they land on emerging wheat leaves, they infect the plant, leading to the appearance of tan-coloured lesions that give the disease its name (Fig. 13.2). This initial infection, caused by the sexual ascospores, is known as primary infection (Fig. 13.2). From this first infection of the leaves, the fungus colonizes the host tissue and forms conidiophores, on which conidia develop. These conidia are then spread by wind or rain, leading to multiple secondary infections within the same growing season.

Disease development requires high humidity and temperatures between 18 and 28°C, although the specific range may vary depending on the host cultivar (Da Luz and Bergstrom, 1986). In addition to wheat stubble, other sources of *P. tritici-repentis* inoculum include infected kernels, volunteer

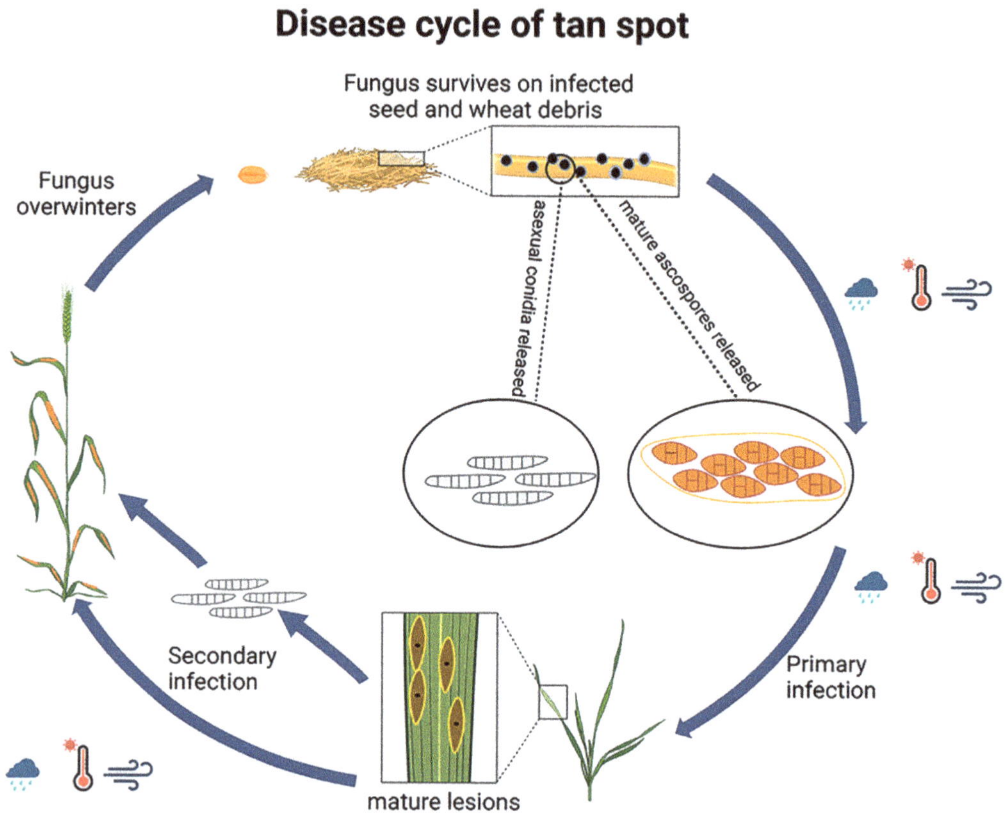

Fig. 13.2. The disease cycle of tan spot of wheat caused by *Pyrenophora tritici-repentis*. The fungus survives on infected crop residues and kernels. In the spring, sexual ascospores produced in pseudothecia and asexual conidia produced on conidiophores are dispersed by wind or water-splash to wheat seedlings, causing primary infection. Disease development requires a combination of high humidity and temperatures between 18 and 28°C. Infected kernels give rise to infected seedlings. Conidiophores bearing conidia are produced on mature lesions, with the conidia acting as secondary inoculum. The infection can then spread vertically up the crop canopy under favourable conditions. Created with BioRender.com.

wheat from the previous cropping season, and infected secondary hosts. Several cycles of conidia may be produced from foliar lesions that develop after the infection of a susceptible host. The conidia act as secondary inoculum, spreading vertically up the canopy under favourable conditions. They can also be carried to other wheat plants and fields by wind or rain-splash. Conidia can be dispersed by wind over long distances, up to 200 km.

Early infection by *P. tritici-repentis* results in the appearance of diamond-shaped lesions with a dark point at the centre, where the fungus has penetrated the leaf. These lesions are tan-coloured and often surrounded by chlorotic halos (Fig. 13.2). As the disease progresses, the lesions coalesce and the leaf becomes necrotic. In severe cases, the pathogen can also infect the ears, causing a condition known as red smudge, which manifests as brownish or bleached discoloration of the glumes, and reddish, shrivelled seeds (See *et al.*, 2020).

Race Classification System

Early studies on tan spot and its causal agent assessed variations in virulence among *P. tritici-repentis* isolates quantitatively, focusing on lesion size or the percentage of foliar necrosis. In the late 1980s, Lamari and Bernier recognized that *P. tritici-repentis* induces two genetically distinct symptoms, necrosis and chlorosis, on different host genotypes. Accordingly, they classified isolates into four pathotypes based on their ability to cause necrosis or chlorosis on selected wheat genotypes. However, the identification of *P. tritici-repentis* isolates that produced similar symptoms on different hosts soon led Lamari *et al.* (1995) to propose a race classification system, which remains widely used today. In this system, fungal isolates are classified into eight races based on their virulence patterns on a differential set of four common wheats ('Glenlea', 6B662, 6B365, 'Salamouni').

The foliar necrosis and chlorosis symptoms characteristic of tan spot disease result from the differential ability of *P. tritici-repentis* isolates to produce necrotrophic effectors (NEs), previously known as host-selective toxins, which interact with specific sensitivity genes in the host (Table 13.1). Consequently, the tan spot pathosystem follows an inverse gene-for-gene

Table 13.1. Relationship between the races and necrotrophic effectors of Pyrenophora tritici-repentis.

Race Host	Race 2	Race 1	Race 8	Race 7	Race 5	Race 6	Race 3	Race 4
'Glenlea' (*Tsn1*)	 ToxA							
6B662 (*Tsc2*)			 ToxB					
6B365 (*Tsc1*)		 ToxC				 ToxC		
'Salamouni'								

Note: Races of *P. tritici-repentis* produce the necrotrophic effectors Ptr ToxA, Ptr ToxB, and Ptr ToxC, either alone or in various combinations. ToxA induces necrosis in the wheat differential cultivar 'Glenlea', which carries the matching sensitivity gene Tsn1. ToxB induces chlorosis in the wheat line 6B662, which carries the matching sensitivity gene *Tsc2*. ToxC induces chlorosis in the wheat line 6B665, which carries the matching sensitivity gene *Tsc1*. Race 4 isolates of *P. tritici-repentis* do not produce any known necrotrophic effectors and are considered non-pathogenic. The wheat cultivar 'Salamouni' is resistant to all described races of the fungus and does not carry any matching sensitivity genes.

model, where the compatible interaction forms the basis for specificity (Lamari *et al.*, 2003). To date, three NEs have been identified: Ptr ToxA, which induces necrosis, and Ptr ToxB and Ptr ToxC, which induce chlorosis in different hosts (Table 13.1). Host sensitivity to these NEs is mediated by the host genes *Tsn1*, *Tsc2*, and *Tsc1*, respectively (Table 13.1). The eight races of *P. tritici-repentis* represent the theoretical maximum number of races that can be distinguished by three loci in the host ($2^3 = 8$), assuming a one-to-one relationship (Lamari *et al.*, 2003).

Discovery of Atypical Isolates and the Need to Expand the Differential Set

The race classification system has been invaluable in characterizing *P. tritici-repentis* isolates and elucidating the role of NEs and host sensitivity genes in tan spot development. However, there have been increasing reports of additional or novel races that do not fit the current race structure. Many of these isolates were recovered from durum wheat in North Africa, a secondary centre of wheat diversity. These isolates, broadly termed 'atypical', exhibited virulence patterns on the tan spot differentials similar to those of existing races but did not possess the expected complement of NEs. For instance, nearly half of *P. tritici-repentis* isolates tested from Tunisia were classified as race 7 but lacked *ToxA* (Kamel *et al.*, 2019). Similarly, Laribi *et al.* (2022a) identified numerous isolates of races 2, 7, and 8 in a different collection from Tunisia that also lacked the *ToxA* gene associated with these races. These studies suggest the involvement of additional NEs that need further characterization and investigation. Wei *et al.* (2021) found that the presence of the ToxA sensitivity gene *Tsn1* in durum wheat did not always result in disease, suggesting that this interaction may not play a significant role in tan spot development in durum wheat and that other factors or NEs might be involved. Supporting this hypothesis, Guo *et al.* (2020) identified numerous race 4 isolates of *P. tritici-repentis* that induced necrosis on durum wheat but not on bread wheat, implying the presence of additional NE(s) that may be specific to durum wheat. Collectively, these findings indicate the need for the inclusion of additional genotypes in the tan spot differential set to enhance the characterization of fungal isolates recovered from durum wheat.

Genetic Diversity and Genomics

In addition to its differential ability to produce various NEs, *P. tritici-repentis* exhibits significant variability in its pathogenicity, genetics, and morphology (Benslimane *et al.*, 2017). This variability can be partly attributed to the homothallic nature of the fungus, which has a life cycle that includes both sexual and asexual reproduction. Earlier studies on the genetic diversity of *P. tritici-repentis*, employing various marker types such as random amplified polymorphic DNA (RAPD) and amplified length fragment polymorphism (AFLP) markers, often produced contradictory results. While some studies identified genetic similarity among isolates based on their geographic origins, others did not observe such relationships (reviewed in Aboukhaddour *et al.*, 2022). However, more recent research has found some differentiation between *P. tritici-repentis* isolates based on the production of NEs or their geographic origin. For example, using a combination of simple sequence repeat (SSR) markers from earlier studies and a global collection of *P. tritici-repentis* representing all eight races, Laribi *et al.* (2022a) found that ToxB-non-producing isolates formed a distinct group. These isolates were genetically distant from the ToxB-producing isolates.

In recent years, the full genome of *P. tritici-repentis* has been investigated more closely, revealing an open pangenome in which the number of accessory genes increases while the number of core genes decreases with each added genome. The tan spot fungus has been shown to possess an open pangenome characterized by a high content of accessory genes and the ability to gain and lose genes. Pangenomic studies have reported a genome ranging in size from 37.56 to 42.19 Mb, demonstrating high plasticity, large-scale structural reorganizations at the chromosomal level, and an extensive content of accessory genes and transposable elements (Gourlie *et al.*, 2022; Moolhuijzen *et al.*, 2022). The NE-encoding genes *ToxA* and *ToxB* are located on different essential chromosomes but have been found to move via large transposons, which have recently been identified as belonging to the starship transposon class (Gourlie *et al.*, 2022).

Disease Management

Several management strategies can be implemented individually or in combination to control the occurrence and severity of tan spot disease. These include cultural practices, effective fungicides, and genetic resistance. Since tan spot is a stubble-borne disease, burning or burying crop debris from previous seasons can significantly reduce infection by minimizing inoculum levels. Additionally, implementing proper crop rotations away from cereals, avoiding cereal monoculture, and planting cultivar mixtures can help reduce disease incidence and severity. Crop rotations away from cereals for at least 2 years, allowing infected wheat residues to decompose, are typically sufficient to reduce *P. tritici-repentis* inoculum in a field. However, fungicides remain among the most commonly used tools for managing tan spot, despite their cost and potential environmental concerns. Strobilurin, carboxamide, and triazole fungicides have been shown to be effective against this disease (reviewed in Sautua and Carmona, 2021). However, concerns about fungicide insensitivity have arisen following reports of strobilurin insensitivity among isolates of *P. tritici-repentis* from various regions (Sautua and Carmona, 2021). One way to reduce the risk of fungicide insensitivity is to apply mixtures with different modes of action and avoid repeated applications. The timing of fungicide application is also crucial and depends on the plant's development stage, host susceptibility, disease spread and severity, and the climatic conditions at the time of application. The frequency of application also depends on the effectiveness of previous treatments and climatic conditions. A particularly effective and practical approach for managing tan spot and other plant diseases is the use of genetic resistance, through the breeding and cultivation of resistant cultivars adapted to particular regions. However, only a few modern cultivars are resistant to tan spot, and additional work is needed in this areaMoolhuijzen *et al.* (2018b).

Host Resistance: Achievements and Challenges

The wheat-*P. tritici-repentis* interaction follows the inverse gene-for-gene model, where each necrotrophic effector produced by the pathogen is matched by a dominant sensitivity gene in the host (Table 13.1). However, most research on tan spot has focused on common wheat, leaving interactions with durum wheat underexplored. Recent studies also suggest that the genetic control of the host's reaction to *P. tritici-repentis* may, in some cases, be more complex than previously thought.

To date, eight recessive genes (*tsr1* to *tsr6*, *TsrHar* and *TsrAri*) and one major gene *Tsr7*, conferring either race-specific or non-race-specific tan spot resistance have been identified. Most of these genes (*tsr2*, *tsr5*, *Tsr7*, and *TsrHar*) are located on chromosome 3B, suggesting that this chromosome is a valuable source of resistance. In breeding for tan spot resistance in common wheat, the interactions between ToxA, ToxB, ToxC, and their corresponding sensitivity genes *Tsn1*, *Tsc2*, and *Tsc1* can play significant roles. However, it has been suggested that the *Tsn1*–ToxA interaction may not be involved in inducing necrotic symptoms in durum wheat (reviewed in Aboukhaddour *et al.*, 2022). This implies the involvement of different or additional recessive genes in the defence mechanisms of durum wheat compared to common wheat, as well as potential additional pathogen virulence factors, as discussed earlier. These genes may play a role in the early stages of resistance to *P. tritici-repentis*, prior to the internalization of ToxA (Aboukhaddour *et al.*, 2022). In addition to *Tsn1*-ToxA, numerous quantitative trait loci may be involved in the *P. tritici-repentis*-durum wheat interaction (reviewed in Faris *et al.*, 2013).

Laribi *et al.* (2023) identified 26 single nucleotide polymorphism (SNP) markers in a genome-wide association analysis of tan spot resistance in a collection of durum wheat accessions and found eight accessions that could serve as possible sources of resistance. Although hundreds of SNPs associated with tan spot resistance have been identified in both common and durum wheat, comparing results across studies has been challenging due to the use of different arrays and reference genomes. This variability may lead to redundancy in results due to the lack of comparative analyses. Moreover, while most durum wheat accessions tested maintained good levels of resistance at both the seedling and adult plant stages, seedling resistance may not always be effective at the adult plant stage (Laribi *et al.*, 2021, 2022b).

This indicates that the resistance mechanisms and genes involved may differ depending on the developmental stage of durum wheat accessionsMoolhuijzen *et al.* (2018a).

Conclusions and Future Perspectives

The emergence of tan spot as a major wheat disease has stimulated research into its biology and management. The recognition of distinct races of *P. tritici-repentis*, along with the role of NEs and matching sensitivity genes in the wheat-*P. tritici-repentis* interaction, has enhanced our understanding of this pathosystem. However, recent studies suggesting the involvement of additional resistance or sensitivity genes and novel NEs, particularly in the durum wheat-*P. tritici-repentis* interaction, highlight the need for further research. A deeper understanding of the genetics and mechanisms underlying tan spot development may facilitate resistance breeding. Cultivars with improved resistance can be combined with other strategies, such as crop rotation and fungicide treatments, for integrated tan spot management in wheat.

Discussion Points

1. What factors have contributed to the emergence of tan spot as a major wheat disease? Could these conditions also favour the development of other diseases?
2. Could genome-editing approaches help in controlling tan spot? For example, by eliminating some of the sensitivity genes in the host?
3. Are there additional strategies used for managing other plant diseases that could be evaluated for tan spot management?
4. Is there potential for discovering novel sources of resistance to tan spot in landraces or wild relatives of wheat from its centre of diversity?

Further Reading

Faris, J.D. and Friesen, T.L. (2020) Plant genes hijacked by necrotrophic fungal pathogens. *Current Opinion in Plant Biology* 56, 74–80.

Faris, J.D., Overlander, M.E., Kariyawasam, G.K., Carter, A., Xu, S.S. *et al.* (2020) Identification of a major dominant gene for race-nonspecific tan spot resistance in wild emmer wheat. *Theoretical and Applied Genetics* 133, 829–841.

Moolhuijzen, P., See, P.T., Hane, J.K., Shi, G., Liu, Z. *et al.* (2018a) Comparative genomics of the wheat fungal pathogen *Pyrenophora tritici-repentis* reveals chromosomal variations and genome plasticity. *BMC Genomics* 19(1), 279.

Moolhuijzen, P.M., See, P.T., Oliver, R.P. and Moffat, C.S. (2018b) Genomic distribution of a novel *Pyrenophora tritici-repentis* ToxA insertion element. *PLOS ONE* 13(10), e0206586.

Moolhuijzen, P., See, P.T. and Moffat, C.S. (2020) PacBio genome sequencing reveals new insights into the genomic organisation of the multi-copy *ToxB* gene of the wheat fungal pathogen *Pyrenophora tritici-repentis*. BMC Genomics 21(1), 645.

References

Aboukhaddour, R., Hafez, M., Strelkov, S.E. and Fernandez, M.R. (2022) Tan spot disease under the lenses of plant pathologists. In: *Achieving Durable Disease Resistance in Cereals*. Burleigh Dodds Science Publishing, Sawston, Cambridge, UK, pp. 589–621.

Benslimane, H., Aouali, S., Khalfi, A., Ali, S. and Bouznad, Z. (2017) In vitro morphological characteristics of *Pyrenophora tritici-repentis* isolates from several Algerian agro-ecological zones. *Plant Pathology Journal* 33(2), 109.

Ciuffetti, L.M., Manning, V.A., Pandelova, I., Faris, J.D., Friesen, T.L. *et al.* (2014) *Pyrenophora tritici-repentis*: A plant pathogenic fungus with global impact. In: Dean, R.A., Lichens-Park, A. and Kole, C. (eds) *Genomics of Plant-Associated Fungi: Monocot Pathogens*. Springer, Berlin, Heidelberg, Germany, pp. 1–39.

Da Luz, W.C. and Bergstrom, G.C. (1986) Effect of temperature on tan spot development in spring wheat cultivars differing in resistance. *Canadian Journal of Plant Pathology* 8(4), 451–454.

De Wolf, E., Effertz, R.J., Ali, S. and Francl, L.J. (1998) Vistas of tan spot research. *Canadian Journal of Plant Pathology* 20(4), 349–370.

Faris, J.D., Liu, Z. and Xu, S.S. (2013) Genetics of tan spot resistance in wheat. *Theoretical and Applied Genetics* 126(9), 2197–2217.

Fernandez, M.R., Clarke, J.M., DePauw, R.M., Irvine, R.B. and Knox, R.E. (1994) Black point and red smudge in irrigated durum wheat in southern Saskatchewan in 1990–1992. *Canadian Journal of Plant Pathology* 16(3), 221–227.

Friesen, T.L., Stukenbrock, E.H., Liu, Z., Meinhardt, S., Ling, H. *et al.* (2006) Emergence of a new disease as a result of interspecific virulence gene transfer. *Nature Genetics* 38(8), 953–956.

Gourlie, R., McDonald, M., Hafez, M., Ortega-Polo, R., Low, K.E. *et al.* (2022) Dissecting the *Pyrenophora tritici-repentis* (tan spot of wheat) pangenome. *BioRxiv* 2022–2003.

Guo, J., Shi, G., Kalil, A., Friskop, A., Ellas, E. *et al.* (2020) *Pyrenophora tritici-repentis* race 4 isolates cause disease on tetraploid wheat. *Phytopathology* 110(11), 1781–1790.

Lamari, L. and Strelkov, S.E. (2010) Minireview/minisynthèse the wheat/*Pyrenophora tritici-repentis* interaction: Progress towards an understanding of tan spot disease. *Canadian Journal of Plant Pathology* 32(1), 4–10.

Kamel, S., Cherif, M., Hafez, M., Despins, T. and Aboukhaddour, R. (2019) *Pyrenophora tritici–repentis* in Tunisia: Race structure and effector genes. *Frontiers in Plant Science* 10, 1562. DOI: 10.3389/fpls.2019.01562.

Lamari, L., Sayoud, R., Boulif, M. and Bernier, C.C. (1995) Identification of a new race in *Pyrenophora tritici-repentis*: Implications for the current pathotype classification system. *Canadian Journal of Plant Pathology* 17(4), 312–318.

Lamari, L., Strelkov, S.E., Yahyaoui, A., Orabi, J. and Smith, R.B. (2003) The identification of two new races of *Pyrenophora tritici-repentis* from the host center of diversity confirms a one-to-one relationship in tan spot of wheat. *Phytopathology* 93(4), 391–396.

Laribi, M., Ben M'Barek, S., Fakhfakh, M., Yahyaoui, A.H. and Sassi, K. (2021) Durum wheat Mediterranean landraces: A valuable source for resistance to tan spot disease. *Agriculture* 11(11), 1148.

Laribi, M., Akhavan, A., Ben M'Barek, S., Yahyaoui, A.H., Strelkov, S.E. *et al.* (2022a) Characterization of *Pyrenophora tritici-repentis* in Tunisia and comparison with a global pathogen population. *Plant Disease* 106(2), 464–474.

Laribi, M., Yahyaoui, A.H., Abdedayem, W., Kouki, H., Sassi, K. *et al.* (2022b) Characterization of Mediterranean durum wheat for resistance to *Pyrenophora tritici-repentis*. *Genes* 13(2), 336.

Laribi, M., Fredua-Agyeman, R., Ben M'Barek, S., Sansaloni, C.P., Dreisigacker, S. *et al.* (2023) Genome-wide association analysis of tan spot disease resistance in durum wheat accessions from Tunisia. *Frontiers in Genetics* 14, 1231027.

Moolhuijzen, P.M., See, P.T., Shi, G., Powell, H.R., Cockram, J. *et al.* (2022) A global pangenome for the wheat fungal pathogen *Pyrenophora tritici-repentis* and prediction of effector protein structural homology. *Microbial Genomics* 8(10), 000872.

Orabi, J., Yahyaoui, A.H., Mamluk, O.F., Lamari, L. and Azmeh, F. (2002) First report on tan spot disease (*Pyrenophora tritici-repentis*) of wheat in Syria. *Phytopathologia Mediterranea* 41(3), 271–274.

Rees, R.G. and Platz, G.J. (1983) Effects of yellow spot on wheat: Comparison of epidemics at different stages of crop development. *Journal of Agricultural Research* 34(1), 589–621.

Sautua, F.J. and Carmona, M.A. (2021) Detection and characterization of QoI resistance in *Pyrenophora tritici-repentis* populations causing tan spot of wheat in Argentina. *Plant Pathology* 70(9), 2125–2136.

See, P.T., Schultz, N. and Moffat, C.S. (2020) Evaluation of *Pyrenophora tritici-repentis* infection of wheat heads. *Agriculture* 10(9), 417.

Wegulo, S.N. (2011) Tan spot of cereals. *Plant Health Instructor* 11. DOI: 10.1094/PHI-I-2011-0426-01.

Wei, B., Despins, T., Fernandez, M.R., Strelkov, S.E., Ruan, Y. *et al.* (2021) Race distribution of *Pyrenophora tritici-repentis* in relation to ploidy level and susceptibility of durum and winter bread wheat. *Canadian Journal of Plant Pathology* 43(4), 582–598.

14 A Global Threat: Epidemiology and Management of Yellow Rust on Wheat

My Livbjerg

Department of Agroecology, Faculty of Science and Technology, Aarhus University, Forsøgsvej 1, 4200 Slagelse, Denmark

Introduction: Yellow Rust as a Global Pathogen

Life cycle and symptoms

P. striiformis (causal organism of yellow rust) is an obligate biotroph with a complex life cycle consisting of five different spore stages on two botanically distinct plant hosts. The life cycle of yellow rust can be categorized into two stages: (1) the asexual stage taking place on the primary host (cereals and grasses) and (2) the sexual stage taking place on the alternate host (*Berberis* spp.). The sexual life stage of yellow rust was not discovered until 2010, when Jin *et al.* (2010) found that several *Berberis* species could serve as alternate hosts for yellow rust. Since then, more than 50 species of *Berberis* and *Mahonia aquifolium* (under controlled conditions) have been reported to be susceptible to the pathogen (Jin *et al.*, 2010; Wang, 2013; Mehmood *et al.*, 2020).

The initial infection of wheat occurs when asexual, dikaryotic urediniospores are deposited on wheat leaf surfaces by wind or raindrops. The urediniospores germinate on the leaf surface and develop a dikaryotic germ tube that penetrates the wheat stoma and subsequently forms infection hyphae inside the leaf, and develops specialized infection structures called haustoria that take up nutrients from infected host cells (Kang *et al.*, 2017). The infection on adult susceptible wheat leaves will be visible 10–18 days after infection, when the wheat epidermal cells break, revealing yellow to orange urediniospores erupting from uredinia (Fig. 14.1d–e), which are arranged in characteristic stripes along the veins of the leaves (Fig. 14.1a–c) (Mehmood *et al.*, 2020; Bouvet *et al.*, 2022). On younger wheat plants and seedlings, uredinia often appear individually or seemingly more at random. On resistant or partially resistant host plants, symptoms will vary from tiny flecks of chlorosis to larger chlorotic/necrotic areas with or without pustules (Fig. 14.2a).

The asexual cycle of yellow rust is often repeated during the growing season and can lead to wide-scale epidemics in areas with susceptible wheat cultivars. Towards the end of the wheat growing season, when the plants begin to senescence, yellow rust may begin to produce two-celled, diploid teliospores that are dark brown to black (Fig. 14.2b). When the teliospores germinate, each cell produces a basidium that carries four haploid basidiospores (Fig. 14.1g) (Liu and Hambleton, 2010). The basidiospores may infect the alternate host, resulting in the production of pycnia, containing haploid pycniospores, on the adaxial side of the leaves. Ultimately, pycniospores will mate with receptive hyphae of another pycnium and develop a mycelium inside the leaf that forms dikaryotic aeciospores capable of infecting the primary host (wheat) and thus complete the life cycle.

Direct evidence of sexual reproduction in yellow rust in nature has, so far, only been found in China; however, it is hypothesized that sexual reproduction occurs in the Himalayas (Nepal and Pakistan) or near-Himalaya regions (China), with yellow rust populations in these areas having a high genetic diversity, recombinant population structures and high sexual reproduction abilities (Zhao *et al.*, 2013; Ali *et al.*, 2014; Cheng *et al.*, 2022; Du *et al.*, 2023). Meanwhile, yellow rust populations in Europe, the Americas and Australia have been characterized previously by their clonal population structures, in which the evolution of virulences is largely shaped by single-step mutations and potentially somatic hybridization (Hovmøller *et al.*, 2016; Ding *et al.*, 2021; Bouvet *et al.*, 2022).

Global dispersion

In the past, yellow rust was primarily an issue in cool climates, such as Northern Europe and North America; however, the emergence of new aggressive isolates, adapted to warmer temperatures led to the disease spreading to all wheat-growing areas in the world (Milus *et al.*, 2009; Hovmøller *et al.*, 2023). The pathogen's international success can be attributed to several factors, one of these being its ability to adapt rapidly via genetic changes, gained through random mutation, selection pressure, sexual recombination or somatic hybridization (Ding *et al.*, 2021). Moreover, the capacity for long-distance migration by urediniospores allows the pathogen to spread between countries, overcome plant resistance and cause recurring local disease outbreaks around the world. Urediniospores of the yellow rust pathogen have been found to travel up to 1700 km with prevailing winds, enabling migration between continents (Bouvet *et al.*, 2022).

While yellow rust has been known in Europe and Asia for hundreds of years, it was not found in North America before the Danish plant pathologist Kølpin Ravn recognized the disease during a visit with the United States Department of Agriculture crop survey team in 1915 (Kang *et al.*, 2017).

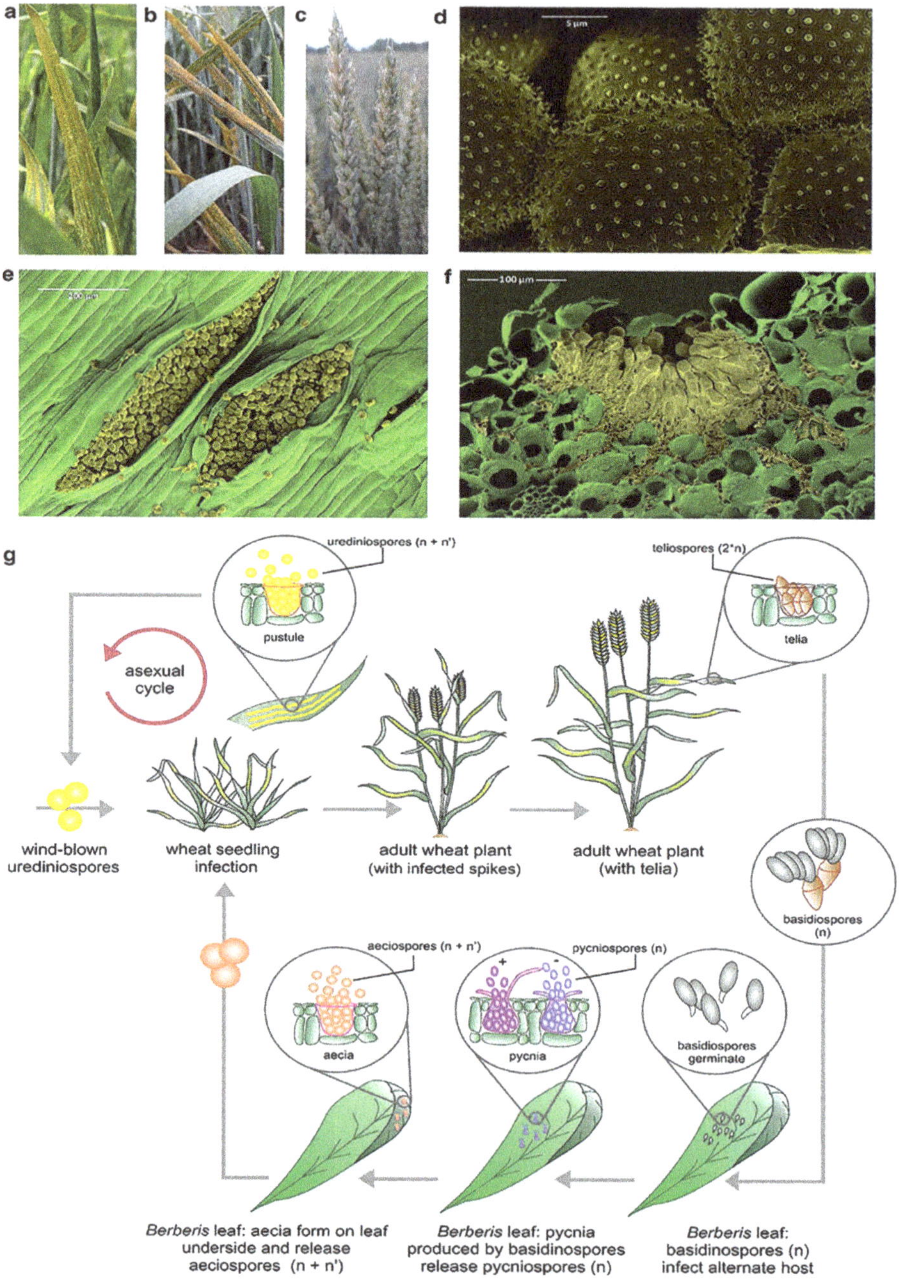

Fig. 14.1. Overview of the symptoms of infection with *Puccinia striiformis* (causal organism of yellow rust) and disease cycle as illustrated by Bouvet *et al.* (2022). (a–c) Symptoms of infection on adult plants in the field. (d) False colour scanning electron microscopy of urediniospores. (e–f) False colour scanning electron microscopy of pustules either on the leaf surface (e) or in a cross-section (f). (g) Illustrated diagram of the disease cycle of yellow rust, with initial infection on the wheat seedlings, followed by subsequent infection on the alternate host (*Berberis* spp).

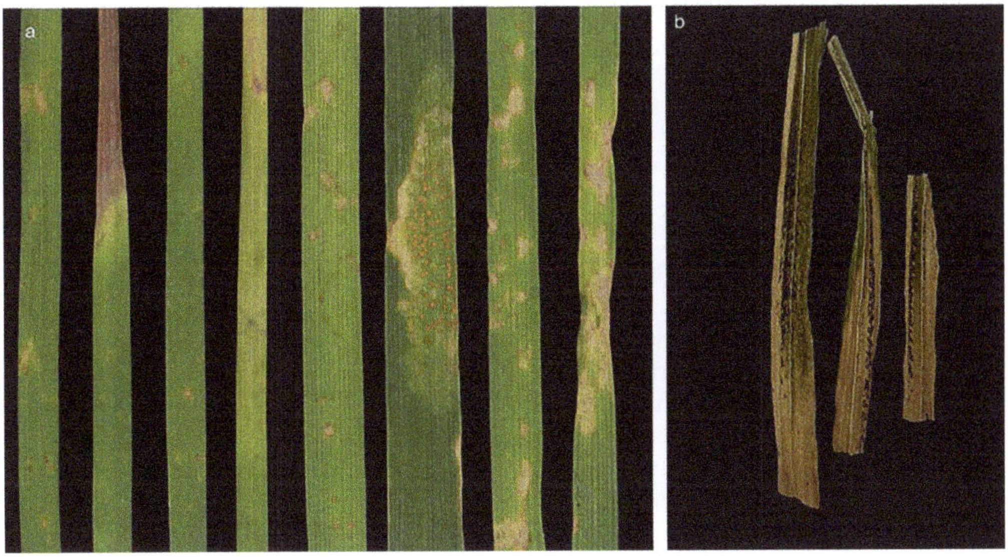

Fig. 14.2. Resistance responses against yellow rust on wheat and black teliospores on wheat leaves. (a) Various plant resistance responses on second leaves of wheat to infection of yellow rust. (b) Black teliospores on senescent wheat leaves.

Moreover, yellow rust did not appear in Australia before 1978, where the sudden incursion has been proposed to be the result of an accidental introduction of the disease by travellers carrying urediniospores from Europe to Australia on their clothing (O'Brien *et al.*, 1980). Likewise, yellow rust emerged for the first time ever in South Africa in 1996, causing epidemic outbreaks throughout most of the wheat-producing areas of the country (Pretorius *et al.*, 1997). It has since been suggested that the disease spread from West or Central Asia, causing this first ever epidemic in South Africa (Hovmøller *et al.*, 2008; Hovmøller *et al.*, 2023).

In addition to the colonization of previously unaffected areas of the world, the long-distance migration capabilities of yellow rust spores have enabled the incursion of exotic races into local populations, which further contributes to the shaping of yellow rust populations. A shining example of the role of exotic incursions on the local population dynamics of yellow rust can be seen in the incursions of the PstS1 genetic group into North America and Australia and closely related PstS2 genetic group into Europe in the early 2000s (Hovmøller *et al.*, 2023). Further examples can be seen specifically in Europe, where the genetic groups PstS7 and PstS8, first observed in 2011, replaced the pre-existing population of yellow rust in Europe in the course of a few years (Hovmøller *et al.*, 2023). These two genetic groups are assumed to have origin in the Himalayan regions (Hovmøller *et al.*, 2016), which is considered the centre of diversity and where sexual recombination seems to also be involved in the shaping of yellow rust populations (Ali *et al.*, 2014).

The abovementioned adaptability, incredible dispersion of yellow rust, both due to wind dispersal and human activities, elucidates the importance of not only thorough disease monitoring but also proper disease control measures and structured field management, in order to limit the damaging effects of emerging virulent races and new incursions.

Disease Management

Plant resistance

Plant resistance breeding is one of the main disease control methods against yellow rust and the use of resistant cultivars is often considered safer, more compatible with other control strategies and environmentally friendly, compared to the use of fungicides. However, the evolutionary arms race that takes place between newly developed resistant wheat varieties and yellow rust puts pressure on researchers to improve the durability and level of resistance in new varieties. This is largely because of the emergence of new virulent races, that can result in resistant cultivars losing their resistance, leading to large yield losses (Chen, 2020).

When it comes to breeding strategies against yellow rust, there are several different types of resistance in wheat used against yellow rust. The two most common types of plant resistance are all-stage resistance (ASR), and adult-plant resistance (APR) (Chen, 2020). ASR can be detected already in the seedling stage and is present through all growth stages, characterized by a hypersensitive response (Wang and Chen, 2017). ASR follows the gene-for-gene hypothesis developed by Flor (1971), which describes that for each R gene present in the host plant, there will be a corresponding avirulence gene in the pathogen. The result of an interaction between the products of the two corresponding genes is a qualitative resistance that is easily recognizable and relatively race-specific, which has made ASR a popular trait for selection in major breeding programs (Bouvet *et al.*, 2022). However, the genes involved in ASR are usually race-specific and are easily overcome by new virulent races of yellow rust, examples of which have been seen in ASR genes such as *Yr2*, *Yr9*, *Yr17* and *Yr27* (Bayles *et al.*, 2000; Hovmøller and Justesen, 2007; Sørensen *et al.*, 2013).

In the case of APR, on the other hand, plants are susceptible in the seedling stage and are characterized by slow rusting or partial resistance manifesting at the adult plant stages (Wang and Chen, 2017; Bouvet *et al.*, 2022). APR genes are generally believed to protect plants from a wide range of pathogen species. Examples of this can be seen in *Yr18/Lr34/Sr67/Pm38* and *Yr46/Lr67/Sr55/Pm46*, whch are effective against yellow rust, leaf rust, stem rust and powdery mildew (Bouvet *et al.*, 2022). APR is considered non-race-specific, although exceptions do exist (Dinglasan *et al.*, 2022; Sørensen *et al.*, 2013). Because APR has a broad-spectrum effect and results in partial resistance, APR genes are considered more durable than ASR genes, since they evoke a lower selection pressure. While the desire to use APR genes in breeding programs has been high, its use has been limited because of the lack of linked molecular markers to support selection, as well as poor knowledge of the gene's phenotypic effects, gene interactions, environmental stability and genetic background effects (Dinglasan *et al.*, 2022). Because of this, efforts are being put into combining ASR and APR to create cultivars with long-lasting and high-level resistance.

As of now, over 300 quantitative trait loci (QTL) conferring yellow rust resistance have been reported, with around 80 of these being permanently named as yellow rust resistance genes and further 63 designated as temporarily designated resistance genes (Bouvet *et al.*, 2022). Most of these are useful in breeding and many have markers developed for marker-assisted selection; however, the continued investigation of resistance genes and QTLs is essential for the future breeding of resistance cultivars control of Pst (Chen, 2020).

Chemical control

Although yellow rust is generally effectively controlled by the use of resistant cultivars, the incursion of new, aggressive strains of yellow rust to various areas globally has overcome many of the long-standing resistance genes and put pressure on farmers, breeders and scientists. Because of this, proper disease monitoring and early warning systems are highly important to alert farmers of new threats. In situations where there is a lack of resistance to wheat cultivars, yellow rust management becomes highly dependent on the use of chemical fungicides. Since the development of systemic fungicides in the 1970s, a wide variety of diverse products with different modes of action have been available for protection against yellow rust on wheat. In general, three classes of fungicides are currently used to control yellow rust on wheat, these being demethylation inhibitors (DMI, FRAC group 3), quinone outside inhibitors (QoI, FRAC code 11) and succinate dehydrogenase inhibitors (SDHI, FRAC code 7) (Carmona *et al.*, 2020; Cook *et al.*, 2021). Of the three classes of chemical fungicides, the most numerous and important group are the triazoles belonging to the DMIs. Refer to Carmona *et al.* (2020) for a list of the most frequently used fungicides to control yellow rust.

It has long been the common perception that most species of the Pucciniales have all been well-controlled with fungicides under field conditions, all belonging to the lowest risk category in regard to fungicide resistance development (Brent and Hollomon, 2007). However, mutations conferring resistance to the three fungicidal classes DMI, QoI and SDHI have been discovered in the Asian soybean rust pathogen *Phakopsora pachyrhizi*, which have been linked to the high-intensity use of fungicides in South America (Cook *et al.*, 2021). Moreover, a study by Cook *et al.* (2021), which screened 363 isolates of *Puccinia striiformis* f.sp. *tritici*, found mutations with the potential to compromise the efficacy of DMI and SDHI fungicides. This highlights the importance of a diverse management strategy against yellow rust, as well as the application of the precautionary principle in the use of fungicides against yellow rust.

Cultural practices

In addition to the use of resistant cultivars and chemical control, yellow rust can be further controlled by the application of proper cultural practices. Farmers can reduce disease pressure from year to year by reducing volunteer plants on and around the fields, on which the pathogen may survive the winter and provide inoculum for the following year (Chen and Kang, 2017). Furthermore, the use of cultivar mixtures to decrease spread of disease and proper management of irrigation to reduce moisture in the field can lead to an overall decrease in the severity of yellow rust epidemics. Moreover, an appropriate timing of sowing is also of importance in the defence against yellow rust. Early sowing of winter wheat can increase the chances of infection during Autumn providing inoculum for further infection in the next season (Chen and Kang, 2017). Therefore, the general recommendation for winter wheat is late planting as far as possible. Meanwhile, the cultivation of spring wheat is highly dependent on the weather conditions and the optimal time for sowing should be determined based on the region and weather conditions.

An integrated disease control strategy with a strong cultural practice that is revised on a yearly basis depending on the resources available, the local environment and weather conditions is instrumental in the sustainable disease control of yellow rust.

Future Research

1. Continued monitoring of yellow rust is of the utmost importance, especially considering the recent tendencies of exotic incursions, in order to increase local preparedness for new disease outbreaks and for the evaluation of evolution of virulence in yellow rust populations around the world. Moreover, a strengthening of global collaboration for early-warning systems would be highly beneficial, especially in areas with limited use of resistant cultivars.
2. Further research in resistance breeding against yellow rust in order to develop more durable and high-level resistant cultivars is needed. The identification and cloning of novel resistance genes, as well as the investigation of the combination and pyramiding of ASR and APR genes, would be highly useful in future breeding programs for the development of long-lasting resistance in new wheat cultivars.
3. Future research into the development of the evolution of fungicide resistance in yellow rust, as well as fungicide resistance monitoring is crucial to ensure the use of current fungicides and their efficacy. Furthermore, research and innovation in fungicide application technology is needed to allow a more timely, accurate and efficient application of fungicides.

Conclusion

While yellow rust is a long-studied disease, known to mankind for many years, the pathogen remains a threat to the global wheat-growing community due to its capacity for long-distance migration and rapid adaptation to new environmental conditions and host resistance. Continued research in the virulence of yellow rust, novel resistance genes and the development of resistant cultivars, as well as global disease monitoring is important in order to reduce future yellow rust epidemics.

Discussion Points

1. List the reasons why yellow rust is a successful global pathogen.
2. Describe which factors play a role in the evolution of virulence of yellow rust – how do these factors vary depending on geographical location?
3. Discuss why all-stage resistance might not always be the most durable resistance mechanism to control yellow rust and suggest how breeders might develop more durable resistance in wheat varieties. Breeders have to select for more than just resistance, consider how they should incorporate durable yellow rust resistance within a much broader context of developing adapted and high-yielding wheat varieties.

Further Reading

Carmona, M., Sautua, F., Pérez-Hérnandez, O. and Reis, E.M. (2020) Role of fungicide applications on the integrated management of wheat stripe rust. In: *Frontiers in Plant Science*. Frontiers Media S.A, p. 11.

Chen, X. and Kang, Z. (2017) *Stripe Rust*, Vol. 1. Springer, Dordrect, The Netherlands, pp. 1–719.

Dinglasan, E., Periyannan, S. and Hickey, L.T. (2022) Harnessing adult-plant resistance genes to deploy durable disease resistance in crops. *Essays in Biochemistry* 66(5), 571–580.

Global Rust Reference Center, Rust tools (n.d.). Available at: https://agro.au.dk/forskning/internationale-platforme/wheatrust (accessed 8 January 2025).

Hovmøller, M.S., Thach, T. and Justesen, A.F. (2023) Global dispersal and diversity of rust fungi in the context of plant health. *Current Opinion in Microbiology* 71, 102243.

References

Ali, S., Gladieux, P., Leconte, M., Gautier, A., Justesen, A.F. *et al.* (2014) Origin, migration routes and worldwide population genetic structure of the wheat yellow rust pathogen *Puccinia striiformis* f.sp. *tritici*. *PLOS Pathogens* 10, e1003903.

Bayles, R.A., Flath, K., Hovmoller, M.S. and Vallavieille-Pope, C. (2000) Breakdown of the Yr17 resistance to yellow rust of wheat in northern Europe. *Agronomie* 20, 805–811.

Bouvet, L., Holdgate, S., James, L., Thomas, J., Mackay, I.J. *et al.* (2022) The evolving battle between yellow rust and wheat: Implications for global food security. *Theoretical and Applied Genetics* 135, 741–753.

Brent, K.J. and Hollomon, D.W. (2007) *Fungicide Reisistance: The Assessment of Risk*. Fungicide Resistance Action Committee.

Carmona, M., Sautua, F., Pérez-Hérnandez, O. and Reis, E.M. (2020) Role of fungicide applications on the integrated management of wheat stripe rust. In: *Frontiers in Plant Science*. Frontiers Media S.A.

Cheng, X.R., Zhuang, H., Zhao, J., Zhan, G.M., Kang, Z.S. *et al.* (2022) Identification of *Mahonia* species as alternate hosts for *Puccinia striiformis* f. sp. *tritici* and determination of existence of sexual propagation of the rust pathogen on *mahonia* under natural conditions in China. *Phytopathology* 112, 1422–1430.

Chen, X. (2020) Pathogens which threaten food security: *Puccinia striiformis*, the wheat stripe rust pathogen. *Food Security* 12, 239–251. DOI: 10.1007/s12571-020-01016-z.

Chen, X. and Kang, Z. (2017) *Integrated Control of Stripe Rust*. Springer Nature, Dordrecht, Netherlands, pp. 559–599.

Cook, N.M., Chng, S., Woodman, T.L., Warren, R., Oliver, R.P. *et al.* (2021) High frequency of fungicide resistance-associated mutations in the wheat yellow rust pathogen *Puccinia striiformis* f. sp. *tritici*. *Pest Manag Sci* 77, sp.

Dinglasan, E., Periyannan, S. and Hickey, L.T. (2022) Harnessing adult-plant resistance genes to deploy durable disease resistance in crops. *Essays in Biochemistry* 66(5), 571–580. DOI: 10.1042/EBC20210096.

Ding, Y., Cuddy, W.S., Wellings, C.R., Zhang, P., Thach, T. *et al.* (2021) Incursions of divergent genotypes, evolution of virulence and host jumps shape a continental clonal population of the stripe rust pathogen *Puccinia striiformis*. *Molecular Ecology* 30, 6566–6584.

Du, Z.M., Peng, Y.L., Zhang, G.S., Chen, L., Jiang, S.C. *et al.* (2023) Direct evidence demonstrates that *Puccinia striiformis* f. sp. *tritici* infects susceptible barberry to complete sexual cycle in autumn. *Plant Disease* 107, 771–783.

Flor, H.H. (1971) Current status of the gene-for-gene concept. *Annual Review of Phytopathology* 9, 275–296. DOI: 10.1146/annurev.py.09.090171.001423.

Hovmøller, M.S. and Justesen, A.F. (2007) Appearance of atypical *Puccinia striiformis* f. sp. *tritici* phenotypes in north-western Europe. *Australian Journal of Agricultural Research* 58, 518–524.

Hovmøller, M.S., Yahyaoui, A.H., Milus, E.A. and Justesen, A.F. (2008) Rapid global spread of two aggressive strains of a wheat rust fungus. *Molecular Ecology* 17, 3818–3826.

Hovmøller, M.S., Walter, S., Bayles, R.A., Hubbard, A., Flath, K. *et al.* (2016) Replacement of the European wheat yellow rust population by new races from the centre of diversity in the near-Himalayan region. *Plant Pathology* 65, 402–411.

Hovmøller, M.S., Thach, T. and Justesen, A.F. (2023) Global dispersal and diversity of rust fungi in the context of plant health. *Current Opinion in Microbiology* 71, 102243.

Jin, Y., Szabo, L.J. and Carson, M. (2010) Century-old mystery of *Puccinia striiformis* life history solved with the identification of berberis as an alternate host. *Phytopathology* 100, 432–435.

Kang, Z., Tang, C., Zhao, J., Cheng, Y., Liu, J. *et al.* (2017) Wheat-Puccinia striiformis interactions. In: Chen, X. and Kang, Z. (eds) *Stripe Rust*. Springer Netherlands, Dordrecht, pp. 155–282.

Liu, M. and Hambleton, S. (2010) Taxonomic study of stripe rust, *Puccinia striiformis* sensu lato, based on molecular and morphological evidence. *Fungal Biology* 114, 881–899.

Mehmood, S., Sajid, M., Zhao, J., Huang, L. and Kang, Z. (2020) Alternate hosts of *Puccinia striiformis* f. sp. *tritici* and their role. In: *Pathogens*. MDPI. Available at: https://www.mdpi.com/2076-0817/9/6/434

Milus, E.A., Kristensen, K. and Hovmøller, M.S. (2009) Evidence for increased aggressiveness in a recent widespread strain of *Puccinia striiformis* f. sp. *tritici* causing stripe rust of wheat. *Phytopathology* 99, 89–94.

O'Brien, L., Brown, J.S., Young, R.M. and Pascoe, I. (1980) Occurrence and distribution of wheat stripe rust in victoria and susceptibility of commercial wheat cultivars. *Australasian Plant Pathology* 9, 14.

Pretorius, Z.A., Boshoff, W.H.P. and Kema, G.H.J. (1997) First report of *Puccinia striiformis* f. sp. *tritici* on wheat in South Africa. *Plant Disease* 81(4), 424–424.

Sørensen, C.K., Justesen, A.F. and Hovmøller, M.S. (2013) Spontaneous loss of *Yr2* avirulence in two lineages of *Puccinia striiformis* did not affect pathogen fitness. *Plant Pathology* 62, 19–27.

Wang, M. (2013) First report of oregon grape (*Mahonia aquifolium*) as an alternate host for the wheat stripe rust pathogen (*Puccinia striiformis* f. sp. *tritici*) under artificial inoculation. *Plant Disease* 97, 839.

Wang, M. and Chen, X. (2017) Stripe rust resistance. In: Chen, X. and Kang, Z. (eds) *Stripe Rust*. Springer, Dordrecht, Netherlands. DOI: 10.1007/978-94-024-1111-9_5.

Zhao, J., Wang, L., Wang, Z., Chen, X., Zhang, H., Yao, J. *et al.* (2013) Identification of eighteen berberis species as alternate hosts of *Puccinia striiformis* f. sp. *tritici* and virulence variation in the pathogen isolates from natural infection of barberry plants in China. *Phytopathology* 103, 927–934.

15 Wheat Leaf Rust, Caused by *Puccinia triticina*, and Mitigation Through Host Genetic Resistance

Brent McCallum[1], Colin Hiebert[1], Xiben Wang[1] and Guus Bakkeren[2]

[1]*Agriculture and Agri-Food Canada, Morden Research and Development Centre, Morden Manitoba, Canada*
[2]*Agriculture and Agri-Food Canada, Summerland Research and Development Centre, Summerland British Columbia, Canada*

Host Range and Origin

Wheat leaf rust is caused by *Puccinia triticina* Erikss (formerly known as *Puccinia recondita* f. sp. *tritici*), a fungal pathogen on bread wheat (*Triticum aestivum*), durum wheat (*Triticum turgidum* ssp. *dicoccoides* and *Triticum turgidum* ssp. *durum*) and related grass species. The pathogen reproduces asexually on these species but can also reproduce sexually on the most abundant and principal alternate host meadow rue (*Thalictrum speciosissimum*) and a few other plant species (Anikster *et al.*, 1997). However, as the alternate hosts are not present in North America, sexual reproduction is therefore rare on this continent (Levine and Hildreth, 1957). Likewise, in most of the other wheat-growing areas of the world, the alternate host is not present. It is present, however, in Europe and the Middle East which could facilitate sexual recombination in those regions (Fellers *et al*, 2021). The pathogen co-evolved with domestication of wheat species 10,000 years ago in the Middle East which includes modern-day Iraq, Syria, Turkey, Jordan, Lebanon, Israel and Palestine(Anikster and Wahl, 1979). This is the centre of evolutionary origin and domestication of many cereal crop species and is also known as the fertile crescent.

Symptoms and Yield Loss

Symptoms of leaf rust infection are initially seen as small circular yellow spots on the upper leaf surface, which over time turn into orange-coloured pustules surrounded by yellow or brown leaf tissue. As the disease progresses the pustules may coalesce to form larger, irregular lesions. The pustules are raised above the leaf surface and contain masses of orange to reddish-brown urediniospores which give the characteristic rust appearance (Fig. 15.1). The urediniospores can be easily dislodged from the pustules and help to distinguish rust diseases from other leaf diseases, such as common wheat leaf spotting diseases such as *Septoria tritici* blotch, *Stagonospora nodorum* blotch and tan spot. There are three common rust diseases found on wheat leaves, leaf rust, stem rust and stripe rust. Stripe rust symptoms are most distinctive with yellow-orange pustules that are elongated running parallel along the leaf veins. Pustules resembling beads on a string along the leaf vein. Stem rust can be distinguished from leaf rust since stem rust pustules are generally larger than leaf rust, have a diamond shape compared to the circular leaf rust pustules, and the urediniospores of stem rust are brick or dark red compared to the orange leaf rust urediniospores (Fig. 15.1).

Colonization of leaves diverts nutrients from the plant to production of fungal mycelium and spores, and dehydration occurs when the epidermis is ruptured by masses of uredia. The affected leaf tissue is unable to photosynthesize leading to stunting of plant growth and reduced grain production. Reports vary considerably on the extent of yield loss caused by leaf rust, but likely range between 5 and 50%, with higher losses in susceptible cultivars, following early on-set of infection, and cool wet weather conditions, with typical losses in the 1 to 10% range in North America (Huerta-Espino *et al.*, 2011). Chai *et al.* (2020) estimated that 94.4% of the world's wheat is vulnerable to leaf rust infection and that global annual yield loss from 2000 to 2050 could reach 8.6–18.3 million tons of grain equivalent to an economic loss of $1.5–$3.3 billion US dollars based on 2016 grain prices. However, their model only attributed an annual crop loss worldwide of 1.2% (baseline scenario) to 2.6% (high-loss scenario) due to leaf rust. Since leaf rust occurs worldwide and spreads by air, it is not a disease that can be controlled by quarantine, particularly between neighbouring countries. Nevertheless, *P. triticina* populations differ significantly among wheat-growing regions. Introduction of races into areas where they do not occur already can cause severe losses to wheat varieties that do not possess appropriate leaf rust resistance. Travellers are therefore advised to avoid accidental long-distance movement of pathogen spores on clothing or footwear for this and other plant and animal diseases.

Fungicide Treatments

Higher precipitation and relative humidity in some areas result in greater yields but also higher incidence of many fungal diseases. Fungicide treatments are common in such areas to control wheat leaf rust. However, fungal

Fig. 15.1. (A) Close view of leaf rust pustules and urediniospores on an infected wheat leaf. (B) Typical symptoms of a heavy wheat leaf rust infection. (C) Stripe rust infection on wheat leaves. (D) Stem rust infection on wheat.

species can develop tolerance to fungicides, which then lose their effectiveness, resulting in increasing yield loss over time even in fungicide-treated fields. Several fungicide products effectively control *P. triticina* and decreased sensitivity has not been reported yet. Nonetheless, fungicide tolerance, due to mutations in the gene targeted by the active ingredient, has been identified in a related fungal rust pathogen, *Puccinia striiformis* f. sp. *tritici*, that causes stripe rust of wheat (Cook *et al.*, 2021). There can sometimes be a temptation to neglect selection for disease resistance in breeding programs, since some resistance genes may initially be associated with lower yield, which seems unacceptable if the disease is not always present. When fungicides are used in breeding programs, selection for leaf rust resistance is no longer possible. Thus, if breeding for leaf rust resistance is not maintained, wheat cultivars may be highly susceptible when fungicides lose their effectiveness.

Life Cycle and Epidemiology

Puccinia triticina is an obligate pathogen thus needing living plant tissue of host species to survive and reproduce (Bolton *et al.*, 2008). The fungus has evolved molecular mechanisms to evade the plant defence mechanisms and depends on the cellular machinery for nutrients to grow, without killing the plant. The complete life cycle of *P. triticina* has five different spore types (Fig. 15.1). Urediniospores (dikaryotic; n+n) are produced during successive rounds of infection and sporulation on wheat. The asexual cycle is common in most wheat-producing regions due to the lack of the alternate host. The urediniospores require 100% humidity for germination and temperatures from 2 to 30°C with an optimum of 20°C (Roelfs *et al.*, 1992). Germ tubes grow along the leaf surface until they encounter a stoma which they recognize because of the shape of the guard cells (Bolton *et al.*, 2008). An appressorium is then formed over the stomatal aperture, a penetration peg originating from the appressorium pushes through the closed stoma to enter the intercellular space, and a substomatal vesicle is formed from which the primary infection hyphae develop (Bolton *et al.*, 2008). These infection hyphae grow outwards until they contact a mesophyll cell, at which point a haustorial mother cell is formed (Bolton *et al.*, 2008). A penetration peg formed from the haustorial mother cell penetrates the host cell wall, followed by the formation of a specialized feeding structure called a haustorium inside the plant cell wall but tightly adjacent to the plasma membrane (Bolton *et al.*, 2008). These haustoria absorb nutrients from plant cells. Additional hyphae continue to grow extracellularly from the haustorial mother cell, to infect other host cells, the fungus ramifies throughout the infected area and after 7 to 10 days after initial infection, uredinia with thousands of urediniospores erupt through the leaf epidermis (Fig. 15.2). The new generation of urediniospores is spread by wind over shorter distance to neighbouring wheat plants or over much longer distances by upper wind currents (Huerta-Espino *et al.*, 2011), and the pathogen continues to propagate asexually in wheat fields throughout the growing season (Samborski, 1985; Bolton *et al.*, 2008). The fungus survives on the host as long as the leaf tissue is alive, and urediniospores can remain viable for weeks without the host as long as they do not encounter high humidity and germinate.

During times of stress such as senescence of the host plant or extreme temperatures, the fungus can switch to form teliospores which mark the initiation of the sexual cycle (Fig. 15.2). A thick, black melanized cell wall enables the teliospores to survive months and even years, during hot summers or cold winters and germinate when conditions become more favourable. In

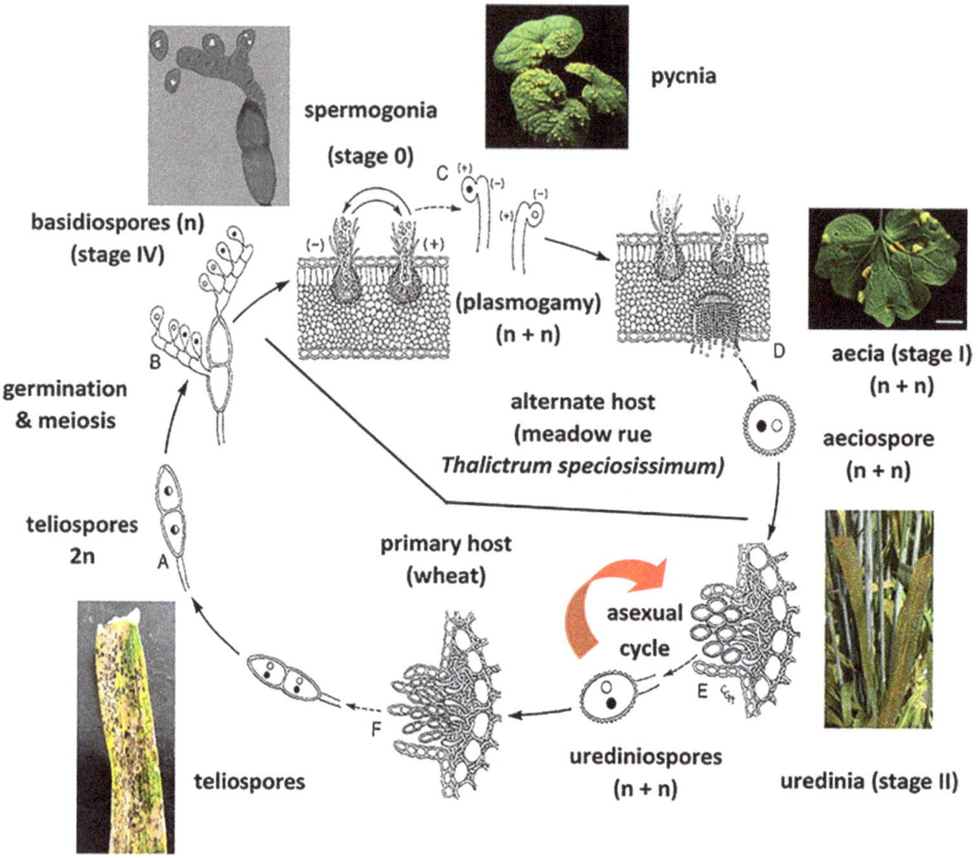

Fig. 15.2. The life cycle of *Puccinia triticina*. Telia with teliospores (A and B) and uredinia with urediniospores (E and F) are produced on the primary wheat host, whereas pycnia with pycniospores (C) and aecia with aeciospore (D) are produced on the upper and lower leaf surfaces on the alternate host, *Thalictrum speciosissimum*. Adapted from1 Alexopoulos *et al.* (1996) by C. Gubbins Hahn (John Wiley & Sons, reproduced by permission). Photos by T. Eilam and J. Kolmer (reproduced by permission).

teliospores, the two haploid nuclei (n + n) perform karyogamy to produce a diploid state (2n). If conditions are conducive for teliospore germination, a basidium is formed in which meiosis occurs. *Puccinia triticina* has a tetrapolar mating system in which the pheromone/pheromone receptor locus (*a* locus) and the homeodomain mating-type transcription regulators (*b* locus) are on different chromosomes; in a dikaryotic cell, each haploid nucleus has different allelic specificities, e.g., *a1b1* + *a2b2*. Since meiosis takes place in the basidium, the opposite mating types segregate, resulting in haploid (n) basidiospores each with either one of the mating types (*a1b1*; *a2b2*; *a1b2*; *a2b1*). An additional mitotic replication after meiosis produces dikaryotic basidiospores in which the two nuclei are identical (e.g., n(*a1b1*) + n(*a1b1*)) (Anikster, 1983). Basidiospores are shot off and can land on leaves of the alternate host whereupon they infect by direct epidermal cell penetration and ramify in the host tissue (Bolton *et al.*, 2008). Pycnia erupts on the upper side of the leaf containing pycniospores and infectious hyphae; each pycnidium represents one mating-type specificity. Insects are attracted to the sweet pycnial fluids and cross-fertilize the fungus by carrying pycniospores from one pycnium to the next. When of different mating-type specificity, fertilization can ensue through cell fusion and one haploid nucleus from a pycniospore combines with another haploid nucleus in the flexious hyphae to produce a new cell type: e.g., n(*a1b2*) + n(*a2b1*). This triggers a new developmental program where the fungus ramifies through the host leaf to produce aecia which erupt on the underside of the leaf, with the production of dikaryotic (n + n) aeciospores. These aeciospores are airborne and can infect wheat which results in the production of uredinia to start the asexual cycle again (Fig. 15.2).

In North America, *P. triticina* overwinters on wheat plants in the southern USA and Mexico, and the initial inoculum, in the form of urediniospores, is spread each year by wind currents into the northern USA and Canada, successively as wheat fields are seeded from south to north (Roelfs, 1985).

Pathogen Diversity and Effectors

Populations of *P. triticina* are genetically diverse as a result of natural mutation during cell divisions, facilitated by large numbers of transposable elements and repetitive DNA regions, which may cause chromosomal recombination. Variation may also be caused by vegetative fusion between different genotypes and genetic reassortment without the sexual cycle (Wang and McCallum, 2009). Consequently, new races that are virulent on prevalent wheat varieties have a selective advantage and increase rapidly (Fellers *et al.*, 2021; McCallum *et al.*, 2021). The rust pathogen produces protein effectors, many of which are recognized by host immune receptors, that often represent the matching leaf rust resistance (*Lr*) genes. This recognition immediately triggers a defence response in the plant, slowing or stopping the pathogen. Mutation and/ or deletion of effectors in the pathogen cause it to become virulent on wheat varieties harbouring the matching *Lr* gene which is no longer able to recognize the mutated effector.

Boom-and-Bust Cycles of *Lr* Genes

The primary leaf rust management strategy is the growing of wheat cultivars with genetic resistance, which has been practiced for over 100 years (Samborski, 1985). Wheat cultivars were developed to have good genetic resistance to the *P. triticina* population at the time of their release. The pathogen population will often evolve to overcome this genetic resistance, particularly when only virulence to a single *Lr* gene is needed to make a cultivar susceptible. Mutations and genetic recombination often allow for the evolution of virulent races that infect and increase these cultivars. This was documented by Samborski (1985) when wheat cultivar Selkirk, with *Lr10* and *Lr16* genes, was introduced in western Canada in 1954, which resulted in a rapid increase of new races that initially became virulent to *Lr10* and later also to *Lr16*. This phenomenon has been called a boom-and-bust cycle, where a new cultivar with resistance to leaf rust has a few successful years of production (the boom), followed years of increasing yield loss because of the development of new virulent races (the bust). A similar situation developed with resistance gene *Lr21*, which was used extensively for many years in both Canada and the United States but became less effective with the appearance of virulent races in both countries. The general concept is illustrated in Fig. 15.3, where a pair of leaves of cultivar Thatcher are susceptible to *P. triticina* isolate 198–1 (left), whereas Thatcher with the *Lr2c* resistance gene confer resistance to isolate 198–1 (middle), but not isolate 196–2 (right). It is commonly stated that resistance has broken down by the development of new races. Sometimes virulent races may already exist in the population at low, nearly undetectable levels at the time of cultivar release, in some cases, new races may have been introduced from other parts of the world.

Types of Leaf Rust Resistance Genes

Genetic analysis of leaf rust resistance has led to identification and genomic mapping of over 80 *Lr* genes, and many more are either unofficial designations or represented by quantitative trait loci (QTL) (Pinto da

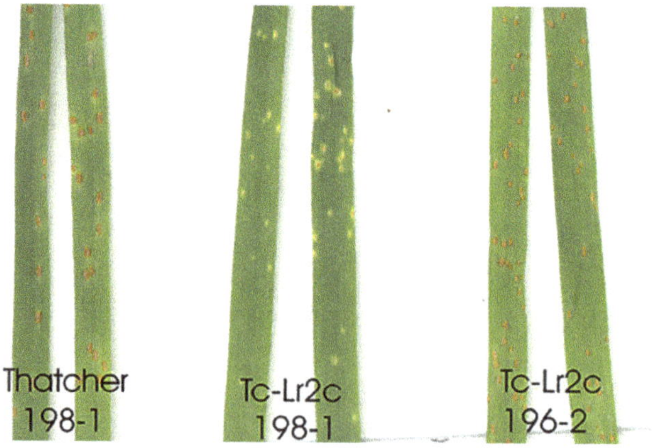

Fig. 15.3. Left: Two leaves of wheat cultivar Thatcher inoculated with *P. triticina* isolate 198–1 resulting in a susceptible reaction seen as large, orange uredinia and extensive sporulation. Middle: Leaves of a near-isogenic line derived from Thatcher with leaf rust resistance gene *Lr2c* inoculated with isolate 198–1, showing a resistant reaction seen as pale, necrotic lesions with reduced sporulation. Right: Leaves of the same near-isogenic line inoculated with isolate 196–2 which has overcome the effect of the *Lr2c* gene, showing the same susceptible reaction as Thatcher to the left.

Silva *et al.*, 2018). The known *Lr* genes confer either seedling or adult plant resistance. Seedling resistance is effective throughout all wheat growth stages, whereas adult plant resistance only becomes effective as the plant matures. Most *Lr* genes, whether classified as seedling or adult are race specific, meaning they are effective against specific races but no other races as illustrated for *Lr2c* in Fig. 15.3. Race non-specific genes have the same level of partial resistance to all races specifically, *Lr34*, *Lr46* and *Lr67*, which are called adult plant slow-rusting genes, since they slow down the development of pustules thereby limiting the spread of the pathogen and damage to the plant (McCallum *et al.*, 2016). Figure 15.4 shows the extent of leaf rust pustules on the wheat cultivar Thatcher with (left) and without (right) the *Lr34* resistance gene. Both *Lr34* and *Lr46* genes have demonstrated durable resistance, since they have been deployed over wide areas for many years without breaking down. The *Lr67* gene is also thought to be durable, although it has yet to be widely deployed. The development of wheat cultivars that combine genes for race-specific resistance with adult plant race non-specific resistance has minimized the negative effect of the boom-and-bust cycles described above.

Fig. 15.4. Effect of the adult plant resistance gene, *Lr34*, on the number and size of *P. triticina* pustules. The two leaves with the *Lr34* gene on the left (Thatcher-*Lr34*) have fewer and smaller pustules compared to the two susceptible leaves on the right (Thatcher) with many large orange pustules. The leaf-tip necrosis on the two Thatcher-*Lr34* leaves is a side-effect of this gene.

Survey of Rust Fungal Races Using Wheat Differentials

Tracking of virulence in *P. triticina* populations is essential for alerting plant breeders and growers to new virulent races that might threaten wheat production, and for determining which *Lr* genes would control leaf rust in the next generation of wheat cultivars. To that end, pathogen populations are being systematicallyassayed annually for the virulence spectrum in many wheat-growing regions of the world. This is done by collecting infected leaf samples from commercial wheat fields or nurseries seeded with leaf rust susceptible lines, during the growing season. Urediniospores are isolated from the leaves and inoculated onto the wheat cultivar Thatcher, which is universally susceptible to all races. A new generation of pustules forms, and urediniospores from a single isolated pustule are collected and stored as a single isolate. A number of isolates are randomly selected, and their urediniospores are inoculated separately onto a set of Thatcher near isogenic lines (NIL), each containing a single different *Lr* resistance gene. When this set of differential plants is rated for rust development, it shows the virulence spectrum of the pathogen population each year (McCallum *et al.*, 2016). The proportion of isolates that are virulent on each *Lr* gene changes annually. Fig. 15.5 shows the proportion of isolates virulent on six *Lr* genes in annual

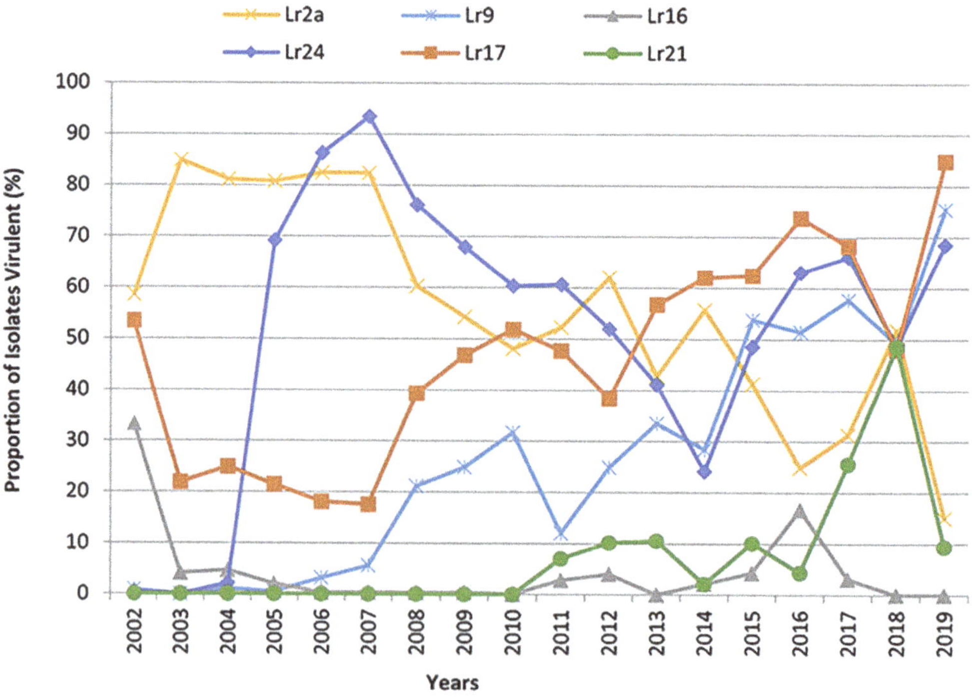

Fig. 15.5. The proportion of *P. triticina* isolates virulent on a set of differential wheat lines having a single different leaf rust resistance gene *Lr2a*, *Lr9*, *Lr16*, *Lr17*, *Lr21* or *Lr24* in a Thatcher genetic background. Isolates were sampled in western Canada from 2002 to 2019 (repinted from McCallum *et al.*, 2021).

surveys from 2002 to 2019 in western Canada. Virulence on *Lr16* has been low and was also low on *Lr21* until 2017 when it was overcome by a higher proportion of virulent isolates, as mentioned earlier. The effectiveness of the other *Lr* genes has fluctuated over the 17 year time period shown here.

Breeding Wheat Cultivars with Leaf Rust Resistance

Decades of breeding of wheat cultivars with resistance to leaf rust have reduced the severity caused by modern-day *P. triticina* populations. A total of 45 wheat cultivars, released in Canada and the USA between 1880 and 2001, were grown side-by-side in field plots between 2007 and 2010. Under relatively high levels of natural *P. triticina* infection, the older cultivars were more susceptible with higher leaf rust severity ratings than more recent cultivars, some of which appeared to be immune to the disease (Fig. 15.6; Martens *et al.*, 2014).

Elucidating the Genetics of Leaf Rust Resistance

Excellent progress has been achieved by breeding wheat cultivars with leaf rust resistance, and many studies have analysed the genetic basis of resistance in cultivars with effective and durable resistance. One recent study focused on the Canadian bread wheat cultivar Carberry (Bokore *et al.*, 2022). Carberry (resistant) was crossed with Thatcher (susceptible) and a population of doubled haploid (DH) lines was generated, so that each progeny line

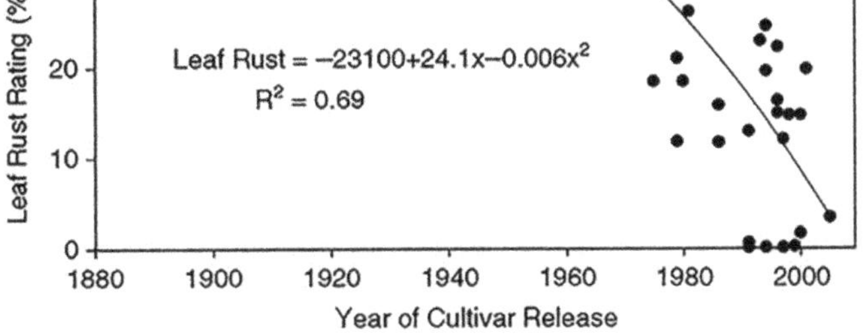

Fig. 15.6. Average leaf rust rating of 45 wheat cultivars, released in Canada and the USA between 1880 and 2001, measured in two field nurseries with a natural *P. triticina* inoculum. The data clearly demonstrate the value of breeding for genetic resistance to leaf rust.

was genetically different, but all genes were homozygous in each line. This population was segregated for two race-specific seedling resistance genes, *Lr2a* and *Lr16*, along with two race-nonspecific adult plant genes, *Lr34* and *Lr46*. The presence of these four genes in each DH line was determined using molecular markers linked to each *Lr* gene. The population was also screened for leaf rust severity over multiple field locations and years. These field plots were inoculated with *P. triticina*, by spraying urediniospores onto rows of susceptible wheat included within the nursery to allow spread to all plots by wind, and irrigated to promote disease development. While the *Lr* genes reduced the level of leaf rust severity individually, combinations of genes, particularly those including *Lr34*, showed lower leaf rust infections (Fig. 15.7). This means that wheat lines carrying *Lr34* together with a race-specific seedling resistance gene were more resistant than lines carrying either gene alone. Carberry has maintained its high level of resistance since its registration in 2009, likely due to stacking of four genes conferring both seedling and adult plant resistance. Other studies have demonstrated that the adult plant resistance genes, *Lr34* and *Lr67*, together with race-specific resistance genes also reduce leaf rust severity in an additive manner (McCallum and Hiebert, 2022). In addition, *Lr34* and *Lr67* confer partial resistance to other diseases of wheat such as Fusarium Head Blight, caused by *Fusarium*

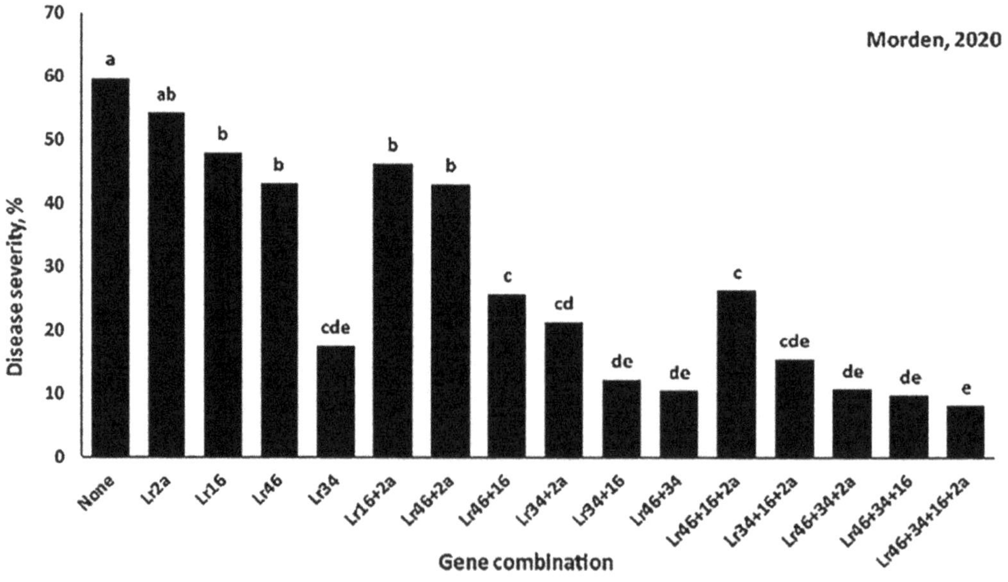

Fig. 15.7. Leaf rust severity, % of flag leaf infected, for doubled haploid lines derived from a cross between wheat cultivars Carberry (resistant with *Lr2a*, *Lr16*, *Lr34*, *Lr46*) and Thatcher (susceptible) in field plots inoculated with *P. triticina*. The presence of different *Lr* genes (x-axis) was determined with molecular markers. Bars represent the average severity for all lines within each genotypic class, and those with different letters were statistically different ($P < 0.05$). This is an excerpt of a larger data set reprinted from Bokore *et al.* (2022).

graminearum, and resulted in lower levels of the mycotoxin deoxynivalenol (DON) in harvested grain samples (McCallum *et al.*, 2024).

Use of Molecular Markers

In wheat, DNA markers linked to a gene of interest can be identified by determining the association of a desired phenotype with a genotypic marker using linkage analysis. Polymorphisms can be identified in the genome either within the actual coding sequence of the gene or in close proximity. Single Nucleotide Polymorphisms (SNPs) are commonly used, while differences in the number of simple sequence repeats (SSRs) in non-coding DNA regions have also been used. Molecular markers are used routinely to genotype large numbers of progeny in wheat breeding programs from which a smaller number of individuals with the appropriate marker can be selected i.e. marker-assisted selection (MAS) (Toth *et al.*, 2018). Situations where the *Lr* gene have been identified, along with the actual DNA sequence, provide the most accurate detection of appropriate progenies and vastly improve the selection process.

It is very valuable to combine the detailed virulence analyses on differential wheat *Lr* genes described above with a genetic analysis of *P. triticina* populations. Understanding the genetic relatedness among isolates helps explain the evolution of populations through mutations within clonal groups, of potential nuclear exchanges creating novel hybrids, and can reveal unexpected alien introductions of novel races. Though initially using SSRs and related small-scale variants in DNA sequences, over the last few years, Genotyping-By-Sequencing (GBS) using high throughput sequencing technologies and mapping to reference genomes, has generated tens of thousands of SNPs which can reveal the genetic relatedness among such genetic lineages at high resolution (Fellers *et al.*, 2021). Such molecular polymorphisms among *P. triticina* isolates can also be exploited to develop markers to quickly identify genetic lineages, and race groups virulent or avirulent against resistance genes deployed in wheat cultivars.

Future Perspectives

Tremendous new insights into the molecular basis of plant-pathogen interaction are happening through the assembly of whole genome sequences of *P. triticina* isolates and wheat varieties, and by comprehensive transcriptome studies of genes and proteins during the infection process. This has allowed the identification of 'susceptibility' genes in wheat shown to be essential for pathogens to gain a foothold in the host. Removing or mutating one or more of these susceptibility genes through genetic engineering using CRISPR/Cas technologies could deprive the pathogen of tools to manipulate the host resulting in less or no infection (Wang *et al.*, 2022). This strategy is believed to lead to more durable resistance, especially if several essential fungal target genes can be engineered without detrimental effects on wheat growth and yield. Large datasets of genomic information from the *P. triticina* – wheat interaction have been used to identify virulence factors and effectors used by the pathogen in the infection process. Inhibition of certain virulence factors through virus-induced gene silencing (VIGS) as well as host-induced gene silencing (HIGS) can reduce rust infection (Panwar *et al.*, 2018). Research in the use of small interfering siRNAs for spray-induced gene silencing (SIGS) is also advancing fast.

Conclusion

A high level of genetic resistance to *P. triticina* can be achieved, but the development of new cultivars must be continuous and well-informed about the prevalent races of the pathogen. Continuous annual surveillance of *P. triticina* populations in important wheat-growing regions of the world generates important information of race structure and changes in their virulence over time. This informs wheat breeders and growers about the effectiveness of the current leaf rust resistance genes. It also directs the production of inoculum representing the current population, which takes place on universal susceptible wheat lines, since *P. triticina* is an obligate pathogen. A combination of adult plant resistance, such as *Lr34*, *Lr46* and *Lr67*, with race-specific *Lr* genes has shown a high level of durable resistance because the pathogen has more difficulty overcoming the different molecular defence pathways. Molecular markers and genomic-assisted selection are important tools to pyramid *Lr* genes in future wheat cultivars. Indeed, this is among the major goals of the 'Expert Working Group on Wheat Pathogens' in the Global Wheat Initiative.

Discussion Points

1. Since *P. triticina* is an obligate pathogen and spreads rapidly via wind currents, what strategies would you use to control this disease?
2. Discuss the characteristics of the two types of leaf rust resistance genes. Which would you use and why?
3. Since *P. triticina* races composition changes over time in specific wheat growing areas, outline the different steps needed to screen wheat lines for resistance to a current leaf rust population.
4. If you were a plant breeder, how would you ensure new wheat cultivars were not overcome by changes in *P. triticina* population similar to boom-and-bust cycles.

Further Reading

Chai, Y., Pardey, P.G., Hurley, T.M., Senay, S.D. and Beddow, J.M. (2020) A probabilistic bioeconomic assessment of the global consequences of wheat leaf rust. *Phytopathology* 110(12), 1886–1896. DOI: 10.1094/PHYTO-02-20-0032-R.

Global Wheat Initiative (n.d.). Available at: https://www.wheatinitiative.org (accessed 30 October 2024).

Huerta-Espino, J., Singh, R.P., Germán, S., McCallum, B.D., Park, R.F. *et al.* (2011) Global status of wheatleaf rust caused by *Puccinia triticina*. *Euphytica* 179(1), 143–160. DOI: 10.1007/s10681-011-0361-x.

McCallum, B.D., Hiebert, C.W., Cloutier, S., Bakkeren, G., Rosa, S.B. *et al.* (2016) A review of wheat leaf rust research and the development of resistant cultivars in Canada. *Canadian Journal of Plant Pathology* 38(1), 1–18. DOI: 10.1080/07060661.2016.1145598.

Samborski, D.J. (1985) 2 – wheat leaf rust. In: Roelfs, A.P. and Bushnell, W.R. (eds) *Diseases, Distribution, Epidemiology, and Control*. Academic Press, pp. 39–59.

References

Alexopoulos, C.J., Mims, C.W. and Blackwell, M.M. (1996) *Introductory Mycology*, 4th edn. John Wiley & Sons, New York.

Anikster, Y. (1983) Binucleate basidiospores – a general rule in rust fungi. *Transactions of the British Mycological Society* 81(3), 624–626.

Anikster, Y. and Wahl, I. (1979) Coevolution of the rust fungi on *Gramineae* and *Liliaceae* and their hosts. *Annual Review Phytopathology* 17(1), 367–403.

Anikster, Y., Bushnell, W.R., Eilam, T., Manisterski, J. and Roelfs, A.P. (1997) *Puccinia recondita* causing leaf rust on cultivated wheats, wild wheats, and rye. *Canadian Journal of Botany* 75, 2082–2096. DOI: 10.1139/b97-919.

Bokore, F.E., Knox, R.E., Hiebert, C.W., Cuthbert, R.D., DePauw, R.M. *et al.* (2022) A combination of leaf rust resistance genes, including *Lr34* and *Lr46*, is the key to the durable resistance of the Canadian wheat cultivar, Carberry. *Frontiers in Plant Science* 12, 2904. DOI: 10.3389/fpls.2021.775383.

Bolton, M.D., Kolmer, J.A. and Garvin, D.F. (2008) Wheat leaf rust caused by *Puccinia triticina*. *Molecular Plant Pathology* 9(5), 563–575. DOI: 10.1111/j.1364-3703.2008.00487.x.

Chai, Y., Pardey, P.G., Hurley, T.M., Senay, S.D. and Beddow, J.M. (2020) A probabilistic bioeconomic assessment of the global consequences of wheat leaf rust. *Phytopathology* 110(12), 1886–1896. DOI: 10.1094/PHYTO-02-20-0032-R.

Cook, N.M., Chng, S., Woodman, T.L., Warren, R., Oliver, R.P. *et al.* (2021) High frequency of fungicide resistance-associated mutations in the wheat yellow rust pathogen *Puccinia striiformis* f. sp. *tritici*. *Pest Management Science*. DOI: 10.1002/ps.6380.

Fellers, J.P., Sakthikumar, S., He, F., McRell, K., Bakkeren, G. *et al.* (2021) Whole-genome sequencing of multiple isolates of *Puccinia triticina* reveals asexual lineages evolving by recurrent mutations. *G3 Genes | Genomes | Genetics* 11(9), jkab219. DOI: 10.1093/g3journal/jkab219.

Huerta-Espino, J., Singh, R.P., Germán, S., McCallum, B.D., Park, R.F. *et al.* (2011) Global status of wheat leaf rust caused by *Puccinia triticina*. *Euphytica* 179(1), 143–160. DOI: 10.1007/s10681-011-0361-x.

Levine, M. and Hildreth, R. (1957) A natural occurrence of the aecial stage of *Puccinia rubigo-vera* var. *tritici* in the United States. *Phytopathology* 47, 110–111.

Liu, J.Q. and Kolmer, J.A. (1998) Molecular and virulence diversity and linkage disequilibria in asexual and sexual populations of the wheat leaf rust fungus, *Puccinia recondita*. *Genome* 41, 832–840. DOI: 10.1139/g98-088.

Martens, G., Lamari, L., Grieger, A., Gulden, R.H. and McCallum, B. (2014) Comparative yield, disease resistance and response to fungicide for forty-five historic Canadian wheat cultivars. *Canadian Journal of Plant Science* 94(2), 371–381. DOI: 10.4141/cjps2013-193.

McCallum, B.D. and Hiebert, C.W. (2022) Interactions between *Lr67* or *Lr34* and other leaf rust resistance genes in wheat (*Triticum aestivum*). *Frontiers in Plant Science* 13. DOI: 10.3389/fpls.2022.871970.

McCallum, B.D., Hiebert, C.W., Cloutier, S., Bakkeren, G., Rosa, S.B. *et al.* (2016) A review of wheat leaf rust research and the development of resistant cultivars in Canada. *Canadian Journal of Plant Pathology* 38(1), 1–18. DOI: 10.1080/07060661.2016.1145598.

McCallum, B.D., Reimer, E., McNabb, W., Foster, A., Rosa, S. *et al.* (2021) Physiologic specializationof *Puccinia triticina*, the causal agent of wheat leaf rust, in Canada in 2015–2019. *Canadian Journal of PlantPathology* 43(sup2), S333–S346. DOI: 10.1080/07060661.2021.1888156.

McCallum, B.D., Hiebert, C.W., McCartney, C.A. and Henriquez, M.A. (2024) The effects of *Lr34* and *Lr67* on Fusarium head blight resistance and deoxynivalenol accumulation in wheat. *Plant Pathology* 73(7), 1901–1911. DOI: 10.1111/ppa.13924.

Panwar, V., Jordan, M., McCallum, B. and Bakkeren, G. (2018) Host-induced silencing of essential genes in *Puccinia triticina* through transgenic expression of rnai sequences reduces severity of leaf rust infection in wheat. *Plant Biotechnology Journal* 16(5), 1013–1023. DOI: 10.1111/pbi.12845.

Pinto da Silva, G.B., Zanella, C.M., Martinelli, J.A., Chaves, M.S., Hiebert, C.W. *et al.* (2018) Quantitative trait loci conferring leaf rust resistance in hexaploid wheat. *Phytopathology* 108(12), 1344–1354. DOI: 10.1094/PHYTO-06-18-0208-RVW.

Roelfs, A.P. (1985) Wheat leaf rust. In: Roelfs, A.P. and Bushnell, W.R. (eds) *The Cereal Rusts*, Vol. 2. Academic Press, NY, pp. 395–424.

Roelfs, A.P., Singh, R.P. and Saari, E.E. (1992) *Rust Diseases of Wheat, Concepts and Methods of Disease Management*. CIMMYT, Mexico.

Samborski, D.J. (1985) 2 – wheat leaf rust. In: Roelfs, A.P. and Bushnell, W.R. (eds) *Diseases, Distribution, Epidemiology, and Control*. Academic Press, pp. 39–59.

Toth, J., Pandurangan, S., Burt, A., Mitchell, F.J. and Kumar, S. (2018) Marker-assisted breeding of hexaploid spring wheat in the Canadian prairies. *Canadian Journal of Plant Science* 99, 111–127. DOI: 10.1139/cjps-2018-0183.

Wang, X. and McCallum, B. (2009) Fusion body formation, germ tube anastomosis, and nuclear migration during the germination of urediniospores of the wheat leaf rust fungus, *Puccinia triticina*. *Phytopathology* 99(12), 1355–1364.

Wang, N., Tang, C., Fan, X., He, M., Gan, P. *et al.* (2022) Inactivation of a wheat protein kinase gene confers broad-spectrum resistance to rust fungi. *Cell* 185(16), 2961–2974. DOI: 10.1016/j.cell.2022.06.027.

16 *Trichoderma* Ear Rot on Maize – When Friends Turn Evil

Annette Pfordt

Georg-August University Goettingen, Institute of Plant Pathology and Crop Protection, Grisebachstrasse 6, 37077, Goettingen, Germany

Introduction

Members of the genus *Trichoderma* (*Ascomycota, Hypocreales*) are commonly considered opportunistic fungi rapidly colonizing diverse organic substrates in various environments. In agricultural habitats, *Trichoderma* spp. are primarily associated with dead wood and contribute to nutrient cycling by decomposing complex organic compounds, thus enhancing soil fertility and promoting overall ecosystem health (Druzhinina *et al.*, 2011). Beyond their ecological functions, *Trichoderma* spp. are used in agriculture as biocontrol agents to promote plant health by directly controlling phytopathogens or indirectly enhancing plant growth (Sood *et al.*, 2020). In maize, *Trichoderma*-based products are used to manage soil-borne and foliar diseases against *Rhizoctonia solanis, Fusarium* spp., or *Pythium* spp. and promote plant growth during drought and temperature stress or increase salt tolerance.

However, in 2018, *Trichoderma* ear rot was reported for the first time in Europe as a disease in maize (Pfordt *et al.*, 2020). In this report, *Trichoderma afroharzianum* was identified as the causal agent of maize ear rot disease in Southern Germany causing severe symptoms and yield reduction. Since then, *Trichoderma* ear rot caused by *T. afroharzianum* has also been reported in France, Austria, Italy (Sanna *et al.*, 2022), and China (Chen *et al.*, 2023) and also described as causal agent of stalk rot disease in India (Harish *et al.*, 2025).

The Genus *Trichoderma*

Species of the genus *Trichoderma* belong to the phylum Ascomycota and are mostly known in their asexual form (*Fungi imperfecti*). *Trichoderma* represents a numerous amount of different species within its genus, with approximately 375 valid species (Cai and Druzhinina, 2021). *Trichoderma* is distributed worldwide and thus inhibiting a wide range of different ecosystems and ubiquitous

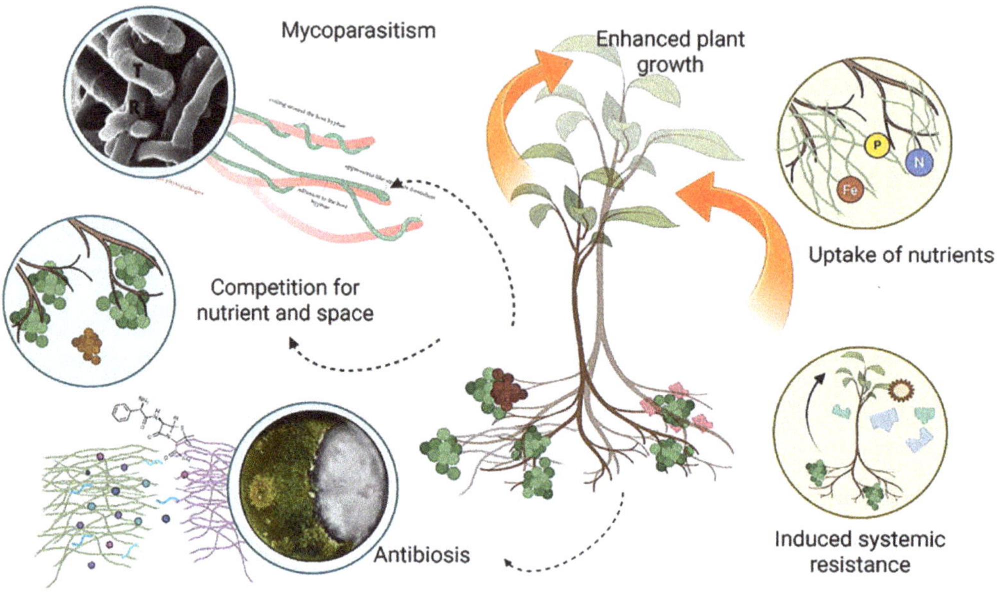

Fig. 16.1. Mode of action of *Trichoderma* spp. as biocontrol agents. *Trichoderma* interacts with plant pathogens and host plants through multiple mechanisms.

in forest or agricultural soils. As a free-living soil-borne fungus, *Trichoderma* is fast-growing and highly competitive and often known for their ability to establish opportunistic symbiotic relationships with plants. *Trichoderma* is known to exhibit antagonistic effects against a variety of plant pathogens, such as fungi, oomycetes, bacteria, and nematodes. Its antagonistic activity against plant pathogens is based on multiple modes of action: fast colonization, competition for nutrients, antibiosis, and mycoparasitism (Harman, 2005). In particular, mycoparasitism—parasitizing other fungi—is a key trait of many *Trichoderma* species. It involves the secretion of cell wall–degrading enzymes and a broad spectrum of secondary metabolites with antimicrobial properties (Woo *et al.*, 2023). In addition, *Trichoderma* can not only induce plant resistance by promoting the plant production of phytohormones but also engage in an endophytic way and support plant growth by triggering various gene expressions (Fig. 16.1) (Gupta and Bar, 2020).

Trichoderma afroharzianum as a Plant Pathogen

Although *Trichoderma* species are widely recognized for their beneficial properties in agriculture, several strains of *T. afroharzianum* have also been reported as pathogens, causing disease and yield loss in maize (*Zea mays*). *T. afroharzianum* belongs to the *T. harzianum* species complex, first separated as an independent species in 2015 by Chaverri *et al.* It reproduces asexually via abundant green conidia formed on branched conidiophores and persists in the soil through chlamydospores and resistant hyphal structures. Under certain conditions, *T. afroharzianum* is also capable of producing secondary metabolites and low-molecular weight compounds with cytotoxic or antimicrobial properties.

Occurrence of *Trichoderma* ear rot

The first report of a *Trichoderma* species associated with maize disease dates back to 1910, when researchers at the University of Kansas observed a greenish-yellow, wet mold growing between the rows of maize kernels. This mold, originally identified as *T. viride*, was linked to premature germination of kernels on the cob and was noted to occur widely in maize fields in Riley County, Kansas. In Ontario, Canada, *Trichoderma* species were found to act as secondary pathogens, colonizing maize roots and stalks after initial infection by *Fusarium oxysporum sensu lato* (Whitney and Mortimore, 1961). In 1972, *T. koningii* was isolated from stunted maize plants in southern Ontario and 3 years later, McFadden and Sutton (1975) identified *T. koningii*, *T. harzianum*, and *T. hamatum* to be the causal agents of lesions on the first internode of maize (McFadden and Sutton, 1975). Several reports have documented their potential role as ear rot pathogens in maize, particularly in the United States. Observations from Iowa, Ohio, and Kentucky describe *Trichoderma*-associated ear rot as presenting dark, blue-green layers of conidia covering and penetrating the kernels. Field evidence further suggests that infections often coincide with physical damage to the ear, such as that caused by bird feeding or mechanical injury, potentially facilitating fungal entry and colonization.

A more recent development in the pathology of maize is the emergence of *T. afroharzianum* as a primary causal agent of maize ear rot in Europe. This species was first identified in association with symptomatic maize cobs in Germany in 2018 (Pfordt *et al.*, 2020). Since then, *T. afroharzianum* infections were recorded in southern Germany, particularly affected from maize-growing regions in Bavaria, Baden-Wuerttemberg along the Rhine and from the drier areas surrounding Magdeburg and Bernburg in Saxony-Anhalt. In France, symptomatic maize cobs were found in the Alsace and Burgundy regions. Northern Italy, particularly the Po Valley and eastern Austria, has reported cases near its borders with Hungary and Slovakia. In addition, confirmed findings came from agricultural sites in Turkey, notably in Salihli (Manisa Province) and Tarsus (Mersin Province) (Fig. 16.2) (Pfordt *et al.*, 2025b).

Symptoms and damage

Trichoderma ear rot is visually characterized by greenish-gray fungal growth between the kernels and on the outer husks, often starting at the ear tip and progressing downward (Pfordt *et al.*, 2020). Due to the similar spore

Fig. 16.2. Geographic distribution of received pathogenic *T. afroharzianum* isolates from maize cobs and agricultural soil samples in Europe.

coloration, visual infection is often mixed up with similar gray–green spore infections caused by *Aspergillus* or *Penicillium*. However, a key distinguishing symptom of *T. afroharzianum* is precocious germination of kernels on the cob, resulting in a watery, soft appearance of the kernels (Video 16.1). This is caused by the production of large amounts of alpha-amylase by the fungus, an enzyme that converts starch into glucose. Since starch constitutes 60–75% of a maize kernel's dry weight, this enzymatic degradation significantly reduces the grain quality and the yield by up to 55%. The result is a soft, sugar-rich kernel with increased moisture content, poor storability, and reduced processing suitability for food and industrial applications. High sugar levels interfere with fermentation and affect flour quality, while excess moisture promotes secondary microbial infections post-harvest. One of the most critical impacts is on seed viability. Infected kernels show severely reduced germination rates—often below 25%—and seedlings that do emerge are typically weak and poorly rooted. This compromises replanting success and can cause serious yield losses (Fig. 16.3) (Pfordt *et al.*, 2024).

Fig. 16.3. Symptoms of Trichoderma ear rot infection on maize.

Disease cycle and infection conditions

The disease cycle of *T. afroharzianum* is not yet fully understood, but recent findings provide important insights. *Trichoderma* species, particularly *T. afroharzianum*, are common soil inhabitants capable of surviving for several years in crop residues and organic matter. At all surveyed field sites where maize ear rot symptoms occurred, *T. afroharzianum* was consistently isolated not only from symptomatic cobs but also from adjacent soil samples. This strong correlation supports the hypothesis that soil-borne populations serve as a primary inoculum source for infections (Pfordt *et al.*, 2025a, b, c). However, the fungus was also recovered from soils in asymptomatic fields, suggesting that it can exist in a latent state and may only cause disease under specific physiological or environmental conditions. These environmental factors play a critical role in disease development. Field observations and inoculation trials suggest that warmer temperatures (20–28°C) combined with dry conditions during silking and early grain filling favor the infection. Infection likely occurs during maize flowering, when the cob is most vulnerable. Prolonged dryness during flowering provides ideal conditions for spores to spread, while heat stress may compromise plant defenses. Maize appears to be the primary host for pathogenic *T. afroharzianum* isolates, although the full host range remains unclear. Preliminary observations indicate that *Trichoderma* may also colonize small grains, such as wheat and barley, leading to brownish discoloration and reduction of thousand kernel weight, though symptoms are less pronounced (Pfordt *et al.*, 2023, 2025c).

These findings raise concerns about the pathogen's broader host range and underline the importance of crop rotation and residue management in controlling its spread.

Pathogenicity of biocontrol products

Biocontrol products based on *Trichoderma* species have gained widespread acceptance as eco-friendly alternatives to chemical pesticides. Their ability to suppress harmful pathogens and promote plant growth makes them valuable tools in sustainable agriculture. However, recent research highlights an important caveat: Not all strains used in these products are entirely harmless. A comprehensive study tested 12 commercial biocontrol products containing *Trichoderma* strains for their pathogenic potential on maize. The results were eye-opening. While many products contained non-pathogenic or only weakly pathogenic strains, three products harbored strains with medium-to-high pathogenicity, causing significant disease symptoms under controlled conditions. Most strikingly, one of these pathogenic strains came from a product officially registered and authorized as a biological plant protection agent for use on maize (Pfordt *et al.*, 2025a, b, c).

This finding raises serious concerns about the current standards and screening procedures for biocontrol agents. This evidence emphasizes the need for more rigorous and standardized protocols in biocontrol product development, including thorough genetic characterization and pathogenicity testing. Reliance solely on species-level identification is insufficient, as even closely related strains can differ drastically in their effects on crops. Without careful screening, the use of biocontrol products risks introducing harmful fungi that may damage crops rather than protect them.

Management Practices

Effective management of *T. afroharzianum* infections in maize remains challenging, partly because early infections often go unnoticed as they are hidden beneath the husk leaves. Chemical fungicides are generally not approved for ear rot control in many European countries, and their efficacy against *T. afroharzianum* is largely untested. In wheat and barley, the application of conventional fungicides at flowering has been shown to reduce infection and maintain grain quality; however, such treatments are not always feasible in organic systems and maize cultivation. Moreover, biological control agents—including *Trichoderma* strains—pose a regulatory dilemma, as their application could potentially introduce or amplify pathogenic strains in the field. Crop rotation and deep ploughing can help reduce inoculum levels in the soil, though the fungus appears to persist naturally in various environments. Currently, no maize varieties show strong resistance or tolerance, underscoring the need for breeding programs focused on tolerance traits.

A major limitation in management is the lack of diagnostic tools to detect pathogenic *T. afroharzianum* strains in the soil before sowing to prevent infection. Although molecular markers such as TEF1-a, RPB2, and CAL are commonly used for species identification, they do not allow differentiation between pathogenic and non-pathogenic (or beneficial) *T. afroharzianum* strains. This diagnostic gap makes it impossible to assess infection risks early or to screen seed lots and soil samples for harmful isolates, significantly hampering the development of monitoring and certification systems.

Further research is needed to clarify infection routes, assess the risk posed by biocontrol applications, and develop integrated disease management protocols.

Discussion Points

1. *Are current safety assessments and regulatory procedures for biocontrol products sufficient to detect potential pathogenic risks of organisms like Trichoderma?*

   Reflect on whether existing frameworks adequately account for the long-term behavior of microbialstrains in diverse field environments. Should regulations include screening for latent pathogenicity?

2. *What environmental and agronomic factors might trigger the shift from endophytic or saprophyticbehavior to pathogenicity in T. afroharzianum?*

   Discuss how stress conditions, host susceptibility, or cropping systems might influence this shift, andhow it compares to other pathogens.

3. *To what extent should plant breeders and crop protection strategies integrate emerging threats likeTrichoderma into their current disease management frameworks?*

   Explore the need for updated resistance screening, monitoring programs, and regulatory guidelines,especially in the context of climate change and evolving pathogen profiles.

Funding statement

Bundesministerium für Ernährung und Landwirtschaft (Grant/Award Number: "FKZ2221NR014").

Further Reading

Pfordt, A., Schiwek, S., Karlovsky, P. and Von Tiedemann, A. (2020) *Trichoderma afroharzianum* ear rot – a new disease on maize in Europe. *Frontiers in Agronomy* 2, 236.

Pfordt, A., Douanla-Meli, C., Schäfer, B.C., Schrader, G., Tannen, E. *et al.* (2025) Phylogenetic analysis of plant-pathogenic and non-pathogenic *Trichoderma* isolates on maize from plants, soil and commercial bio products. *Applied and Environmental Microbiology* 91(3), e0193124. DOI: 10.1128/aem.01931-24.

References

Cai, F. and Druzhinina, I.S. (2021) In honor of John Bissett: Authoritative guidelines on molecular identificationof *Trichoderma*. *Fungal Diversity* 107(1), 1–69. DOI: 10.1007/s13225-020-00464-4.

Chen, Y., Han, J., Yang, H., Qi, X., Guo, H. *et al.* (2023) Different maize ear rot fungi deter theoviposition of yellow peach moth (*Conogethes punctiferalis* (*Guenée*)) by maize volatile organic compounds. *Agronomy* 13(1), 251. DOI: 10.3390/agronomy13010251.

Druzhinina, I.S., Seidl-Seiboth, V., Herrera-Estrella, A., Horwitz, B.A., Kenerley, C.M. *et al.* (2011) *Trichoderma*: The genomics of opportunistic success. *Nature Reviews Microbiology* 9(10), 749–759. DOI: 10.1038/nrmicro2637.

Gupta, R. and Bar, M. (2020) Plant immunity, priming, and systemic resistance as mechanisms for*Trichoderma* spp. biocontrol. In: Sharma, A.K. and Sharma, P. (eds) *Trichoderma*. Springer Singapore (Rhizosphere Biology), Singapore, pp. 81–110.

Harish, J., Prasannakumar, M.K., Karan, R., Venkateshbabu, G., Vamsidharreddy, N. *et al.* (2025) *Trichoderma* spp. as a novel pathogen for maize post-flowering stalk rot in India. *Crop Protection* 195, 107255. DOI: 10.1016/j.cropro.2025.107255.

Harman, G.E. (2005) Overview of mechanisms and uses of *Trichoderma* spp. *Phytopathology* 96, 190–194.

McFadden, A.G. and Sutton, J.C. (1975) Relationships of populations of *Trichoderma* spp. in soil to disease in maize. *Canadian Journal of Plant Science* 55(2), 579–586. DOI: 10.4141/cjps75-085.

Pfordt, A., Schiwek, S., Karlovsky, P. and Von Tiedemann, A. (2020) *Trichoderma afroharzianum* ear rot-a new disease on maize in Europe. *Frontiers in Agronomy* 2, 1–7. DOI: 10.3389/fagro.2020.547758.

Pfordt, A., Gaumann, P. and Von Tiedemann, A. (2023) Pathogenicity of *Trichoderma afroharzianum* incereal crops. *Pathogens* 12, 936. DOI: 10.3390/pathogens12070936.

Pfordt, A., Steffen, L., Raz, T. and Naumann, M. (2024) Impact of *Trichoderma afroharzianum* infection on fresh matter content and grain quality in maize. *Frontiers in Plant Science* 15, 1–15.

Pfordt, A., Douanla-Meli, C., Schäfer, B.C., Schrader, G., Tannen, E. *et al.* (2025a) Phylogenetic analysis of plant-pathogenic and non-pathogenic *Trichoderma* isolates on maize from plants, soil and commercial bio products. *Applied and Environmental Microbiology* 91(3), e0193124. DOI: 10.1128/aem.01931-24.

Pfordt, A., Douanla-Meli, C., Voll, W.F., Schäfer, B.C., Schrader, G. *et al.* (2025b) Effectof temperature and precipitation on the occurrence of *Trichoderma* ear rot in Europe, phytopathology research, research square. PrePrint. DOI: 10.21203/rs.3.rs-6871003/v1.

Pfordt, A., Pruszynski-Beck, S. and Von Tiedemann, A. (2025c) Susceptibility of small grain cereals to *T. afroharzianum* ear infection in the field related to fungicide treatment and crop cultivar. *Frontiers inAgronomy*. PrePrint bioRxiv. DOI: 10.1101/2025.10.06.680620.

Sanna, M., Pugliese, M., Gullino, M.L. and Mezzalama, M. (2022) First report of *Trichoderma afroharzianumcausing* seed rot on maize in Italy. *Plant Disease* 106(7), 1773–2006. DOI: 10.1101/2025.10.06.680620.

Sood, M., Kapoor, D., Kumar, V., Sheteiwy, M.S., Ramakrishnan, M. *et al.* (2020) *Trichoderma*: The "secrets" of a multitalented biocontrol agent. *Plants (Basel, Switzerland* 9(6), 1–25. DOI: 10.3390/plants9060762.

Whitney, N.J. and Mortimore, C.G. (1961) Root and stalk rot of field corn in southwestern Ontario: ii. Development of the disease and isolation of organisms. *Canadian Journal of Plant Science* 41(4), S854–S861. DOI: 10.4141/cjps61-125.

Woo, S.L., Hermosa, R., Lorito, M. and Monte, E. (2023) *Trichoderma*: A multipurpose, plant-beneficialmicroorganism for eco-sustainable agriculture. *Nature Reviews Microbiology* 21(5), S312–S326. DOI: 10.1038/s41579-022-00819-5.

17 Phoma Stem Canker (Blackleg) of Oilseed Rape (Canola) and Other Brassica Crops

Jonathan S. West and Kevin M. King

Rothamsted Research, Harpenden, AL5 2JQ, UK

Introduction

This article explains the current taxonomy, identification methods, disease symptoms, epidemiology and control of phoma stem canker or blackleg of Brassicas, which is caused by two species, *Plenodomus lingam* or *P. biglobosus* (formerly known as *Leptosphaeria maculans* and *L. biglobosa* and they may be known by the asexual stage, *Phoma lingam*). These are important plant pathogens of oilseed rape (canola; *Brassica napus*), vegetable Brassicas (e.g. *B. oleracea*: cabbage, broccoli and kale; *B. rapa*: turnip and pak choi) and some other herbaceous plants, mainly in temperate areas, but *P. lingam* has not been found in China, nor in most parts of both South America and Africa. In addition, *P. dezfulensis* (a close relative of *P. biglobosus*) causes phoma leaf spot in Iran. Phoma stem canker can cause yield losses exceeding 50% in some individual fields, but usually losses are 0–15%. The fungi survive saprophytically on crop debris and spread by two types of spores—ascospores, which are wind dispersed and can travel hundreds of metres and potentially longer, and conidia, which are rain-splashed short distances (e.g. <2 m) but occasionally a few hundred metres as a fine spray of splashed rain in strong wind. Spread to new locations can also occur through infected seed or crop debris. Infection of leaves requires high humidity and is via stomata or wounds, and the fungi grow down the leaf vascular tissue to infect the stem, where they cause stem canker, a type of dry stem rot. Fungicides are not effective once the fungus has reached the stem, so prevention of leaf spots is the best control strategy, aided by good crop rotation and cultivar resistance. Identification methods by PCR and DNA sequencing are also described.

Taxonomy and Nomenclature

Phoma stem canker is caused by two closely related species, *Plenodomus lingam* (Tode) Höhnel and *P. biglobosus* (formerly *Leptosphaeria maculans* and *L. biglobosa*), which are ascomycetes in the class Dothideomycetes and family Leptosphaeriaceae. The asexual stage of *P. lingam* is *Phoma lingam*. Many sources of literature refer to them as *L. maculans* and *L. biglobosa*, and before 2001, *P. biglobosus* (*L. biglobosa*) was known as the weakly aggressive, B type or Tox^0 type of *L. maculans*.

Currently, there are subclades of both species, which are *P. lingam* subclades "brassicae" and "lepidii" and *P. biglobosus* subclades "americensis", "australensis", "brassicae", "canadensis", "erysimii", "occiaustralensis" and "thlaspii", see Mendes-Pereira *et al.* (2003). In addition, a very close relative of *P. biglobosus* has been named from samples in Iran and is termed *Plenodomus dezfulensis*.

Description and Morphological Identification

Ascospores of both species are identical physically although the pattern of initial hyphae produced by germinating ascospores is different. *P. lingam* hyphae emerge initially from central cells of the ascospore and branch profusely with side branches angled almost perpendicularly from the initial hypha, while *P. biglobosus* produces germ tubes initially from the terminal cells at each end of the ascospore, and these grow relatively long and straight hyphae, branching by dichotomising at relatively shallow angles (Huang *et al.*, 2001). Other physical differences between the two species are the production of a dark yellow pigment in culture by *P. biglobosus* and production of pink to violet conidial ooze (cirri) by *P. biglobosus* compared to pale cream coloured cirri in *P. lingam*, but there can be some overlapping characteristics in some isolates, so molecular diagnostics are needed for complete identification (see later).

Ascomata (pseudothecia) are flattened black spheres 200–250 µm high × 300–500 µm in diameter with a single pore on the top surface. The ascomata occur alone or clustered in groups initially immersed or partly immersed in crop debris but becoming raised above lignified material as softer dead plant tissues degrade away. The mature ascocarp can be hydrated in water and squashed or sectioned to show the outer dark pigmented thick-walled cells of the outer layer, an inner subhyaline or light brown layer of angular cells and the internal contents of sterile hairs (periphyses) and sacs (asci) containing eight ascospores. The asci are 90–110 × 10–12 µm hyaline, bitunicate (double-walled) structures sausage-shaped apart from tapering towards the base.

The ascospores are 35–70 × 5–8 µm, pale tan coloured, ellipsoids, sometimes slightly curved and comprising six cells divided by five septae, and

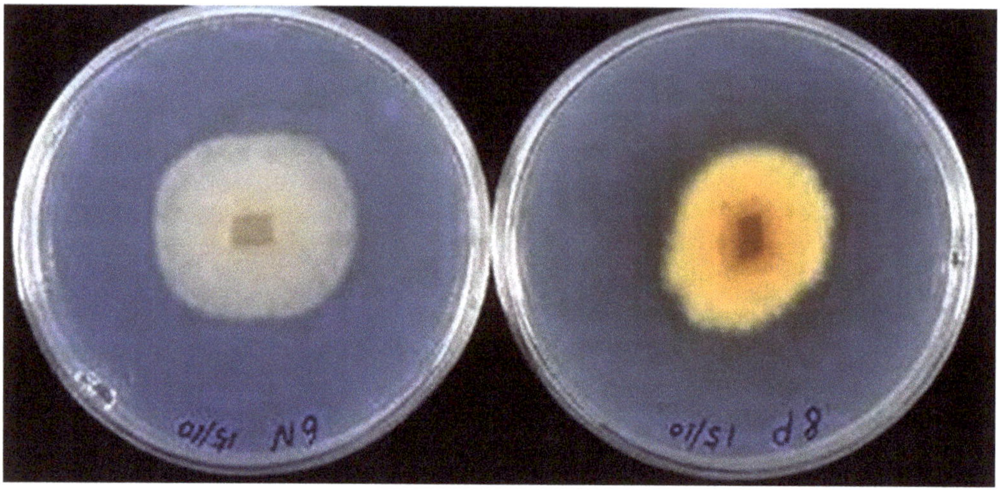

Fig. 17.1. Cultures of *P. lingam* (left) and *P. biglobosus* (right) growing on potato dextrose agar and showing yellow pigment production by *P. biglobosus*.

sometimes slightly constricted at the central-most septum. Each cell may contain one or more guttules (vesicles) (Fig. 17.2).

The asexual stage is produced in pycnidia, which are embedded in leaf lesions and are 150–350 µm in diameter, dark-brown to black flattened spheres with a small pore at the apex of a raised neck or spout in the centre of the upper surface. The pycnidia occur singly or more normally in groups within the leaf lesion. Alternatively, larger pycnidia occur on the surface of crop debris and are 300–700 µm in diameter and slightly more spherical that the type of pycnidia occurring in leaf lesions. In both cases, the pycnidium comprises several layers of pseudoparenchymatous cells pigmented dark brown. Conidia produced in the pycnidia are 3–5 × 1–2 µm, hyaline, aseptate, elongated spheres, occasionally with two small guttules near each end. The mass of conidia produced from the ostiole or pore of the pycnidium appear as a pale white, cream or pink-coloured ooze (cirrus), which readily disperses into water.

In culture, both *P. lingam* and *P. biglobosus* grow on common media like PDA or V8-agar at 15–25°C in the dark, producing a mat of mycelium ranging from white to pale grey in colour, sometimes with other pigments and particularly most isolates of *P. biglobosus* produce a dull yellow-orange pigment, which is easily seen on PDA (Fig. 17.1). Both also produce pycnidia in culture. *P. biglobosus* tends to stale in culture and loses viability if kept for a long time.

Symptoms and Disease Cycle

The fungus survives on stem and root debris for a few years or more unless the debris is immersed by flood water, which tends to cause it to be replaced

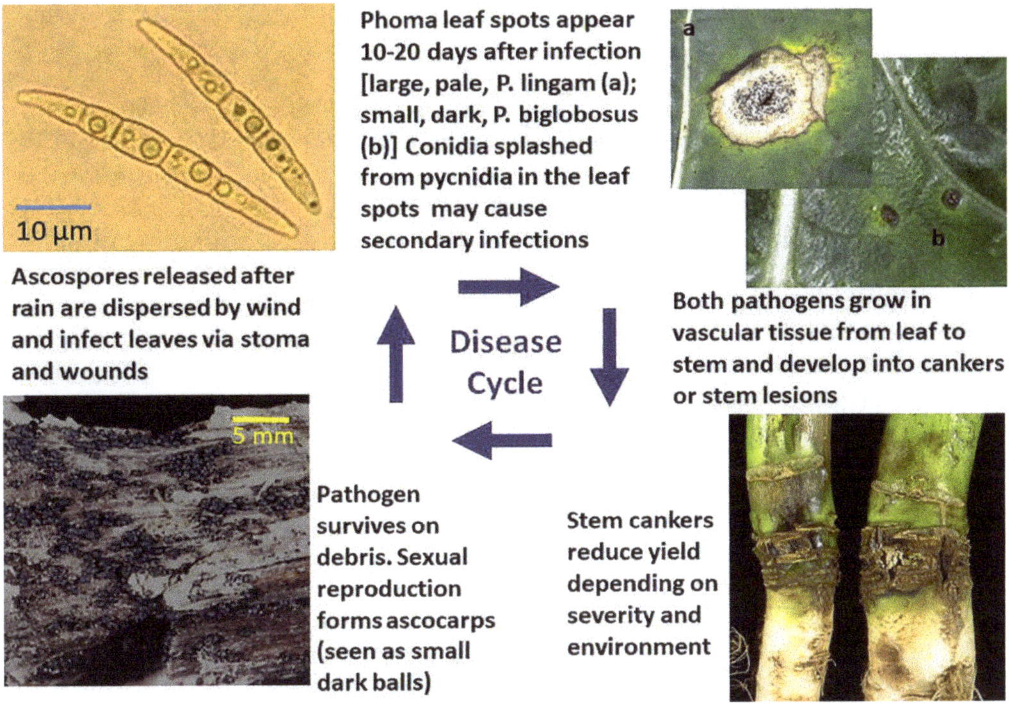

Fig. 17.2. Disease cycle of phoma stem canker, showing air dispersed ascospores approximately 50 µm long (top left), Phoma leaf spots of both types (top right), stem cankers (bottom right) and ascocarps on the surface of dead stem debris. All photos by Jon West except ascocarps which was by Rothamsted's photography group.

by other microbes (Peluola *et al.*, 2013). Both pycnidia and ascocarps can be produced on the surface of crop debris. Pycnidia tend to survive on crop debris only short periods, and as they empty, further fruiting body production tends to be ascocarps. In temperate locations, these mature and release wind-dispersed ascospores in the autumn and to a lesser extent into the winter and spring. In Canada, due to longer, cold winters, ascospores are released from the previous season's crop debris the following spring, while in Australia, they can be released to infect autumn-winter-sown crops of spring type canola crops and can continue to be released the following year. Factors affecting maturation and timing of ascospore release are discussed in a later section. Ascospores infect leaves in wetness films by producing hyphae that penetrate stoma, hydathodes or wounds.

Leaf spots develop about 7–20 days after infection but vary in appearance according to host resistance and leaf age. *P. lingam* causes pale lesions approximately 5–20 mm in diameter which develop multiple small dark specks in them (Fig. 17.2a). These are pycnidia, which produce a type of asexually produced spores called conidia (described earlier). Conidia can cause secondary infections by splashing short distances to other leaves and nearby plants. In contrast, *P. biglobosus* tends to cause small dark brown spots which may be surrounded by chlorotic tissue (Fig. 17.2b). The centre of some spots may degrade or drop out completely. In humid conditions, aerial mycelium may be produced, particularly at the margin of lesions.

Stem cankers and stem lesions result from the fungus growing down vascular tissue of the petiole to reach the stem. Cankers appear from flowering onwards in oilseed rape at stem positions where the infected leaf had been attached. Green stem tissue darkens to almost black immediately around the edge of cankers leading to the alternative name, blackleg, but the centre of cankers appear as a dry pale brown lesion, which with maturation towards harvest, may crack open and also develops pycnidia and later still ascocarps. The severity of canker is a product of the extent of girdling and internal stem infection and may result from several cankers coalescing. In severe cases, upper parts of the plant become drought-stressed due to insufficient water transport, leading to reduced yield and premature senescence. This is exacerbated by hot and windy weather due to increased water demand. In very severe cases, the plant may lodge.

Infections of later leaves (after the rosette leaves) and bracts can lead to stem lesions, whereas stem cankers occur at the stem base or crown of the plant. Upper stem lesions tend to spread up and down the softer stem tissues more readily, so the black or dark discolouration becomes more apparent.

Pod lesions can occur from late spore releases after flowering and are similar to leaf spots but occurring on the pods [siliques] and peduncles, causing some constriction and reduced development of the affected pod. These can be important by infecting seed in the pods, which is one of the main ways for the disease to be introduced to completely new locations.

Root infection also occurs either by growth of the fungus down the stem from basal stem cankers or directly from the soil where spores may have been washed or where the fungus may have survived on crop debris.

Tap roots and the root collar persist longest after harvest due to their lignified, woody composition and on affected plants, these may appear to have blackened areas encrusted with melanised ascocarps or their empty shells.

Diagnostic Tools

Morphological discrimination of *P. lingam* and *P. biglobosus*, for instance by growing the isolated fungus on artificial agar media such as potato dextrose agar (PDA) and examining cultures for the production of pigment (see *Description*), can be a cost-effective method for species identification, particularly in resource-poor laboratories, but can also be time-consuming and not always reliable (King and West, 2021).West *et al.* (2002b) Identification of the pathogens can also be attempted via isolation from infected seeds using the blotter test described by Andreoli and Maguire (1995), although again this approach can be time-consuming and give inconclusive results. Thus molecular PCR-based diagnostic approaches are often used, including species-specific quantitative PCR (qPCR) assays that can determine amounts of DNA of each pathogen species in environmental samples. These can be simple SYBR-green qPCR assays that are relatively cheap but rely on a non-specific fluorescent dye binding to amplified DNA, and thus, results might be affected by any spurious non-target PCR amplification. Alternatively, Taqman qPCR diagnostics are available for *Plenodomus* spp. (Jacques *et al.*, 2021) that are considered more reliable given the inclusion of an additional but expensive internal target-sequence specific probe, with fluorescence only detectable upon successful probe binding to the target DNA sequence. For all molecular-based diagnostics, however, a good understanding of *Plenodomus* population biology and genetics is essential to ensure the required target is being correctly amplified and identified. For example, in Europe, both *P. biglobosus* "brassicae" and the rare "canadensis" subclades are present West *et al.* (2002a)(King and West, 2021), but the SYBR green assay of Liu *et al.* (2006) does not detect the "canadensis" subclade (Kaczmarek *et al.*, 2023). Other molecular diagnostics, such as loop-mediated isothermal amplification (LAMP), utilising six primers for each target species and operating at a single constant temperature, have also been developed for detection and discrimination of *P. lingam*/*P. biglobosus* (Omer and Wallenhammar, 2020). LAMP offers some advantages over PCR-based approaches, including purportedly improved speed, specificity and sensitivity and the potential for applied use of this technology by end-users such as growers in the field using a portable device. Last, genomic approaches are now being used for monitoring and quantifying *Plenodomus* spp. in field populations. For example, a work at Rothamsted Research (UK) is currently utilising minION third-generation nanopore sequencing to investigate ratios of *P. lingam*/*P. biglobosus* in air and plant samples to explore how pathogen populations might be influenced by agronomic practices (e.g. application of fungicides).

Discussion Points on Epidemiology and Control

Ascospores can be dispersed many kilometres but tend to decline to the regional background level within a few hundred metres of a source. Conidia have been shown to cause secondary infections on new leaves and adjacent plants, and in some cases in Canada, they were shown to blow in a fine spray combined with wind to infect plants in adjacent fields. Management of crop debris and separation of subsequent crops to avoid close proximity to debris of the previous crop are important cultural control options to reduce disease pressure.

Plant breeders working with plant pathologists in pre-breeding schemes have developed a number of varieties of oilseed rape with resistance to *P. lingam* in recent years. Less is known about host resistance in vegetable brassicas, and little is known about host resistance to *P. biglobosus*. About 20 different host resistance genes are now known in oilseed rape against *P. lingam* (*L. maculans*) (Borhan *et al.*, 2022). At the time of writing (2024), all are named according to the former pathogen name of *Leptosphaeria maculans*, e.g. Rlm1 or LepR3.

In 2024, UK commercially available oilseed rape varieties have resistance ratings of 4 to 9 on a 1–9 scale. Current varieties that have excellent resistance to the disease include Vegas (Hybrid variety rated 9) and Murray (rating 8), and others are Dolphin (7), Tennyson (7), Matrix CL (7) and Miraculix CL (7). In many cases, the most durable resistance comes from a good background level of quantitative (polygenic or partial) resistance which does not stop infection but reduces the severity of disease.

Cold weather slows down disease development resulting in reduced canker severity by harvest and a delay in release of ascospores the following autumn. This is why the disease is generally not a problem in cooler locations (e.g. in the UK north of Lincolnshire). However, climate change predictions suggest that warmer weather will exacerbate this disease.

Fungicides are often used in combination with host resistance, particularly in locations where the yield potential of oilseed rape is high. Fungicides should be targeted at preventing leaf infections in initial growth stages as this prevents early infection of the stem. Although some fungicides can have a negative growth regulatory effect if applied to very young plants, in parts of Australia, where leaf infections can often occur on very young plants, growers sometimes coat fertiliser pellets with fungicide, and this is drilled below the seed to boost any short-term fungicide effect from seed coatings. In Europe, it is more normal to apply one or two autumn fungicides to oilseed rape, and this is done because the yield potential is relatively high due to the temperate conditions. Indeed, one or two fungicide applications, by controlling this disease, were found to reduce the greenhouse gas emissions per ton of grain produced by 15%. Fungicides have very little effect once the fungus has reached the stem. The best timing of fungicide application is after the first significant release of ascospores in the autumn, ideally before more than 10% of plants have phoma leaf spots. The timing of spore release varies each year due to the weather – in Europe, wet summers promote an early release of spores to infect autumn-sown crops (Huang *et al.*, 2007). Ascospores in Europe are mainly from the previous season's debris, while in North America, spring-sown plants are infected from spores produced on debris of the previous or preceding crop while in Australia, autumn-sown varieties are infected by ascospores released from debris 6–18 months old. In the UK, a weather-based forecasting tool is available, which forecasts the date when 10% of plants will have phoma leaf spot – see Evans *et al.* (2007) and https://ahdb.org.uk/provision-of-oilseed-rape-decision-support-systems-to-the-uk-arable-industry (accessed 29 May 2024).

Due to improvements in host resistance, locations that routinely used two autumn fungicides have now reduced applications to a single autumn spray unless phoma leaf spotting is very early or if the variety grown is known to be very susceptible. On more resistant varieties of oilseed rape, fungicide applications can even be omitted completely (against phoma stem canker) and particularly if phoma leaf spotting is late. Fungicide options are to use at least two modes of actions such as a DMI (FRAC code 3, e.g. mefentrifluconazole) plus an SDHI (FRAC code 7, e.g. bixafen) or a QoI (FRAC code 3, e.g. pyraclostrobin).

Further Reading

Bingol, E., Qi, A.M., Karandeni-Dewage, C., Ritchie, F., Fitt, B.D.L. *et al.* (2023) Co-inoculation timing affects the interspecific interactions between phoma stem canker pathogens *Leptosphaeria maculans* and *Leptosphaeria biglobosa*. *Pest Management Science.* DOI: 10.1002/ps.7799.

Fungicide Resistance Action Committee (FRAC) (n.d.). Available at: https://www.frac.info/

Stachowiak, A., Olechnowicz, J., Jedryczka, M., Rouxel, T., Balesdent, M.H. *et al.* (2006) Frequency of avirulence alleles in field populations of *Leptosphaeria maculans* in Europe. *European Journal of Plant Pathology* 114, 67–75. DOI: 10.1007/s10658-005-2931-z.

van de Wouw, A.P., Scanlan, J.L., Al-Mamun, H.A., Balesdent, M.-H., Bousset, L. *et al.* (2023) A new set of international *Leptosphaeria maculans* isolates as a resource for elucidation of the basis and evolution of blackleg disease on *Brassica napus*. *Plant Pathology.* DOI: 10.1111/ppa.13801.

West, J.S., Kharbanda, P.D., Barbetti, M.J. and Fitt, B.D.L. (2001) Epidemiology and management of *Leptosphaeria maculans* (phoma stem canker) on oilseed rape in Australia, Canada and Europe. *Plant Pathology* 50, 10–27.

West, J.S., Fitt, B.D.L., Leech, P.K., Biddulph, J.E., Huang, Y.-J. *et al.* (2002a) Effects of timing of *Leptosphaeria maculans* ascospore release and fungicide regime on phoma leaf spot and phoma stem canker development on winter oilseed rape (*Brassica napus*) in Southern England. *Plant Pathology* 51, 454–463.

West, J.S., Balesdent, M.-H., Rouxel, T., Narcy, J.P., Huang, Y.-J. *et al.* (2002b) Colonisation of winter oilseed rape tissues by A/Tox+ and B/Tox0 *Leptosphaeria maculans* (phoma stem canker) in France and England. *Plant Pathology* 51, 311–321.

References

Andreoli, C. and Maguire, J.D. (1995) Preliminary investigations on the viability of the use of abscisic-acid (aba) in crucifer seed health test for detecting Phoma-lingam. *Pesquisa Agropecuaria Brasileira* 30, 291–294.

Borhan, M.H., Wouw, A.P. and Larkan, N.J. (2022) Molecular interactions between *Leptosphaeria maculans* and *Brassica* species. *Annual Review of Phytopathology* 60, 237–257. DOI: 10.1146/annurev-phyto-021621-120602.

Evans, N., Baierl, A., Semenov, M.A., Gladders, P. and Fitt, B.D.L. (2007) Range and severity of a plant disease increased by global warming. *Journal of the Royal Society Interface* 5, 525–531. DOI: 10.1098/rsif.2007.1136.

Huang, Y.J., Underwood, C.T., Fitt, B.D.L., Todd, A.D., West, J.S. *et al.* (2001) Effects of temperature on germination and hyphal growth from ascospores of A-group and B-group *Leptosphaeria maculans* (phoma stem canker of oilseed rape). *Annals of Applied Biology* 139, 193–207. DOI: 10.1111/j.1744-7348.2001.tb00396.x.

Huang, Y.J., Liu, Z., West, J.S., Todd, A.D., Hall, A.M. *et al.* (2007) Effects of temperature and rainfall on date of release of ascospores of *Leptosphaeria maculans* (phoma stem canker) from winter oilseed rape (*Brassica napus*) debris in the UK. *Annals of Applied Biology* 151, 99–111. DOI: 10.1111/j.1744-7348.2007.00157.x.

Jacques, N., Balesdent, M.-H., Rouxel, T. and Laval, V. (2021) New specific quantitative real-time PCR assays shed light on the epidemiology of two species of the *Leptosphaeria maculans-Leptosphaeria biglobosa* species complex. *Plant Pathology* 70, 643–654.

Kaczmarek, J., West, J.S., King, K.M., Canning, G.G.M., Latunde-Dada, A.O. *et al.* (2023) Efficient qpcr estimation and discrimination of airborne inoculum of *Leptosphaeria maculans* and *L.*

biglobosa, the causal organisms of phoma leaf spotting and stem canker of oilseed rape. *Annual Review of Phytopathology*. DOI: 10.1002/ps.7800.

King, K.M. and West, J.S. (2021) Detection of the *Phoma pathogens Plenodomus biglobosus* subclades 'brassicae' and 'canadensis' on wasabi, and 'canadensis' in Europe. *European Journal of Plant Pathology*. DOI: 10.1007/s10658-021-02428-z.

Liu, S.Y., Liu, Z., Fitt, B.D.L., Evans, N., Foster, S.J. *et al.* (2006) Resistance to *Leptosphaeria maculans* (phoma stem canker) in *Brassica napus* (oilseed rape) induced by *L. biglobosa* and chemical defence activators in field and controlled environments. *Plant Pathology* 55, 401–412. DOI: 10.1111/j.1365-3059.2006.01354.x.

Mendes-Pereira, E., Balesdent, M.-H., Brun, H. and Rouxel, T. (2003) Molecular phylogeny of the *Leptosphaeria maculans–L. biglobosa* species complex. *Mycological Research* 107, 1287–1304. DOI: 10.1017/S0953756203008554.

Omer, Z.S. and Wallenhammar, A.C. (2020) Development of loop-mediated isothermal amplification assays for rapid detection of blackleg pathogens in Swedish winter oil seed rape. *European Journal of Plant Pathology* 157, 353–365. DOI: 10.1007/s10658-020-02004-x.

Peluola, C., Fernando, W.G.D., Huvenaars, C., Kutcher, H.R., Lahlali, R. *et al.* (2013) Effect of flooding on the survival of *Leptosphaeria* spp. in canola stubble. *Plant Pathology* 62, 1350–1356. DOI: 10.1111/ppa.12036.

Part IV: Diseases of Horticultural and Plantation Crops

Horticultural and plantation crops hold a unique place in global agriculture, comprising high-value commodities such as grapevine, coffee, oil palm, and berry species that drive important economic sectors while also posing distinct challenges in disease management. These perennial systems require significant capital investment and involve long periods between establishment and profit, making outbreaks of disease particularly damaging. Unlike annual crops, which can be rotated, perennial crops serve as continuous hosts for pathogens, allowing their establishment and evolution over many decades.

This section explores both historically significant diseases and modern emerging threats through seven case studies. Coffee leaf rust, caused by *Hemileia vastatrix*, exemplifies historical impact, having devastated early coffee cultivation and influenced global trade for over 150 years. Despite extensive research, this pathogen still threatens production due to breakdowns in genetic resistance and ongoing questions about optimal management strategies within evolving systems like organic and agroforestry methods. Emerging threats include *Xylella fastidiosa*, recently found in Mediterranean Europe, capable of infecting over 600 host species and adapting to various disease environments, making its spread hard to predict. Similarly, *Neoscytalidium dimidiatum* has emerged as a threat to pineapple production, causing internal black rot that is undetectable before fruit is shipped, demonstrating how opportunistic pathogens can exploit intensive cultivation systems.

Disease management in perennial crops is challenging. Grapevine downy mildew caused by *Plasmopara viticola* and grey mould from *Botrytis cinerea* highlight the reliance on fungicides, but both pathogens develop resistance, necessitating rotation of chemicals. Developing resistant cultivars through grapevine breeding with wild Vitis species shows promise, but long breeding cycles and the need to maintain quality limit progress. For oil palm with basal stem rot caused by *Ganoderma boninense*, no single control method exists, so integrated approaches are required to reduce disease during replanting.

Modern management practices are increasingly focusing on biological control and integrated strategies as alternatives to heavy chemical use. The challenge of applying such integrated management in perennial systems, especially when involving biocontrol agents and organic certification standards, highlights the complex decision-making process. This process involves balancing pest control with market demands, sustainability objectives, and economic limitations.

The case studies provide comprehensive insights into pathogen biology, historical impacts, current management strategies, and emerging research areas including resistance breeding, biocontrol development, and microbiome manipulation.

18 Management of *Plasmopara viticola*: From the Tradition to the Innovation

Marco Cesarini[1,2], Oscar Giovannini[2], Sabrina Sarrocco[3] and Puopolo Gerardo[1,2]

[1]*Center Agriculture Food Environment (C3A), University of Trento, San Michele all'Adige, Italy*
[2]*Research and Innovation Centre, Fondazione Edmund Mach, San Michele all'Adige, Italy*
[3]*Department of Agriculture, Food and Environment, University of Pisa, Pisa, Italy*

Introduction

The grapevine (*Vitis vinifera L.*) in 2021 produced 64.0 million tons of grapes globally, with 33.9 million dedicated to the production of wine (Available at: https://www.oiv.int/it). Millions of hectares are necessary for cultivation and production at this scale. Because of its importance as a crop and scale of cultivation, we must be able to deal with its high susceptibility to several diseases, including downy mildew, caused by *Plasmopara viticola* (Berk. & M.A. Curtis). This plant pathogen is not a fungus but an oomycete. The disease, originally endemic to American *Vitis* species, causes considerable yield and grape quality losses. It was introduced in Europe in the 19th century and quickly proved its impact in France by destroying about half of the grape production. More than a century later, grapevine downy mildew still regularly causes significant damage to vineyards, resulting in the need for frequent and precise chemical fungicide applications to control *P. viticola*. As the use of chemical plant protection products requires a substantial cost for producers and causes environmental damage, research is moving towards alternative methods for controlling *P. viticola* more sustainably, efficiently, and economically.

The Plant Pathogen: What is *Plasmopara Viticola*?

Plasmopara viticola is a biotrophic organism that can only develop when associated with its host plant (*V. vinifera*) and has a life cycle that includes both sexual and asexual phases. During the sexual phase, the antheridium (male sexual structure) fuses with the wall of oogonium (female sexual structure) inside the plant host tissues, forming the resting spore named oospore (Fig. 18.1).

Oospores are dispersed via dead leaf tissues into leaf litter, entering a dormancy stage. Once mature, oospores will end dormancy and germinate in spring (optimum 25°C) after periods of rain by producing and releasing macrosporangia, responsible for the primary infection of grapevine green tissues. Once they arrive on grapevine plant tissues, macrosporangia will release biflagellate zoospores that can swim on the plant tissue surfaces and encyst into stomata in the presence of a layer of water. Under favourable conditions, encysted zoospores produce germinative tubes that penetrate plant tissues and form mycelium at the intercellular level. Once the network is established, the plant pathogen establishes a biotrophic interaction with the host cell by absorbing plant nutrients from the cytoplasm (Figs 18.2 and 18.3). It takes six to ten days from initial infection for sporangiophores to emerge from stomata to release sporangia for secondary infection cycles (Fig. 18.1). The sporulation is visible as a dense cottony-white mold on the abaxial surface of grapevine leaves, green shoots, and young berries (Figs 18.2 and 18.3). In case of sporulation on grapevine leaves, oily spots occur on the adaxial surface region corresponding to the sporulation site (Fig. 18.2b).

The most impactful infections of *P. viticola* occur on grapevine bunches. The symptoms that can be visualized on grapevine bunches are the presence of sporulation (white mold) on shoots and young grapevine berries, the deformation of shoots and the rachis, and berry death (Fig. 18.3, Rumbolz *et al.*, 2002; Palliotti *et al.*, 2010; Koledenkova *et al.*, 2022).

Breeding Approach to Select Grapevine Varieties Resistant to *Plasmopara Viticola*

European *V. vinifera* cultivars are highly susceptible to *P. viticola*. At the same time, several *Vitis* species from America (*e.g. Vitis rotundifolia*, *Vitis rupestris*, *Vitis labrusca*, *Vitis riparia*, and *Vitis cinerea*) and Asia (*e.g. Vitis amurensis*, *Vitis piasezkii*, and *Vitis coignetiae*) show different levels of resistance to this plant pathogen (Reynolds, 2015; Sargolzaei *et al.*, 2020). The selection of new grapevine varieties resistant to downy mildew helps manage the disease and the economic impact. Indeed, it reduces the need for fungicide treatments for each growing season, decreases labour, and increases grape yields. Despite these advantages, consumers perceive wines from grapevine-resistant varieties as inferior quality. Additionally, several of these grapevine varieties are often excluded from the production regulations of many protected

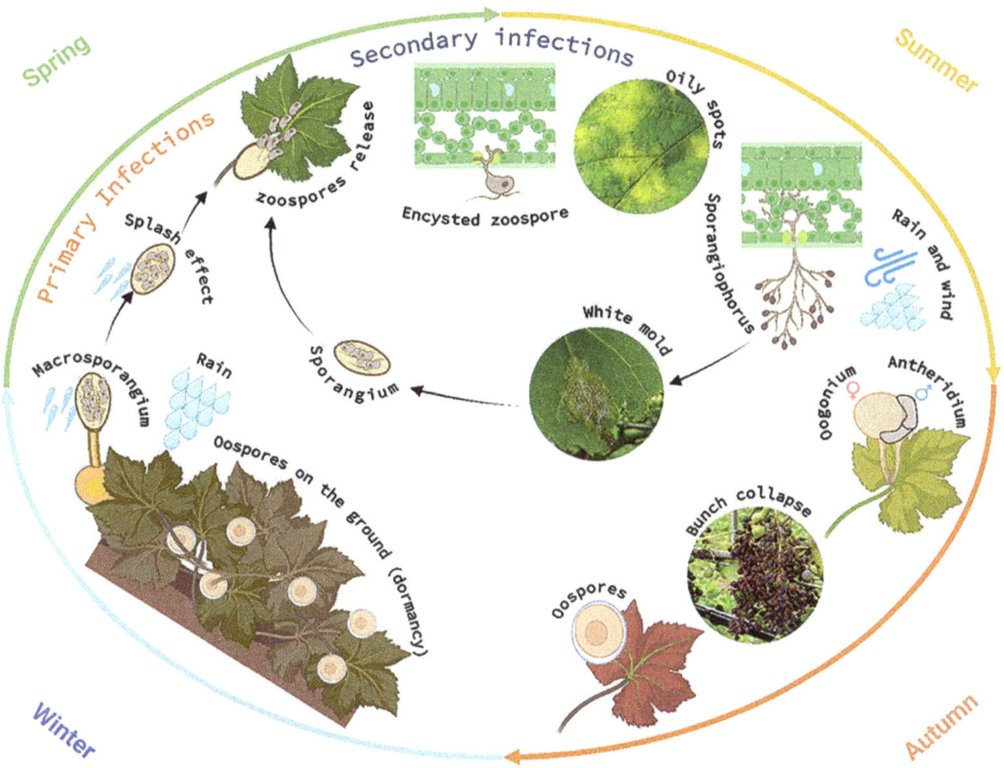

Fig. 18.1. Disease cycle of *Plasmopara viticola* following the seasonality. Primary infections occurring in spring are linked to *macrosporangia* produced by *oospores*. These spores are projected from the ground to young grapevine leaves and/or grapevine green tissues through the *rain splash* in spring. Once on the leaves, macrosporangia release *biflagellate zoospores* that swim on the abaxial surface of grapevine toward the stomata where they lose flagella and *encyst*. The cysts produce the *germinative tubes* that colonise leaf parenchyma and suck the cellular nutrients producing haustoria. The intercellular colonization and the sucking of nutrients will result in the typical *oily spot symptom*. With favourable environmental conditions, *P. viticola* evades from grapevine leaves through stomata producing *sporangiophores* that bear *sporangia*. The evasion corresponds to the classical *cottony-white mold symptom* on the abaxial grapevine leaf surface. Secondary infections that occur from spring to summer are caused by the sporangia that are transported through the rain and wind on other grapevine green tissues. Once arrived, sporangia release biflagellate zoospores that will infect the grapevine green tissues as reported above. Infections on grapevine bunches grapevine bunches produce *shriveled berries* and *bunch death*. At the end of summer and beginning of autumns, *sexual reproduction* occurs on old grapevine leaves, and the final product will be represented by oospores that overwinter in plant debris fallen on the ground (Created in BioRender).

denominations of origin, a key feature for quality wine production (Fuller *et al.*, 2014; Espinoza *et al.*, 2018). New Plant Breeding Technologies (NBTs) could be applied to grapevine to modify grapevine varieties included in protected denominations of origin (Purnhagen and Wesseler, 2021). Despite

Fig. 18.2. Grapevine downy mildew symptoms on grapevine leaves. Symptoms caused by *Plasmopara viticola* are characterized by a *cottony-white mold* on the abaxial surface of grapevine leaves (a); *yellow blotches*, defined as *oil spots*, on the adaxial grapevine leaf surface occurring on the same spot of the sporulation (b).

their potential, NBTs face several constraints, mainly due to juridical aspects, especially in Europe. Indeed, NPBTs are defined by multiple legislative documents – Directive 2001/18 (n.d.); Regulation (EC) 1829 (2003); Regulation (EC) 1830 (2003); Directive (EU) 2015/412 (n.d.) that make the procedures for producing and selling Genetically Modified Organism products complicated and long in Europe. Thus, the development of new cultivars developed by NBTs for cultivation must go through an approval process (long, costly, and uncertain), which results in a reduced number of farmers and industries that take this risk (Purnhagen and Wesseler, 2021).

The Tradition in Grapevine Downy Mildew Management: Prevention of *Plasmopara viticola* Infections Using Chemical Plant Protection Products

Despite the potential of grapevine breeding, the restriction of the market and consumer choice makes the use of chemical plant protection products the most reliable strategy to control *P. viticola* in grapevine-producing countries. For instance, a significant portion of chemical plant protection products used in Europe—around 70%—is used to control *P. viticola* through multiple preventive applications during a single growing season, from bud break to the onset of the ripening of the grapes.

In organic viticulture, the use of copper-based plant protection products is currently unavoidable due to their preventive properties. However, the

Fig. 18.3. Grapevine downy mildew symptoms occurring on grapevine bunches and shoots. Occurrence of sporulation on *purplish-brown grapevine* and *infected distorted inflorescences* covered with sporulation. (a) *Sporulation* on the surface of young grapevine berries. (b) Bunch partially attacked by *P. viticola* showing *shriveled berries*. (c) grapevine bunches with *shriveled berries* after a heavy infection by *P. viticola*.

environmental and health impacts of copper, including soil and groundwater accumulation and potential human toxicity, underscore the importance of reducing its application in agriculture (Stern *et al.*, 2007). Predictive systems utilizing microclimate data have been developed for this purpose, which can improve application efficiency and reduce unnecessary use (Fedele *et al.*, 2024). A promising recent innovation is the control of downy mildew using a novel leaf-fertilizer Aqua-Hort® which contains a low amount of copper, reducing the pollution deriving from the use of copper-based fungicide (Behrendt *et al.*, 2024).

Several multi- and single-site synthetic plant protection products offer long-lasting control of grapevine downy mildew. However, the use of synthetic plant protection products has some limitations. For instance, the mode of action of single-site synthetic plant protection products increases the risk of the development of resistance in *P. viticola* populations, making their applications unsuccessful (Massi *et al.*, 2023). Moreover, the negative impact of synthetic plant protection products on human health and the environment led to a constant withdrawal of main active ingredients from the market. Nonetheless, these negative aspects, such as chemical plant protection products, remain a cornerstone in the management of downy mildew in vineyards.

The Innovation in Grapevine Downy Mildew Management: Development of Bioproducts Alternative to Chemicals

Due to the impact of the chemical input in viticulture, it is envisaged that wine growers will face the necessity to have the availability of new products that could be employed in the management of grapevine downy mildew. The challenge to find alternatives to chemicals has been embraced by several research groups and companies. However, alternative products registered for the control of *P. viticola* are not available yet in some grapevine-growing areas, such as Europe. The lack of registered products makes the management of grapevine downy mildew a case study, especially in Europe.

The following paragraphs will be dedicated to providing evidence of potential candidates ready to be developed and registered as new bioproducts that can replace chemicals in the control of *P. viticola*.

Use of plant extracts to control *Plasmopara viticola*

Plant extracts are receiving substantial interest for their use as potential plant protectants. They generally exhibit a significant interference activity in disease development for several plant pathogens, including *P. viticola*. For example, Chen *et al.* (2002) evaluated 58 plant extracts derived from *Chloris vigata*, *Dalbergia hupeana*, *Pinus massoniana*, *Paeonia suffruticosa*, and *Robinia pseudoacacia* that showed an inhibitory effect on *P. viticola* sporangia germination higher than 80%. Similarly, Andreu *et al.* (2018) reported ethanol extracts of *Artemisia absinthium*, *Artemisia vulgaris*, *Equisetum arvense*, and *Salix alba* as inhibitors of sporangial germination and described a just visible sporulation on leaf discs when treated after infection. *Glycyrrhiza glabra* leaf extract was effective in controlling *P. viticola* on grapevie leaf discs, while *Yucca schidigera* and *Abies sibirica* extracts were slightly less effective (La Torre *et al.*, 2019).

Finally, species belonging to the Pinaceae family attracted interest in plant protection for their characteristics. Compounds derived from the bark of *Larix decidua* showed a 60% efficacy when used independently, and this efficacy increased to 80% when combined with a low-copper strategy. Additionally, extracts from *Picea abies* drastically reduced *P. viticola* attacks on grapevine leaf discs, while *Pinus pinaster* node extracts effectively impeded the mobility and development of the pathogen's zoospores (Thuerig *et al.*, 2018).

Data on the efficacy of plant extracts against *P. viticola* indicate they can be considered a valuable source of new active ingredients to be developed as an alternative to copper and synthetic chemical plant protection products. However, there are still crucial points that need to be resolved to achieve efficient disease control using plant extracts. Overall, results refer to controlled experimental conditions that still have high production costs, limited availability in large quantities, and, above all, poor persistence and tolerance to rain (Dagostin *et al.*, 2011). However, it is still necessary to stabilize the plant extracts for their use in the field, providing sufficient plant protection at a reasonable cost to the farmers.

Development of microbial biocontrol agents for the control of *Plasmopara viticola*

The use of plant-beneficial fungi and bacteria in agriculture has captured significant interest due to their potential to serve as effective alternatives to chemical plant protection products. In particular, bacterial and fungal biocontrol agents have demonstrated their ability to restrict plant pathogens through various modes of action, including mycoparasitism, competition, antibiosis, and induction of resistance in plants (Sarrocco *et al.*, 2015a).

Fungal biocontrol agents, encompassing both filamentous fungi and yeasts, stand out as a significant group of organisms with the potential to curb plant pathogens' growth and enhance plant health. Their unique characteristics make them a compelling choice as potential bioactive ingredients of commercial plant protection bioproducts (Sarrocco *et al.*, 2015b).

Among the most promising candidates, using endophytic fungi such as *Acremonium persicinum* and *Acremonium sclerotigenum* is noteworthy. The latter actively colonizes grapevine tissues, and parasitizes *P. viticola*, and inhibits sporangia germination (Burruano *et al.*, 2008; Lo Piccolo *et al.*, 2015).

Several *Trichoderma* spp. have shown promising results as biocontrol agents. Notably, *T. harzianum* T39 has demonstrated an impressive suppression of *P. viticola* by inducing systemic resistance in grapevine plants (Perazzolli *et al.*, 2011). Other *Trichoderma* species, such as *T. asperellum* and *T. atroviride*, produce volatile organic compounds that significantly reduce the severity of *P. viticola* infection by inducing callose deposition and enhancing the modulation of defence genes (Lazazzara *et al.*, 2021). *Trichoderma harzianum* HL1 and *T. viride* HL5 were found to reduce downy mildew severity under field conditions while promoting plant growth and berry quality (El-Sharkawy *et al.*, 2018).

In the case of bacterial biocontrol agent, several strains from different species have also shown a relevant efficacy against *P. viticola*. For example, *Bacillus subtilis* GLB191 and *B. pumilus* GLB197 reduced disease incidence by over 50% on grapevine leaves (Zhang *et al.*, 2017). Some bacteria have shown ability to control *P. viticola* indirectly. Thus *Ochrobacterium* sp. SY286 reduced disease severity by 90% by applying its fermentation liquid to leaves (Zang *et al.*, 2020). Finally, in a field study conducted by the application of *Streptomyces* spp. isolates resulted in a nearly 70% reduction in grapevine downy mildew severity (El-Sharkawy *et al.*, 2018).

Among bacterial biocontrol strains effective in controlling *P. viticola*, *Lysobacter capsici* AZ78 is one of the most promising, as it is possible to combine it with low doses of copper-based plant protection products (Puopolo *et al.*, 2014). At a genome level, *L. capsici* AZ78 has genes involved in producing antibiotics, lytic enzymes, and siderophores. From an ecological point of view, it can tolerate UV light irradiation, thermal shock, and extended periods without nutrients (Puopolo *et al.*, 2016). These elements render it an ideal candidate for developing a plant protection bioproduct against *P. viticola*. A first prototype formulation of *L. capsici* AZ78 was designed and evaluated under controlled conditions and in small-scale field trials (Segarra *et al.*, 2015, Segarra *et al.*, 2016). Markellou *et al.* (2022) recently showed that the application of *L. capsici* AZ78 formulation had the same effectiveness as a copper-based plant protection product. Moreover, the application of *L. capsici* AZ78 had no impact on the viability of *Saccharomyces cerevisiae* and the fermentation of red and white wine, thus maintaining the qualitative parameters unchanged. Importantly, for future registration of *L. capsici* AZ78 as the active ingredient of a commercial plant protection bioproduct, the bacterial strain did not show any impact or mortality on non-target arthropod and crustacean species, indicating that it does not affect the ecosystem's non-target organisms (Markellou *et al.*, 2022).

Discussion

Plasmopara viticola still represents one of the most serious threats for wine growers worldwide despite all the efforts made to develop strategies to reduce its negative impact on grapevine production (Koledenkova *et al.*, 2022). Reduction of the copper input in agriculture (Stern *et al.*, 2007), the occurrence of *P. viticola* populations resistant to chemical plant protection products (Ghule *et al.*, 2020; Massi *et al.*, 2023), the withdrawal from the market of chemical plant protection products (Massi *et al.*, 2023), and constraints in the application of NBTs for the generation of grapevine varieties resistant to downy (Purnhagen and Wesseler, 2021) make the development of plant protection bioproducts a case study that needs attention of the scientific community.

Several plant extracts and microbial biocontrol agents have been evaluated for their efficacy in controlling *P. viticola*, achieving satisfactory results, especially in controlled conditions (Dagostin *et al.*, 2011). Some of the tested plant extracts and microbial biocontrol agents have been evaluated in the vineyards and showed efficacy levels similar to the reference chemical plant protection products (Koledenkova *et al.*, 2022). Despite the relevant efforts made by companies and the research community in developing new plant protection bioproducts, few and/or no (as in the case of Europe) commercial product is registered for the management of grapevine downy mildew.

The path to developing new innovative plant protection bioproducts is still long. However, the work carried out to find new potential candidates for controlling *P. viticola* is a sound basis for the future management of this plant pathogen requiring a lower input of chemical plant protection products.

Group Discussion Questions

1. When should plant protection products be applied to control *Plasmopara viticola* effectively?
2. What features of the *Plasmopara viticola* life cycle suggest intervention points for controlling or managing this disease?
3. What are the constraints of the application of gene editing to make grapevine varieties resistant to grapevine downy mildew?
4. Would chemical plant protection products be sufficiently available to control *Plasmopara viticola* in the future?
5. Why has no plant extract and or biocontrol agent been registered in Europe for the control of *Plasmopara viticola*?

Further Readings

Collinge, D.B., Jensen, D.F., Rabiey, M., Sarrocco, S., Shaw, M.W. *et al.* (2022) Biological control of plant diseases – What has been achieved and what is the direction? *Plant Pathology* 71, 1024–1047.

Markellou, E., Kapaxidi, E., Karamaouna, F., Samara, M., Kyriakopoulou, K. *et al.* (2022) Recently showed that the application of *L. capsici* AZ78 formulation had the same effectiveness as copper based fungicides in a field trial carried out in Greece. *Biocontrol Science and Technology* 32(8), 930–951. DOI: 10.1080/09583157.2022.2064431.

Salotti, I., Caffi, T., Fedele, G. and Rossi, V. (2024) Resistant grapevine varieties to Downy Mildew. *Plant Health Cases.* DOI: 10.1079/planthealthcases.2024.0008.

Tronsmo, A.M., Collinge, D.B., Djurle, A., Munk, L., Yuen, J. *et al.* (2020) *Plant pathology and plant diseases.* CAB International, Wallingford, UK.

References

Andreu, V., Levert, A., Amiot, A., Cousin, A., Aveline, N. *et al.* (2018) Chemical composition and antifungal activity of plant extracts traditionally used in organic and biodynamic farming. *Environmental Science and Pollution Research International* 25(30), 29971–29982.

Behrendt, R.A., Toldam-Andersen, T.B., Arneborg, N., Collinge, D.B. and Jensen, B. (2024) Wine yeast as potential biological control agent against downy mildew of grapevine

(*Plasmopara viticola*). In: *Proceedings of the 21st International Conference on Organic Fruit-Growing, Filderstadt*, Germany, pp. 51–57.

Burruano, S., Alfonzo, A., Lo Piccolo, S., Conigliaro, G., Mondello, V. *et al.* (2008) Interaction between *Acremonium byssoides* and *Plasmopara viticola* in *Vitis vinifera*. *Plant Phytopathologia Mediterranea* 47(2), 122–131.

Chen, J., Dai, G., Gu, Z. and Miao, Y. (2002) Inhibition effect of 58 plant extracts against grape downy mildew (*Plasmopara viticola*). *Natural Product Research and Development* 14(5), 9–13.

Dagostin, S., Schirer, H.J., Pertot, I. and Tamm, L. (2011) Are there alternatives to copper for controlling grapevine downy mildew in organic viticulture? *Crop Protection* 30(7), 776–788.

Directive (EU) 2001/18 (n.d.) Council of 12 March 2001 amending Directive 2001/18/EC on the deliberate release into the environment of genetically modified organisms and repealing Council Directive 90/220/ EEC.

Directive (EU) 2015/412 (n.d.) Council of 11 March 2015 amending Directive 2001/18/EC as regards the possibility for the Member States to restrict or prohibit the cultivation of genetically modified organisms (GMOs) in their territory.

El-Sharkawy, H.H.A., Abo-El-Wafa, T. and Ibrahim, S. (2018) Biological control agents improve the productivity and induce the resistance against downy mildew of grapevine. *Journal of Plant Pathology* 100(1), 33–42.

Espinoza, A.F., Hurbert, A., Raineau, Y., Franc, C. and Giraud-Héraud, É. (2018) Resistant grape varieties and market acceptance: An evaluation based on experimental economics. *OENO One* 52(3), 247–263.

Fedele, G., Salotti, I., Caffi, T. and Rossi, V. (2024) Plant health cases, CABI, multi-modelling approach to support decision making in crop protection. *Plant Health Cases* 2024, phcs20240004.

Fuller, K.B., Alston, J.M. and Sambucci, O.S. (2014) The value of powdery mildew resistance in grapes: Evidence from California. *Wine Economics and Policy* 3(2), 90–107.

Ghule, M.R., Sawant, I.S., Sawant, S.D. and Saha, S. (2020) Resistance of *Plasmopara viticola* to multiple fungicides in vineyards of Maharashtra. India. *Journal of Environmental Biology* 41(5), 1026–1033.

Koledenkova, K., Esmaeel, Q., Jacquard, C., Nowak, J., Clément, C. *et al.* (2022) *Plasmopara viticola* the causal agent of downy mildew of grapevine: From its taxonomy to disease management. *Frontiers in Microbiology* 13, 889472.

La Torre, A., Righi, L., Tovino, V. and Battaglia, V. (2019) Evaluation of copper alternative products to control grape downy mildew in organic farming. *Journal of Plant Pathology* 101(4), 1005–1012.

Lazazzara, V., Vicelli, B., Bueschi, C., Parich, A., Pertot, L. *et al.* (2021) *Trichoderma* spp. volatile organic compounds protect grapevine plants by activating defense-related processes against downy mildew. *Physiologia Plantarum* 172(4), 1950–1965.

Lo Piccolo, S., Alfonzo, A., Giambra, S., Conigliaro, G., Lopez-Llorca, L.V. *et al.* (2015) Identification of acremonium isolates from grapevines and evaluation of their antagonism towards *Plasmopara viticola*. *Annals of Microbiology* 65(4), 2393–2403.

Markellou, E., Kapaxidi, E., Karamaouna, F., Samara, M., Kyriakopoulou, K. *et al.* (2022) Evaluation of plant protection efficacy in field conditions and side effects of *Lysobacter capsici* AZ78, a biocontrol agent of plasmopara viticola. *Biocontrol Science and Technology* 32(8), 930–951.

Massi, F., Torriani, S.F., Waldner-Zulauf, M., Bianco, P.A., Coatti, M. *et al.* (2023) Characterization of Italian *Plasmopara viticola* populations for resistance to oxathiapiprolin. *Pest Management Science* 79(3), 1243–1250.

Palliotti, A., Poni, S., Berrios, J.G. and Bernizzoni, F. (2010) Vine performance and grape composition as affected by early-season source limitation induced with anti-transpirants in two red *Vitis vinifera* L. cultivars. Grape and Wine Research 16(3), 426–433.

Perazzolli, M., Roatti, B., Bozza, E. and Pertot, I. (2011) Trichoderma harzianum T39 induces resistance against downy mildew by priming for defense without costs for grapevine. Biological Control 58, 1–7482.

Puopolo, G., Giovannini, O. and Pertot, I. (2014) *Lysobacter capsici* AZ78 can be combined with copper to effectively control *Plasmopara viticola* on grapevine. *Microbiological Research* 169(7–8), 633–642.

Puopolo, G., Tomada, S., Sonego, P., Moretto, M., Engelen, K. *et al.* (2016) The *Lysobacter capsici* AZ78 genome has a gene pool enabling it to interact successfully with phytopathogenic microorganisms and environmental factors. *Frontiers in Microbiology* 7, 96.

Purnhagen, K. and Wesseler, J. (2021) EU regulation of new plant breeding technologies and their possible economic implications for the EU and beyond. *Applied Economic Perspectives and Policy* 43, 1621–1637.

Regulation (EC) 1829 (2003) EC (European Commission) of the European Parliament and of the Council of 22 September 2003 on Genetically Modified Food and Feed.

Regulation (EC) 1830 (2003) EC (European Commission) of the European Parliament and of the Council of 22 September 2003 concerning the traceability and labelling of genetically modified organisms and the traceability of food and feed products produced from genetically modified organisms and amending Directive 2001/18/EC.

Reynolds, A. (2015) *Grapevine Breeding Programs for the Wine Industry*. 1st edn. Elsevier, Amsterdam, The Netherlands, pp. 1–466.

Rumbolz, J., Wirtz, S., Kassemeyer, H.-H., Guggenheim, R., Schäfer, E. *et al.* (2002) Sporulation of *Plasmopara viticola*: Differentiation and light regulation. *Plant Biology* 4(3), 413–422.

Sargolzaei, M., Maddalena, G., Bitsadze, N., Maghradze, D., Bianco, P.A. *et al.* (2020) *Rpv29, Rpv30* and *Rpv31*: Three novel genomic loci associated with resistance to *Plasmopara viticola* in *Vitis vinifera*. *Frontiers in Plant Science* 11, 562432.

Sarrocco, S., Diquattro, S., Avolio, F., Cimmino, A., Puntoni, G. *et al.* (2015a) Bioactive metabolites from new or rare fimicolous fungi with antifungal activity against pathogenic fungi. *European Journal of Plant Pathology* 142(1), 61–71.

Sarrocco, S., Diquattro, S., Baroncelli, R., Cimmino, A., Evidente, A. *et al.* (2015b) A polyphasic contribution to the knowledge of auxarthron (onygenaceae). *Mycological Progress* 14(11), 112.

Segarra, G., Puopolo, G., Giovannini, O. and Pertot, I. (2015) Stepwise flow diagram for the development of formulations of non-spore-forming bacteria against foliar pathogens: The case of *Lysobacter capsici* AZ78. *Journal of Biotechnology* 216, 56–64.

Segarra, G., Puopolo, G., Porcel-Rodríguez, E., Giovannini, O. and Pertot, I. (2016) Monitoring *Lysobacter capsici* AZ78 using strain specific qpcr reveals the importance of the formulation for its survival in vineyards. *FEMS Microbiology Letters* 363(3), fnv243.

Stern, B.R., Solioz, M., Krewski, D., Aggett, P., Aw, T.C. *et al.* (2007) Copper and human health: Biochemistry, genetics, and strategies for modeling dose-response relationships. *Journal of Toxicology and Environmental Health* 10(3), 157–222.

Thuerig, B., James, E.E., Schärer, H.-J., Langat, M.K., Mulholland, D.A. *et al.* (2018) Reducing copper use in the environment: The use of larixol and larixyl acetate to treat downy mildew caused by *Plasmopara viticola* in viticulture. *Pest Management Science* 74, 477–488.

Zang, C., Lin, Q., Xie, J., Lin, Y., Zhao, K. *et al.* (2020) The biological control of the grapevine downy mildew disease using *Ochrobactrum* sp. *Plant Protection Science* 56(1), 52–61.

Zhang, X., Zhou, Y., Li, Y., Fu, X. and Wang, Q. (2017) Screening and characterization of endophytic *Bacillus* for biocontrol of grapevine downy mildew. *Crop Protection* 96, 173–179.

19 Coffee Leaf Rust: Wreaking Havoc in Coffee Production Areas Across the Tropics

ATHINA KOUTOULEAS

Department of Geosciences and Natural Resource Management, University of Copenhagen, Copenhagen, Denmark

Introduction

"Civilisation has produced only three important non-alcoholic beverages—the extract of the tea plant, the cocoa bean, and the coffee bean." – William Ukers, 1921

The appropriately titled book *All about Coffee* (Ukers, 1922) proposed that the lure and universal appeal of coffee can be explained by a so-called twofold effect. First, the enjoyable, aromatic sensations of drinking coffee and second, the increased efficiency which provides its recipient via caffeine stimulation. It is no wonder that coffee is virtually ubiquitous in kitchens, cafes and in the hands of pedestrians on just about every street corner. Ethiopia may be the widely accepted origin of the oldest known species of coffee, *Coffea arabica*, but Yemen is thought to be the birthplace of consuming coffee in the beverage form we know today (Friis, 2015). Early observations from Ethiopia described the invigorating effects of chewing roasted coffee beans as "better than a loaf of bread or a meal of meat" (Friis, 2015). In Yemen, coffee as a hot drink went by at least three Arabic names: "bunna", "gahve" and "geschr" and was said to be drunk by people at any time of day (Friis, 2015). Nowadays, most coffee drinkers could not imagine going a day without their cup of "liquid gold". This means that coffee is a big business; annual global earnings exceed US$200 billion. Consequently, any pest and disease which impact coffee yield are of keen interest to the farmer and the scientific community at large.

Coffee leaf rust (CLR) is one of the most significant diseases to strike coffee production. Due to its widespread prevalence across all global coffee-growing regions and the calamity it brings in terms of yield reduction, CLR is by far the most studied coffee disease. CLR damages coffee plants by causing heavy leaf defoliation, leading to dramatic yield losses (up to 75%) and in

Fig. 19.1. Severe leaf defoliation caused by coffee leaf rust can lead to plant death. Courtesy of Dr Melanie Bordeaux, director of NICAFRANCE Foundation's research centre, Nicaragua.

some cases plant death (Fig. 19.1). Countries plagued by CLR suffer financial losses at the national level, which have trickle-down effects on poor farmers and, in extreme circumstances, has led to the mass abandonment of coffee farms. In fact, estimates suggest that the most recent epidemic of CLR cost up to US$1 billion in losses and cut 250,000 jobs, mainly on farms.

CLR was first officially discovered on uncultivated, wild coffee off the shores of Lake Victoria (Kenya) in 1861 by a British explorer (Koutouleas

Fig. 19.2. Tea, coffee and chocolate were once considered as luxury imports by the British. Courtesy of British Museum (1750–1760). Obtained from the British Museum, Public domain, via Wikimedia Commons. https://commons.wikimedia.org/wiki/File:Print,_trade-card_(BM_Heal,68.45).jpg

et al., 2019). Yet it was not until 1869 that the disease reached the public eye after it reaped havoc across several colonial coffee plantations in Sri Lanka (then called Ceylon). The British colonialists were so taken aback by CLR in their fields that they resorted to tearing out the coffee plants and replacing them with tea bushes instead.

Through this case study, the longstanding history of CLR will be considered and linked to key colonial trade activities (Fig. 19.2). The relevance of the disease will be highlighted briefly in terms of modern coffee farming practices using testimonies and data from Central American and East African coffee farms. Future prospects of coffee farming in new ways to combat disease pressure by *H. vastatrix* will also be explored.

Causal Agent: *Hemileia Vastatrix*

Hemileia vastatrix Berk. & Broome is an obligate parasitic fungus in family Pucciniaceae (Koutouleas *et al.*, 2019). This means that the fungus cannot survive in the absence of its plant host. Optimal environmental conditions for the development of *H. vastatrix* are 21–25°C air temperature with 100%

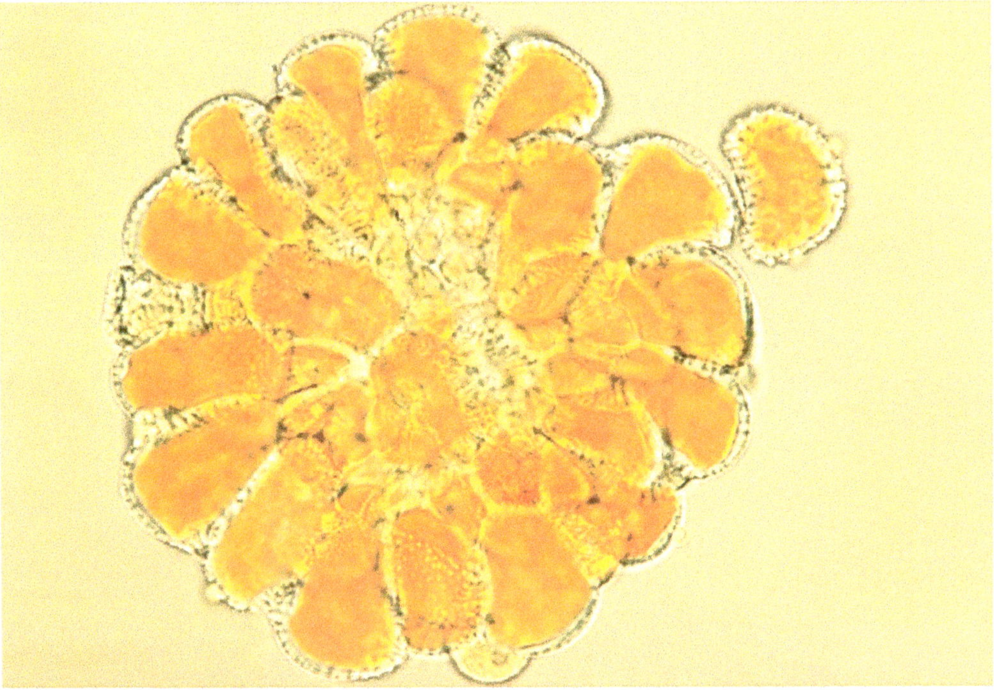

Fig. 19.3. Uredinium of *H. vastatrix* budding orange rust-coloured urediniopores (bar = 20 µm). Courtesy of Carvalho *et al.* (2011), CC BY 2.5 via Wikimedia Commons. https://commons.wikimedia.org/wiki/File:Hemileia_vastatrix_Uredinium_02.png

relative humidity. CLR infection peaks, therefore, tend to occur immediately after the peak of the rainy season across tropical and sub-tropical regions. Under optimal environmental conditions, a fungal germ tube of *H. vastatrix* penetrates via stomatal openings on the underside of the coffee leaf. Upon entry, *H. vastatrix* grows nutrient-absorbing mycelium within the leaf mesophyll cells using haustoria to invaginate the mesophyll cells to obtain nutrients. Vibrant bouquet-shaped, orange uredinia and telia sporulate from the stomata (Figs 19.3 and 19.4).

Uredinia give rise to urediniospores, dikaryotic cells (containing two unfused nuclei), which can survive up to six weeks on detached coffee leaf tissue. Under cool, dry conditions, the telia give rise to two more spore types: teliospores and subsequently basidiospores. Teliospores are two-celled, thick-walled and produce basidia; which then develop four haploid basidiospores—the sexual spores of *H. vastatrix*.

By producing both asexual and sexual spore types, rust fungi such as *H. vastatrix* increase their chances of transmission via multiple hosts and across the seasons. For this reason, many rusts are observed to have complex disease cycles with different spore types or reproductive structures (Koutouleas *et al.*, 2019).

Fig. 19.4. Inspecting the underside of coffee leaves for early signs of infection on an experimental coffee farm in Costa Rica (Alajuela). Photo courtesy of the author.

The covert and complex nature of the fungal attack lies in the fact that urediniospores adhere to and enter via the underside of the leaf (where stomata are most numerous). This means that the early onset of the disease can go largely undetected to the naked eye. When traversing a coffee field infected by CLR, yellow pin spots on the upper leaf surface can be seen as a sign of disease. However, by the time this visual indicator catches the attention of a coffee worker, the fungus has already established its nutrient supply from the leaf tissues via penetration of the epidermal cells through the stomata.

Discovery of "Devastating Emily" (Ceylon, 1869)

During the 15th century, coffee plants were transported from their centre of origin in Africa, across the Red Sea to southern Arabia (Yemen) and meticulously cultivated along sloped terraces. Until the end of the 17th century, the world's limited supply of coffee was almost exclusively distributed by Yemeni merchants and traders (Friis, 2015). Ethnobotanists mark this time as the great Yemen coffee monopoly era, given that there was a careful watch on the export of any green coffee beans by Arabians (Ukers, 1922). In fact, any precious coffee beans leaving Yemen at the time were ordered to be steeped in boiling water or parched, in order to destroy any possibility of foreign germination of the cash crop. Yemen's coffee supply monopoly came to its end by the early 17th century, at which point the Dutch drove the Portuguese out of Ceylon (now Sri Lanka) and began coffee cultivation. The British then entered the coffee scene through their occupation of this tropical island in 1796. However, they only began formally cultivating coffee in 1825. In the 50 years thereafter, coffee was one of the island's leading export products. *H. vastatrix* first appeared in coffee fields in 1869. At first, the disease was dubbed by coffee planters and field workers as "Devastating Emily" due to the destruction it brought to the coffee fields. Little did they know that this plant disease would single-handedly ruin the coffee industry in Ceylon.

For more than three centuries coffee was cultivated by both the Arabian and Dutch colonialists without any reports of CLR. The onset of extensive cultivation methods (introduced by the British) provided the perfect substrate for the fungal pathogen *H. vastatrix* to manifest a serious epidemic, destroying coffee productivity and profitability. Enter the British tea industry.

At the time of the first major CLR outbreak, exporters and researchers, under secondment by the British government to Ceylon, were trying to figure out what was causing the coffee yield losses and plant death. Coffee planters and field staff surmised the onset of "Devastating Emily" was due to poor plant growth and humid environments. Cultural practices with the aim of bolstering plant vigour and growth were used as a result of these evaluations. When these efforts failed, a simple experiment by a tea planter James Taylor to helped re-write the history books of early coffee production. Taylor demonstrated that tea plants were completely "resistant" to the foliar disease and henceforth, the majority of coffee plants were ripped out of the ground and replaced with tea plants. These days, it is widely known that *H. vastatrix* only infects coffee (as its primary plant host) and no alternate plant host has been discovered – yet. Given this, Taylor's findings were somewhat simplistic, as just about any plant (other than coffee) would have been deemed "resistant" to CLR at that time. It is often said that this piece of history explains why the British drinks more tea than coffee, even today. Although this is considered to be more of a fable than a fact, since any drop in coffee supply could have easily been recouped from other British colonial territories growing coffee during the same period (i.e. Jamaica, Tanzania, northeastern Australia, etc.).

Scholars are still uncertain why "Devastating Emily" became so severe in Ceylon during the first major outbreak. One explanation could be that other outbreaks may have occurred but either went undetected or unreported by local plantation managers at the time. In fact, British colonial activities represent some of the most detailed records that we have today concerning trade and agriculture in the period of the first outbreak. Another explanation could be a change in wind patterns which helped to spread fungal spores across long distances. However, this explanation does not explain why CLR outbreaks were not detected across other Asian coffee-producing countries at the time. Intensive monoculture farming of the coffee crop in the Ceylon region no doubt led to a concentration in coffee plant substrate for *H. vastatrix*, further increasing the likelihood of such a significant outbreak (compared to coffee gardens and small-scale production employed in other regions). The increase in transportation routes of people and goods during the late 19th century is another likely entry mechanism for the primary inoculum of rust spores into coffee plantations.

Worldwide Spread of the "Big Rust"

After the debut of CLR in Ceylon, the fungus made its way across Asian and Oceania coffee-producing countries such as Australia, Fiji, India, Indonesia, Mauritius, Pakistan, Papua New Guinea, the Philippines and Vietnam. The coffee disease was then reported to have made a return to the mother continent of Africa with widespread outbreaks first in the East (e.g. Uganda, Kenya and Mozambique) and then in the West (e.g. Ivory Coast). Latin American coffee production was free from CLR until the 1970s when the first report was made in Brazil. The disease was coined the "big rust" after the significant outbreak period starting in 2008 in Colombia (McCook and Vandermeer, 2015). The epidemic spread north to Central America and Mexico by 2012–2013 and since 2014 has begun impacting coffee farms in Ecuador and Peru. In 2020, *H. vastatrix* reached Hawaii, the only country growing coffee that avoided CLR for more than a century. Today, CLR plagues all coffee-producing countries, albeit to varying degrees depending on climatic conditions, on-farm resources, de-forestation activities (in which trees acted as buffer zones between farms) and general coffee plant health. A group of smallholder coffee farmers in Nicaragua has recently declared that CLR continues to cause heavy disease pressure in the major coffee-growing region of Matagalpa. Irregular rainfall patterns coupled with unusually high average air temperatures could be leading to more sporadic and localized outbreaks, which are being largely controlled through chemical fungicide application (when accessible to the farmer). Recent studies from Ethiopia, the motherland of coffee, have shown that the fungal disease is increasing in severity over time across major coffee-growing areas of the Southern districts. A surge in CLR has been reported across Ethiopia, with an average incidence rate of 35.3%. Moreover, the disease has shown to be spread upward from lowland to mid and highland coffee-growing areas. In the neighbouring coffee growing country, Uganda, recent reports suggest that CLR has reduced average annual yield per hectare from 855 to 432 kg, representing a 49.5% decrease in productivity and income, as well as a gradual loss of *C. arabica* plants (due to high susceptibility to the disease). These are just a few developing countries that rely on coffee as a key export commodity and are therefore at the mercy of this devastating pathogen.

Current Best Management Practices for *Hemileia Vastatrix*

Carbohydrate production by photosynthesis in leaves ensures the maintenance, growth and reproduction of the coffee plant. When *H. vastatrix* sporulates heavily, coffee leaves are damaged by both the intracellular growth and the nutrient leaching of the fungus. In order to protect coffee leaves from CLR, some form of disease control should be considered. This is especially relevant during the fruit development phase which demands the presence of healthy plant foliage for carbohydrate supply (De Filho and Domian, 2019). Both chemical and cultural practices are recommended for the control and containment of CLR (De Filho and Domian, 2019). These include:

- Protective and systemic fungicides
- Coffee plant and shade tree pruning
- Use of resistant cultivars
- Renovation of old coffee farms with new plants
- Biological control agents

The use of multiple lines of defence against plant pests or pathogens is called an Integrated Pest-Disease Management system (IPDM) and coffee farmers benefit from employing such an approach when faced with CLR. Today, the primary mode of control recommended is a mixture of fungicides (triazoles and strobilurin) to control and prevent the disease (Table 19.1). Resistant and/more tolerant coffee varieties and hybrids have been identified and selectively bred, based on nine known resistance genes (called SH genes). Resistance genes SH1, SH2, SH4 and SH5 were discovered in the *C. arabica* species; SH6, SH7, SH8 and SH9 come from the *C. canephora* species and the SH3 gene originates from the less cultivated *C. liberica* species (De Filho and

Table 19.1. Systemic and protective fungicides recommended for control of CLR (adapted from De Filho and Domian, 2019). The dose should be applied to every 0.5 ha of coffee land area.

Technical name	Type	Grams or ml/liter
Copper oxide	Protective	2.5–3.5 g
Copper hydroxide	Protective	2.5–3.5 g
Copper oxychloride	Protective	4.5–5.0 g
Propiconazole	Systemic	1.25 ml
Triadimenol	Systemic	1.25 ml
Tebuconazol & Triadimenol	Systemic	1.75 ml
Cyproconazol & Carbendacin	Systemic	1.25 ml
Cyproconazol & Pyraclostrobin	Systemic	2.50 ml
Cyproconazol & Trifoxistrobin	Systemic	1.50 ml

Domian, 2019). Combined use of resistant coffee varieties or hybrids, with these SH genes, along with fungicide application can provide even better protection from severe CLR outbreaks. However, *H. vastatrix* has evolved to overcome these forms of control to varying degrees. Given this, other cultural practices are also recommended in an IPDM system for a coffee farm.

By pruning both the coffee plant and the shade trees which surround them, the coffee field benefits from better air circulation, lowering the inter-canopy relative humidity levels, which can hinder the sporulation of *H. vastatrix* urediniospores. The mere act of pruning also removes the possible build-up of primary spore inoculum on the coffee plant itself.

Selection of locally adapted and/or improved varieties, cultivars, or coffee hybrids which exhibit known resistance to CLR is considered as frontline protection from the pathogen. Locally adapted coffee types can exhibit better plant vigour, which, in turn, bolsters their overall defence when encountered with a disease such as CLR. Complete farm renovation or rehabilitation of old coffee plants with these vigorous and/or resistant varieties is a decision that coffee farmers can make only a few times in the long production cycles of the coffee crop. Therefore, the choice of the coffee variety grown is a critical decision to make and requires a cost-benefit analysis weighing up the suitability of the variety to the local environment, yield potential, niche market value of the beans, and pest and disease resistance.

Coffee plants that have high production of fruits tend to be more severely attacked in the subsequent year by CLR. This has been attributed to the transport of protective phenolic compounds from the leaves to the beans, leaving the evergreen leaves vulnerable to pathogen attack (De Filho and Domian, 2019). Given these findings, a holistic approach to developing an IPDM against CLR involves the fine balance of the coffee plant yielding capacity with the age of the plant (i.e. older plants will be less vigorous and therefore more prone to attack) along with the planting density in the field (i.e. high-density planting will lead to higher in-field relative humidity thereby promoting disease outbreaks). New biological control agents (BCAs) are being investigated (Table 19.2) which are environmentally friendly and show good potential for control of CLR (Koutouleas *et al.*, 2022). However, these BCAs still require large-scale testing for efficacy in the coffee field. Asides from the application of chemical fungicides, most practices recommended for the control of CLR are compatible with farming systems such as organic, bird-friendly and agroforestry.

Discussion

Can natural enemies control *H. vastatrix*?

Habitat complexity in farming coffee can lead to enhanced ecological pest-disease control and overall better plant health (compared to monoculture farming). By farming coffee together with different plant and tree species, a higher number of predators and natural enemies may be present in the coffee

Table 19.2. Summary of biological control agents (BCAs) tested against coffee leaf rust (adapted from Koutouleas *et al.*, 2022).

Known BCAs	Tested BCAs
Fungus: *Calonectria hemileiae, Lecanicillium lecanii, Mycodiplosis hemileiae* and *Trichoderma harzianum* Insect: *Ricoseius loxocheles* Gastropod: *Bradybaena similaris*	Bacterial isolates: *Bacillus cereus, Bacillus lentimorbus*, *Bacillus megaterium, Bacillus subtilis, Brevibacillus choshinensis, Cedecea davisae*, *Microbacterium testaceum Pectobacterium carotovorum, Pseudomonas sp*. and *Salmonella enterica* Fungal isolates: *Acremonium sp, Aspergillus sp, Cladosporium sp., Fusarium spp*. and *Penicillium sp*. Natural Polymer: Chitosan[a] (an amino polysaccharide obtained from crustaceans and insects)

[a]Chitosan is not technically a BCA but is rather considered a biopesticide.

Fig. 19.5. The hyperparasite *Lecanicilium lecanii* (white spores) attacking *H. vastatrix* (orange spores) on a coffee leaf. Courtesy of Dr Melanie Bordeaux, director of NICAFRANCE Foundation's research centre, Nicaragua.

field. Control of certain pests or pathogens can also be achieved through the timely introduction of microorganisms or insects which act as BCAs in the coffee field (Koutouleas *et al.*, 2022). BCAs may act through various mechanisms to control a pest or disease. These mechanisms include competition, antibiosis, hyperparasitism, or induced resistance. Table 19.2 highlights some of the BCAs which are known to target *H. vastatrix*. These organisms range from other fungi, bacteria and insects which can have an inhibitory effect on the growth, survival and/ or reproduction of *H. vastatrix*. However, there are challenges associated with disease control with BCAs such as off-target effects on other organisms in the coffee field or a lack of efficacy due to the presence of other enemies which may feed on or compete with the BCA (Fig. 19.5).

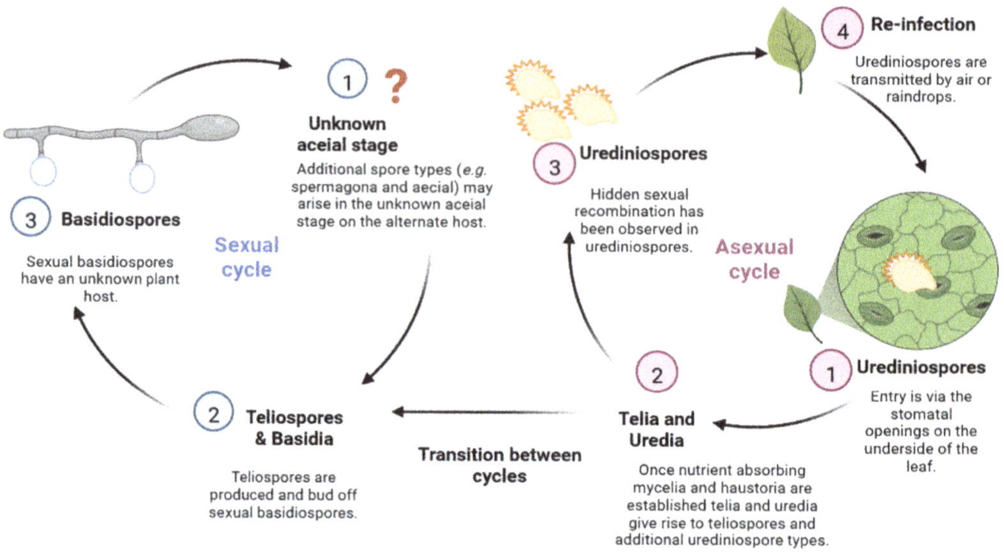

Fig. 19.6. Disease cycle of *Hemileia vastatrix*. The asexual cycle has been well documented, while the sexual cycle has been hypothesised to take place on an unknown alternate plant host. Courtesy of Koutouleas and Collinge (2022).

An alternate host or secret sex life?

H. vastatrix is considered a macrocyclic rust, which means it produces different spore types. Thus far, the role of the sexual basidiospores in the fungal life cycle remains unknown. Basidiospores are very small (around 5 microns). The baffling thing is, no one knows the target host these spores are intended for. It is hypothesised that *H. vastatrix* uses basidiospores to alternate from its primary host (the coffee plant) to an alternate plant host (unknown) (Fig. 19.6). This lifestyle pattern allows the fungus to "overwinter" and complete its sexual cycle on an alternate plant host when the season changes. However, in tropical coffee-growing regions, there is no real winter season, and since the first outbreak of coffee leaf rust in 1869, no alternate host has ever been detected in or around coffee plantations. With no known host to complete its sexual cycle, it is a mystery how the fungus can possibly create new, more virulent versions of itself, which are today plaguing our coffee production and therefore causing huge economic losses. A similar situation is known for stripe (yellow) rust of cereal crops and grasses (caused by *Puccinia striiformis*). The rust fungus was discovered in the mid-18th century, yet the alternate host plant (*Berberis* spp.) was not detected until 2010.

Some researchers postulate the basidiospores of *H. vastatrix* are remnants of an earlier rust ancestor and no longer utilised by the fungus. Others argue that the preservation of the basidiospores in the observed disease cycle is enough evidence for the possibility of a viable, alternate host. Recent scholars have uncovered a so-called "crypto-sexuality", in which *H. vastatrix* can asexually produce new versions of itself without the need for an alternate host (Carvalho *et al.*, 2011).

Despite apparently not completing its sexual cycle, the fungus is capable of producing new virulent variants. These new "races" hinder plant-breeding efforts. In addition, the fungus can develop resistance to fungicides.

Future coffee farming under agroforestry systems

Simply put, agroforestry means "farming with trees". However, this farming system is far from simple and the ecological interactions which lie within are largely unknown both from an agronomical and ecological point of view. A coffee agroforestry system is a farming approach that incorporates the growth of the coffee crop along with other shade trees and/or shrub species. The shade tree species are often fast-growing timber, fuel and/or fruit trees which can provide an alternate means of income for coffee farmers. Since *C. arabica* has its evolutionary origins in the Afromontane forest understorey, many coffee scientists share the opinion that the plant is shade-loving (or shade-tolerant to say the least). With global reforestation being a top priority, the use of agroforestry systems in coffee production has recently gained increased traction. Some would argue that traditional African farmers intuitively practiced different forms of agroforestry together with small-scale coffee production. However, since the expansion of coffee production across sub-tropical growing regions (e.g. Vietnam, Bolivia, etc.), coffee agroforestry systems have decreased in popularity.

Monoculture coffee farming in open fields with full sun exposure has been shown to increase short-term yields and easily scale-up production (compatible with modern agricultural technologies). But how does infection by "Devastating Emily" compare between a monoculture versus agroforestry coffee production system? Researchers have been elucidating the interactions between shade and CLR. Fig. 19.7 illustrates the key dynamics that have been identified to play a role in disease development in shaded coffee. Under the shade, coffee tends to produce a lower fruit load which in turn reduces the likelihood of rust infection. However, dense shade trees promote a higher overall relative humidity which allows for more leaf moisture and the potential for optimal water conditions for the sporulation of *H. vastatrix* urediniospores. The habitats under the shaded coffee system can also play a role in the disease development of CLR. More diverse plant species can support the presence of natural enemies of *H. vastatrix* such as the hyperparasitic fungus *Lecanicillium lecanii*. Given these dynamics, it is easy to see how the use of shade is a contentious issue for coffee farming.

Coffee leaf rust and climate change

Interactions between *H. vastatrix* and weather patterns have been linked to different stages of the disease cycle. However, predictions concerning

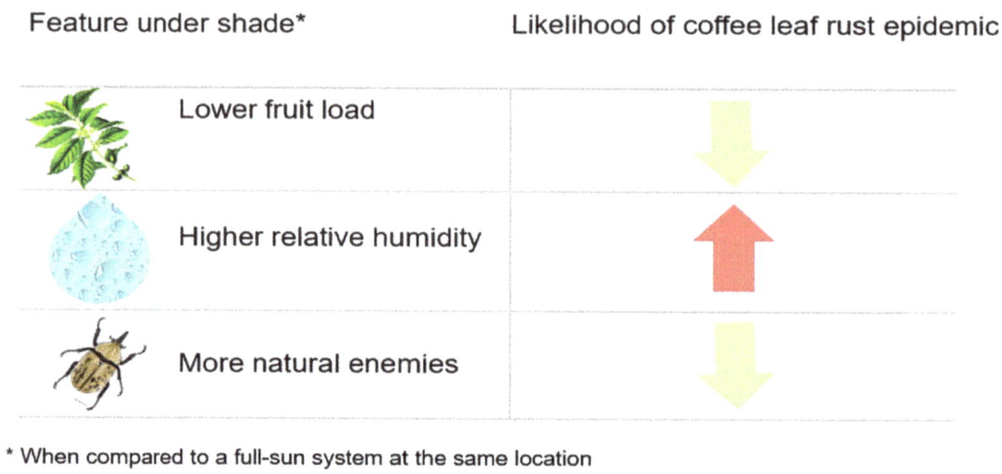

Fig. 19.7. The key features under a shaded coffee production system that can influence the likelihood of a coffee leaf rust epidemic. Figure courtesy of the author.

future climate changes and disease pressure are difficult to unravel. Like most fungal pathogens, the disease cycle of *H. vastatrix* is tightly regulated to the atmospheric temperature and humidity. Provided that the coffee plant leaf is moist for 6–24 hours and air temperatures are maintained between the optimal of 21 and 25°C, then *H. vastatrix* can germinate in most coffee-growing environments. If future coffee-growing climates are warmer and drier then CLR prevalence may be reduced. On the other hand, if excessive rainfall occurs in the dry seasons then there could be more opportunity for additional CLR outbreaks to occur. Nevertheless, recent modelling efforts in coffee-growing areas of Brazil have failed to predict connections between CLR outbreaks and climate forecasting.

Conclusion

There are many plant diseases that can impact our daily lives; CLR impacts the reliability of our future coffee supply. To some, this is a serious threat to a daily habit, natural stimulant and source of culinary pleasure. To small-holder coffee farmers, this disease poses a danger to farm productivity and in turn the livelihoods of those who depend on coffee beans as a main source of household income. By understanding the causal fungal pathogen and the best methods to prevent and control CLR, we can ensure the health of coffee plants across the tropics which supply the nearly 10 billion kg of beans needed annually for our cups of "liquid gold".

Discussion Points

1. What might be some of the knock-on effects of releasing BCAs into the coffee field?
2. What is the evolutionary advantage of *H. vastatrix* producing basidiospores if there is no alternate host present to be infected?
3. Do the benefits outweigh the costs of farming coffee with other trees?
4. What is the likely impact of climate change on coffee production in relation to coffee leaf rustJin *et al.* (2010)?

Further Reading

Arneson, P.A. (2011) Coffee rust. Available at: https://www.cabdirect.org/cabdirect/abstract/20137804111

Avelino, J., Zelaya, H., Merlo, A., Pineda, A., Ordóñez, M. *et al.* (2006) The intensity of a coffee rust epidemic is dependent on production situations. *Ecological Modelling* 197(3–4), 431–447.

Avelino, J., Cabut, S., Barboza, B., Barquero, M., Alfaro, R. *et al.* (2007) Topography and crop management are key factors for the development of American leaf spot epidemics on coffee in Costa Rica. *Phytopathology* 97(12), 1532–1542.

Avelino, J., Cristancho, M., Georgiou, S., Imbach, P., Aguilar, L. *et al.* (2015) The coffee rust crises in Colombia and Central America (2008–2013): Impacts, plausible causes and proposed solutions. *Food Security* 7(2), 303–321.

Berihun, G. and Alemu, K. (2023) Status of coffee leaf rust (*Hemileia vastatrix*) and its management in Ethiopia: A review. *Archives of Phytopathology and Plant Protection.*

Fungicide Testing for Coffee Leaf Rust Control in Hawaii Presented by Dr. Lisa Keith – Feb, 8, 2022 (n.d.). Available at: https://youtu.be/LJEJEOCDvz8

Jackson, D.W., Zemenick, K. and Huerta, G. (2012) Occurrence in the soil and dispersal of *Lecanicillium lecanii*, a fungal pathogen of the green coffee scale (*Coccus viridis*) and coffee rust (*Hemileia vastatrix*). *Tropical and Subtropical Agroecosystems* 15(2).

Jin, Y., Szabo, L.J. and Carson, M. (2010) Century-old mystery of *Puccinia striiformis* life history solved with the identification of Berberis as an alternate host. *Phytopathology* 100(5), 432–435.

Motisi, N., Bommel, P., Leclerc, G., Robin, M.H., Aubertot, J.N. *et al.* (2022) Improved forecasting of coffee leaf rust by qualitative modeling: Design and expert validation of the ExpeRoya model. *Agricultural Systems* 197, 103352.

Rey Mallén, P.R. (2014) Coffee rust has claimed $1B in losses and 250,000 jobs in Central America, US fears increase of migration. *International Business Times.*

Silva, M.D., Várzea, V., Guerra-Guimarães, L., Azinheira, H.G., Fernandez, D. *et al.* (2006) Coffee resistance to the main diseases: Leaf rust and coffee berry disease. *Brazilian Journal of Plant Physiology* 18, 119–147.

Silva, M.D., Guerra-Guimarães, L., Diniz, I., Loureiro, A., Azinheira, H. *et al.* (2022) An overview of the mechanisms involved in Coffee-Hemileia vastatrix interactions: Plant and pathogen perspectives. *Agronomy* 12(2), 326.

Talhinhas, P., Batista, D., Diniz, I., Vieira, A., Silva, D.N. *et al.* (2017) The coffee leaf rust pathogen Hemileia vastatrix: One and a half centuries around the tropics. *Molecular Plant Pathology* 18(8), 1039.

Understanding La Roya: The Coffee Rust Disease (n.d.). Available at: https://youtu.be/twWfTCUpfRw

Wariyo, A., Gebreselassie, H., Gerbatsedik, W. and Belachew, K. (2021) Current status on coffee leaf rust (*Hemileia vastatrix*) in sidama and gedeo zone, Southern Ethiopia. *International Journal of Agricultural Extension* 9(1), 01–11.

References

Carvalho, C.R., Fernandes, R.C., Carvalho, G.M.A., Barreto, R.W. and Evans, H.C. (2011) Cryptosexuality and the genetic diversity paradox in coffee rust, *Hemileia vastatrix*. *PLoS ONE* 6(11), e26387.

De Filho, E.V. and Domian, C.A. (2019) *Prevention and Control of Coffee Leaf Rust: Handbook of Best Practices for Extension Agents and Facilitators*. Tropical Agricultural Research and Higher Education Center (CATIE), Turrialba, Costa Rica. 2019.

Friis, I. (2015) Coffee and QAT on the Royal Danish expedition to Arabia–botanical, ethnobotanical and commercial observations made in Yemen 1762–1763. *Archives of Natural History* 42(1), 101–112.

Koutouleas, A. and Collinge, D.B. (2022) Coffee Leaf Rust. DOI: 10.13140/RG.2.2.28685.41444.

Koutouleas, A., Jørgen Lyngs Jørgensen, H., Jensen, B., Lillesø, J.P., Junge, A. *et al.* (2019) On the hunt for the alternate host of *Hemileia vastatrix*. *Ecology and Evolution* 9(23), 13619–13631.

Koutouleas, A., Collinge, B.D. and Ræbild, A. (2022) Alternative plant protection strategies for tomorrow's coffee. *Plant Pathology*. DOI: 10.1111/ppa.13676.

McCook, S. and Vandermeer, J. (2015) The big rust and the red queen: Long-term perspectives on coffee rust research. *Phytopathology* 105(9), 1164–1173.

Ukers, W.H. (1922) *All about coffee*. The Tea and Coffee Trade Journal Company, New York, USA. Available at: https://www.gutenberg.org/files/28500/28500-h/28500-h.htm

20 Management Strategies of Grey Mould Disease in Berries

MOHAMMED RADI[1] AND RACHID LAHLALI[2]

[1]*Department of Plant Protection, Phytopathology Unit, Ecole Nationale d'Agriculture de Meknès, Km 10, Rte Haj Kaddour, BPS/40, Meknès, 50001, Morocco*
[2]*Laboratory of Environment and Valorization of Microbial and Plant Resources, Faculty of Sciences, Moulay Ismail University, PO Box 11201, Zitoune, Meknès, Morocco*

Introduction

Grey mould, induced by the fungus *Botrytis cinerea*, is a highly destructive disease impacting berry crops globally, including strawberries, raspberries, and blackberries. *B. cinerea* is placed second in the global Top 10 list of fungal plant diseases based on scientific and economic significance (Dean *et al.*, 2012). *B. cinerea* is a significant post-harvest pathogen due to the favourable conditions present throughout the post-harvest handling process, such as injuries, elevated humidity, deteriorating plant tissue, and high sugar concentration (Romanazzi and Feliziani, 2014). Ambient temperature and humidity are the primary determinants of the occurrence and severity of grey mould disease. Relative humidity above 80% and temperatures ranging from 15 to 25°C are the optimal conditions for the proliferation of grey mould disease (Ciliberti *et al.*, 2015). *B. cinerea* can infect all plant parts, like leaves, petals, crowns, and flower stems, due to its capacity to thrive on both living and dead tissues. These infections can be severe at all stages of fruit development, even post-harvest (Maung *et al.*, 2021). Furthermore, the application of chemical fungicides is extensively employed to regulate *B. cinerea* in both greenhouse and field crops (Geng *et al.*, 2022). *B. cinerea* frequently acquires resistance to fungicides, highlighting a significant constraint of pesticide application in the management of grey mould (Diao *et al.*, 2020). The extensive application of chemical fungicides results in major pollution of the environment. To address these challenges, the invention of bio-fungicides and biological control has emerged as a pivotal focus in the research of plant disease management (Lahlali *et al.*, 2022).

Importance and Economic Losses of *Botrytis cinerea*

Botrytis cinerea is a highly studied necrotrophic fungal pathogen responsible for grey mould rot in over 500 species of plants, including important crops such as tomato (Dean *et al.*, 2012), roses (Elad *et al.*, 2007), apple, kiwi, and pomegranate (Droby and Lichter, 2007; Romanazzi and Feliziani, 2014), as well as berries (Williamson *et al.*, 2007). These host variations reflect its ability to adapt to diverse environments and highlight its economic and ecological importance. This necrotrophic infection damages multiple berry crops, such as strawberries, raspberries, and blueberries, causing degradation both before and after harvest. The pathogen can be found within the leaves, stems, fruit, flowers, or seeds. It may manifest evident signs and symptoms during the pre-harvest phase or remain dormant until the post-harvest phase (Fillinger and Elad, 2016). *Botrytis cinerea* is considered one of the most significant post-harvest diseases affecting fresh vegetables and fruits (Zhang *et al.*, 2014). The yearly revenue losses attributed to *B. cinerea* surpass US$10 billion globally (Rahman *et al.*, 2024). According to its scientific and commercial significance, *B. cinerea* is currently designated as the second most important plant pathogen (Poveda *et al.*, 2020).

Taxonomy

The designation *Botrytis cinerea* is directly sourced from the morphology of the fungus: "Botrytis" originates from the Greek term for "bunch of grape berries," which characterizes the grape-like structure of conidiophores, while "cinerea" denotes the grey colour of sporulation (Williamson *et al.*, 2007; Petrasch *et al.*, 2019).

B. cinerea represents the asexual phase (anamorph) of the teleomorph Botryotinia fuckeliana. The global genus Botrytis has 22 acknowledged species and a single hybrid (Staats *et al.*, 2005), every one of which causes diseases of plants. *B. cinerea* is the sole species within its genus that exhibits a wide host range, while the other Botrytis species are considered specialized to a single plant species (Staats *et al.*, 2005).

The primary identification of *B. cinerea* is based on its distinctive spore and macroconidiophore morphology (Alfonso *et al.*, 2000; Williamson *et al.*, 2007). The following is a specialized structure that comprises a terminal cluster of sequentially generated hydrophobic conidia, supported by a robust macroconidiogenous hypha, like a cluster of grapes (Williamson *et al.*, 2007; Avenot *et al.*, 2020; Notte *et al.*, 2021). The hyphae develop via a cushion-shaped stroma located immediately below the host surface (Boff *et al.*, 2001; Williamson *et al.*, 2007; Feliziani and Romanazzi, 2016). Macroconidia have been significant, specialized blastoconidia with dimensions of 8–14 × 6–9 μm (Williamson *et al.*, 2007; Feliziani and Romanazzi, 2016). Macroconidia are round, desiccated, and hydrophobic, formed individually on lateral branches for efficient dissemination. After maturation, conidia are readily displaced from the ampullae by

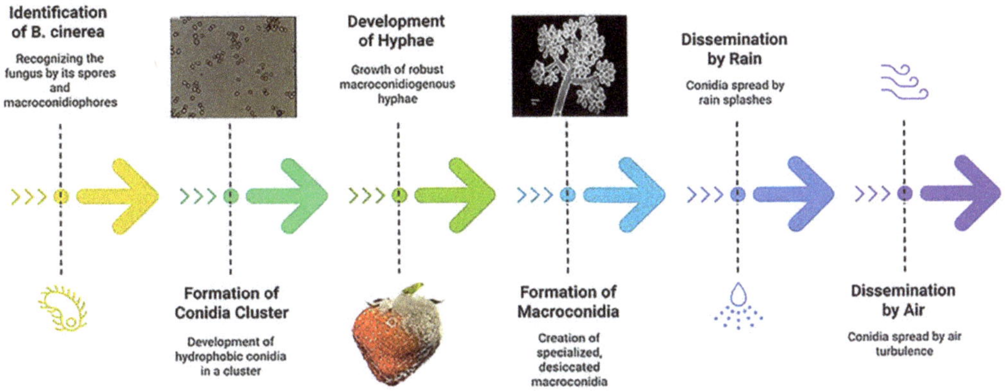

Fig. 20.1. Illustration of key criteria of identification of *Botrytis cinerea*.

rain splashes and/or air turbulence, which also facilitate their dissemination (Fig. 20.1) (Williamson *et al.*, 2007; Avenot *et al.*, 2020; Notte *et al.*, 2021).

Disease Cycle

Botrytis cinerea survives the winter as sclerotia or latent mycelium, predominantly in decayed strawberry debris, as well as in mummified berries and plants. During spring, under optimal conditions, the fungus forms mycelium and generates substantial amounts of conidia, which serve as the primary inoculum source. Senescent petals, calyces, and stamens may promote primary infections in strawberries (Petrasch *et al.*, 2019).

B. cinerea produces many specialized metabolites, toxins, and enzymes to aid in its invasion of the strawberry plant. *Botrytis cinerea* can invade strawberry tissues directly by the action of pectinases, chitinases, and proteases, or by utilizing natural openings such as stomata or existing wounds (Brauna-Morževska *et al.*, 2019).

Conidia are emitted by conidiophores in reaction to vibrations or a swift decline in relative humidity and are disseminated by circulation of air and splashes of water (Lumînare, 2023). The germ tubes of conidia readily penetrate the strawberry epidermis beneath the petals, pistils, and stamens. After establishment in the floral structures, *B. cinerea* infiltrates the developing fruit, generally lying dormant until the fruit sustains damage. Spore production commences rapidly as the berries mature and persists throughout the period of fruiting. This functions as a supplementary source of inoculum. *B. cinerea* commences the necrotrophic phase during another infection without entering dormancy (Fig. 20.2) (Rhouma *et al.*, 2022).

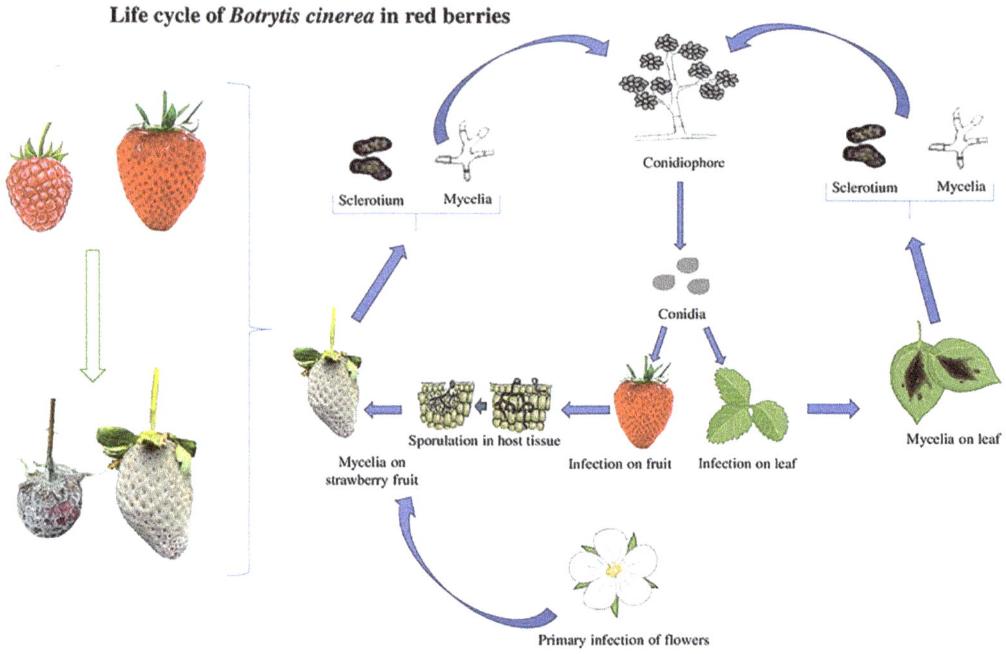

Fig. 20.2. Disease cycle of *Botrytis cinerea* in red berries.

Symptoms

Immature leaves are susceptible to infection by this pathogen, although they exhibit no signs as the fungus keeps dormant within these tissues (D'ercole, 1986). As the leaves develop and commence decomposition, *B. cinerea* may activate and generate the distinctive grey mould on the necrotic regions of the leaf (Williamson *et al.*, 2007).

Affected flowers exhibit brown and discoloured lesions on the petals on them, sepals, along with receptacles. If the pathogen progresses into the flower, *B. cinerea* will necrotize the pedicel, resulting in the withering and demise of the entire bloom (Alfonso *et al.*, 2000).

During the development of fruits, *B. cinerea* activates and induces light brown rot at the calyx end of the immature fruit (Rhouma *et al.*, 2022). Ripe red fruits are especially vulnerable to *B. cinerea* infection after sustaining physical injury, since the airborne pathogen can rapidly invade the compromised tissues and disseminate throughout the fruit (Petrasch *et al.*, 2019).

Completely decayed fruits become desiccated, arid, and mummified. Immature fruits may distort and perish before maturation. Strawberries maintained in chilly temperatures postharvest may exhibit a whiter appearance of the disease, since the pathogen often requires light for the complete development of grey spores (Fig. 20.3) (Fillinger and Elad, 2015; Rhouma *et al.*, 2022).

Fig. 20.3. Symptoms of *Botrytis cinerea* on strawberry fruit at different stages of development.

Management Strategies

In response to the substantial economic losses caused by *B. cinerea*, significant measures have been implemented to manage Botrytis-induced diseases, including cultural practices, chemical control, resistance inducers, and biological management (Hua *et al*., 2018).

Cultural practices

Cultural practices are an important aspect of reducing incidents of grey mould in berry crops by minimizing conditions favourable to disease development. These methods include proper air circulation around the plants; plant spacing and pruning must be carried out in such a way as to ensure sufficient air circulation around the plants. Consequently, reduced humidity offers conditions that are not conducive to infection, since *B. cinerea* proliferates in humid conditions (Koike *et al*., 2007). Irrigation management is another very significant cultural aspect. Sprinkler irrigation creates a humid environment that favours the growth of fungi. It is, therefore, possible to reduce humidity on berries and leaves by switching from sprinkler irrigation to drip irrigation, thereby reducing the risk of grey mould infection. In addition, irrigation should be carried out early in the morning to give plant surfaces time to dry out during the day (Elad and Shtienberg, 1995). Similarly, optimal sanitation is important. Regular removal of diseased plant material, including infected berries and leaves, can greatly reduce the inoculum source for the disease. This practice minimizes spore spread within the crop as a whole (Williamson *et al.*, 2007). In addition, rotating crops, in addition to not planting berries in the same location year after year, can further reduce the buildup of grey mould inoculum in the soil itself (Bulger *et al.*, 1987).

Chemical control

The primary strategy for managing postharvest diseases caused by *B. cinerea* is the application of synthetic fungicides. Fungicides are classified into five types based on their mechanisms: those affecting the process of respiration, osmoregulation, microtubule assembly, sterol biosynthesis, and those whose effects can be reversed by methionine (Leroux, 2007). Among these, dicarboximides have historically been widely utilized as botryticides, while their principal target site remains unidentified. They exhibit activity directed at both conidia and mycelium by influencing sensitivity to osmotic stress. Resistance to this class of drugs was attributed to a singular polymorphic gene, Daf1 (Faretra and Pollastro, 1991). In addition, the anilinopyrimidines serve as effective botryticides, which are inhibited by methionine and some other amino acids. These fungicides can inhibit the release of hydrolytic enzymes involved in disease, including lipases, cutinases, proteases, and cellulases (Miura *et al.*, 1994). Another key class, the pyridine-carboxamides, includes boscalid, a member of the pyridine-carboxamides class, which inhibits energy generation by acting as a succinate dehydrogenase inhibitor. It inhibits the production of vital cellular components, hence obstructing fungal growth and adversely affecting spore germination, germ tube elongation, mycelial development, and sporulation (Romanazzi and Feliziani, 2014). Similarly, fluopyram, a pyridinyl-ethyl-benzamide fungicide, exerts its effect by inhibiting succinate dehydrogenase, disrupting the fungal respiratory chain complex II (Romanazzi and Feliziani, 2014). Further contributing to the range of fungicides is fenhexamid, a fungicide that inhibits sterol production and is the latest and most efficient agent against *B. cinerea* (Rosslenbroich and Stuebler, 2000). Specific isolates from a particular *B. cinerea* subpopulation exhibit varying levels of resistance to this specific fungicide in the laboratory (Albertini *et al.*, 2002; Fournier *et al.*, 2003). The following Table 20.1 provides an overview of common chemical and non-chemical control agents used against *Botrytis cinerea* in major berry-producing regions.

Despite the wide application of synthetic fungicides, their use is often costly, as managing *B. cinerea* typically requires greater dosage rates compared to other fungal diseases. The costs for the management of Botrytis and associated species were approximately 8% of the global fungicide market

Table 20.1. Common chemical and non-chemical control agents for Botrytis cinerea

Country/Region	Chemical control agents	Non-chemical control methods	References
United States	*Boscalid*, *Cyprodinil*, *Fludioxonil*, *Fenhexamid*	Biological agents (*Trichoderma* spp., *Bacillus subtilis*), Drip irrigation, Sanitation practices	Romanazzi and Feliziani (2014); Fillinger and Elad (2016); Köhl *et al.* (2019)
European Union	*Fenhexamid*, *Pyrimethanil*, *Iprodione*, Fluopyram	*Aureobasidium pullulans* (e.g., Botector®), Crop rotation, Optimal air circulation, Mulching	Köhl *et al.* (2019); Petrasch *et al.* (2019)
China	*Mancozeb*, *Carbendazim*, *Difenoconazole*	*Gliocladium catenulatum* (e.g., Prestop®), Soil solarization, Controlled humidity	Hua *et al.* (2018);Diao *et al.* (2020)
Australia	*Captan*, *Iprodione*, *Azoxystrobin*	*Bacillus subtilis* (e.g., Serenade®), Removal of infected debris, Adjusted irrigation timing	Elad *et al.* (2007); Köhl *et al.* (2019)
South America (Chile)	*Fludioxonil*, *Cyprodinil*, *Boscalid*	*Pseudomonas fluorescens*, Biological fungicides, Greenhouse humidity management	Bonaterra *et al.* (2022)
India	*Carbendazim*, *Mancozeb*	*Trichoderma harzianum*, Compost-based soil amendments, Resistant cultivars	Hua *et al.* (2018); Sharma (2018); Köhl *et al.* (2019)

(Fira *et al.*, 2018). In addition, chemical fungicides are detrimental to both the environment and human health, particularly due to their toxicological residues. Consequently, numerous regulatory constraints are imposed on the use of botryticides (Droby *et al.*, 2009).

Resistance inducers

Systemic plant resistance to *B. cinerea* can be induced not only by post-pathogen attack but it can also be pre-activated through the administration of several exogenous chemical substances and biological elicitors. The external application of protective plant hormones on fresh produce over postharvest has demonstrated the ability to activate plant resistance against various pathogens, particularly through the use of compounds derived from jasmonic acid, such as methyl jasmonate (MeJA) (Poveda, 2020). In plants, multiple chemical compounds that can induce a priming-type systemic resistance against *B. cinerea* are currently identified, including BABA in *A. thaliana*

(Zimmerli *et al.*, 2001) and tomato (Wilkinson *et al.*, 2018), benzothiadiazole (BTH) in poinsettia (Kułek and Floryszak-Wieczorek, 2002), hexanoic acid (Vicedo *et al.*, 2009) and riboflavin in tomato plants (Azami-Sardooei *et al.*, 2010), as well as elicitors like chitosan (Trotel-Aziz *et al.*, 2006) and laminarin in grapevine (Aziz *et al.*, 2003). In this sense, mechanical damage and injuries can also trigger a plant's systemic resistance to *B. cinerea* (Tomas-Grau *et al.*, 2018), along with unfavourable climatic conditions including elevated temperatures (Widiastuti *et al.*, 2011) and UV radiation (Mercier *et al.*, 2000).

MeJA treatment generated resistance in Chinese bayberries against fungal pathogens by priming defence mechanisms, stimulating the hydrogen peroxide burst, enhancing the translation levels of defence-related proteins, and increasing the concentrations of antimicrobial substances (Wang *et al.*, 2014). Cinnamic acid has been demonstrated to efficiently manage grey mould rot caused by *B. cinerea* in table grapes. Its mechanisms of action involve reducing mycelial growth by compromising plasma membrane integrity and enhancing host fruit resistance by inducing the activity of proteins like polyphenol oxidase and peroxidase (Zhang *et al.*, 2015).

Furthermore, culture circumstances may activate a plant's systemic resistance to *B. cinerea*. The priming process against infections in hydroponic crops was documented (Chinta *et al.*, 2015) following the incorporation of biochar (Mehari *et al.*, 2015) and olive mark compost (Segarra *et al.*, 2013). The induction of systemic defence responses against *B. cinerea*, facilitated by favourable fungi and bacteria, was previously documented.

Biological control

The biological management of plant diseases has reached notable advancements, especially in the creation of commercially viable biocontrol agents (BCAs) that specifically target critical pathogens such as *Botrytis cinerea*, responsible for grey mould. Research highlighted by Collinge *et al.* (2022) demonstrates successful applications of fungal and bacterial antagonists, such as *Trichoderma* spp., *Bacillus subtilis*, and Coniothyrium minitans, which suppress pathogens through antibiosis, competition, and induced systemic resistance. These particular strains have demonstrated efficacy in numerous studies, leading to their utilization as industrial and commercial bioproducts. For example, *Serenade®* (*Bacillus subtilis*) (Robinson-Boyer *et al.*, 2009), *Botector®* (*Aureobasidium pullulans*)(Fabbro *et al.*, 2016) , and *Prestop®* (*Gliocladium catenulatum*) (Caproni *et al.*, 2021) have been shown to effectively inhibit pathogen proliferation or prevent infections. However, adoption is limited, with about 10–15% of European cultivators employing these techniques. Principal obstacles encompass elevated expenses and inconsistent effectiveness across different conditions. Incorporating biopesticides into comprehensive management systems promotes acceptance and diminishes chemical residues and environmental effects (Köhl *et al.*, 2019).

The bacterial genus Bacillus comprises species extensively studied and utilized as biological controls against the phytopathogenic fungi in the agricultural sector, due to their multiple specialized metabolism and capacity to synthesize a broad range of structurally distinct antagonistic compounds via a mechanism termed antibiosis (Fira *et al.*, 2018). Inhibition rates of 75 to 90% were attained when grey mould in strawberries was treated with B. subtilis and B. licheniformis (Poveda *et al.*, 2020). Furthermore, B. subtilis and B. amyloliquefaciens are recognized as species capable of synthesizing and emitting antifungal volatile organic compounds (VOCs) against *B. cinerea*; these compounds assist in managing the fungi by blocking spore germination and hyphal growth, both in vitro and planta (Paz *et al.*, 2018). For instance, effective antagonism has been characterized by the emission of VOCs in strawberries by Galactomyces candidum (Chen *et al.*, 2018).

In addition, Pseudomonas spp. are recognized as effective biological control agents with several mechanisms of action (Bonaterra *et al.*, 2022). Wafaa and Mostafa (2012) demonstrated that under natural conditions, a P. fluorescens formulation effectively reduced *B. cinerea* disease in strawberry foliage and fruit. Pre-harvest treatment methods preserved fruits from Botrytis postharvest disease more successfully compared to fungicides. Studies demonstrated that bacterial treatment significantly improved growth indices, drying weights, and yield.

However, the application of yeasts as antagonistic microorganisms substantially mitigated postharvest deterioration induced by *B. cinerea* in several fruits. Antagonistic yeasts are favoured due to their safety, as no risky specialized metabolites have been identified in their interactions with pathogens (Hua *et al.*, 2018). Aureobasidium pullulans is the predominant yeast species employed against *B. cinerea*, thanks to its proficiency in competing for space and nutrients on plant surfaces and in wounds, as well as its production of many antifungal chemicals (Wang *et al.*, 2018).

Furthermore, the integration of fungi as biocontrol agents represents a potential and sustainable strategy for managing plant diseases (Thambugala *et al.*, 2020). The genus *Trichoderma* is acknowledged as the principal biological control agent against *B. cinerea*. A number of species of this genus are frequently used for biological control agents in agriculture due to their mechanisms of action, which include antibiosis, mycoparasitism, and competition for space and nutrients in the rhizosphere (Sood *et al.*, 2020). Research has demonstrated the effectiveness of *Trichoderma* spp. in reducing berry diseases. Jalali *et al.* (2016) indicate that the volatile organic compounds (VOCs) of *T. asperellum* considerably reduced the incidence and severity of rot induced by *B. cinerea* on strawberries. Furthermore, Savas *et al.* (2021) observed that the volatile metabolites of T. atroviride EGE-K-71 resulted in a 71.8% reduction of *B. cinerea* growth. In addition, these volatile metabolites exhibited the greatest percentage of *B. cinerea* growth suppression in vitro relative to various evaluated fungicides.

Future directions in biological control emphasize enhancing efficacy through microbiome engineering, CRISPR-based microbial consortia, and synergies with emerging technologies. Collinge *et al.* (2022) highlight the necessity of optimizing biocontrol agent combinations with host resistance features, precision agricultural technologies (such as sensors for microclimate monitoring), and reduced-risk chemicals to manage grey mould in berries. Research aims to elucidate the tripartite interactions among biocontrol agents, plants, and pathogens during changing conditions, alternatively utilizing genetic insights to engineer biocontrol agents with multifunctional characteristics (e.g. stress resilience and biofilm development). In the future, innovations in berry production may include phage therapy against Botrytis or RNAi-based biocontrols that can be designed to inhibit fungal virulence pathways. Policy frameworks that facilitate expedited BCA registration and enhance farmer education on application timing will be essential for the sustainable scaling of these solutions.

Biotechnological Approaches

Recent advances in biotechnological approaches for grey mould management focus on precision-targeted strategies that disrupt fungal virulence while minimizing environmental impact. RNA interference (RNAi) technology shows promise through spray-induced gene silencing (SIGS), where double-stranded RNA (dsRNA) molecules silence critical fungal genes like BcBmp1 (involved in MAP kinase signalling) or BcPls1 (essential for host penetration) (Spada *et al.*, 2023). Engineered biocontrol microbes, such as *Trichoderma harzianum* strains modified to overexpress chitinases or p-1,3-glucanases, enhance mycoparasitism of Botrytis hyphae while inducing systemic resistance in berry plants (Cortes *et al.*, 2023). CRISPR-Cas9 editing enables the design of *Bacillus subtilis* variants with optimized biosynthetic gene clusters for antifungal lipopeptides (e.g. surfactin, iturin), improving suppression of Botrytis conidial germination on flower surfaces (Wang *et al.*, 2019).

Emerging directions integrate phage therapy with microbiome engineering to create synergistic control systems. Lytic bacteriophages (e.g. Botrytis phage cpBc1) targeting fungal cell wall synthesis genes can be combined with phage-resistant *Pseudomonas fluorescens* auxotrophs that secrete botryticidal cyclic lipopeptides (Huss and Raman, 2020; Lenneman *et al.*, 2021). Advances in synthetic biology enable engineered microbial consortia with division-of-labor strategies, for example, Saccharomyces cerevisiae modified to secrete p-glucanase enzymes that weaken Botrytis cell walls, paired with Aureobasidium pullulans strains producing antifungal aureobasidins (Chen *et al.*, 2023). These systems are being optimized for berry production environments through nanoencapsulation in pH-responsive polymers, ensuring controlled release during critical infection windows like flowering and fruit ripening.

Challenges and Future Perspectives in the Management of *Botrytis cinerea*

One of the issues in berry cultivators is fungicide resistance, since the repetitive use of chemical treatments has resulted in resistant strains, hence reducing the efficacy of commonly utilized fungicides like QoI and SDHI (Fernández-Ortuño *et al.*, 2017). In addition, irregular environmental conditions, aggravated by climate change, favour the proliferation of the disease, preventing control measures (Elad and Pertot, 2014). Although advances have been made in the further development of resistant berry varieties, they remain limited and not widespread in availability (Petrasch *et al.*, 2019). However, the prospects appear promising for the future. Biopesticides represent some ecological alternatives that include *Trichoderma* spp. and helpful microorganisms that would contribute to combating fungicide resistance of crops (Sharma, 2018). Advances in Smart sensors and AI technologies may afford the possibility to fully revolutionize early detection, providing the possibility of intervention before disease manifestation (Behmann *et al.*, 2014). Moreover, new gene-editing technologies, including CRISPR, could accelerate the development of resistant varieties, thanks to more specific targeting of the genes responsible for susceptibility. In addition, precision agriculture could minimize the risk of Botrytis, using tools and techniques such as drones or advanced irrigation systems to optimize the environment. Eventually, it will be necessary to develop a more viable approach that will integrate biological, cultural, and technical measures in order to lessen the influence that disease control has on the environment (Elmer and Reglinski, 2006).

Conclusion

Grey mould caused by *Botrytis cinerea* is one of the main problems faced by berry growers in general, with significant economic losses due to reduced fruit quality. Current disease management practices, such as the application of Cultural Practices, chemical treatments, and biological control, are effective to some extent; however, there are many challenges in terms of fungicide resistance and other environmental issues. The future of *Botrytis cinerea* management now depends on the development of integrated and sustainable approaches, combining new crop protection technologies with more precise farming. Continued research into resistant berry varieties, biopesticides, and the adaptation of agriculture to the climate will also be crucial in reducing crop susceptibility to grey mould. Ultimately, it will be a matter of striking a delicate balance between preserving the environment and maintaining crop yield and quality, so as to overcome this disease in the long term.

Discussion Points

1. List potential management strategies that can be employed to control grey mould caused by *Botrytis cinerea*.
2. What are the challenges associated with integrated management of *Botrytis cinerea*, especially when combining chemical, biological, and cultural controls?
3. What are the limitations of biological control in managing grey mould, and how can these be addressed?
4. If biocontrol agents and resistant varieties are successful against *Botrytis cinerea*, how can they be scaled for commercial use in large-scale berry production?

Funding

Core organic co-fund/MESRI, (Grant/Award Number: "Resberry Project").

Further Reading

Dean, R., Kan, J.A.L., Pretorius, Z.A., Hammond-Kosack, K.E., Pietro, A. *et al.* (2012) The top 10 fungal pathogens in molecular plant pathology. *Molecular Plant Pathology* 13(4), 414–430.

Elad, Y., Williamson, B., Tudzynski, P. and Delen, N. (2007) *Botrytis: Biology, Pathology, and Control*. Springer.

Fillinger, S. and Elad, Y. (2016) *Botrytis: The Fungus, the Pathogen, and its Management in Agricultural Systems*. Springer.

Petrasch, S., Knapp, S.J., Van Kan, J.A.L. and Blanco-Ulate, B. (2019) Grey mould of strawberry: A devastating disease caused by the ubiquitous necrotrophic fungal pathogen *Botrytis cinerea*. *Molecular Plant Pathology* 20(6), 877–892.

Romanazzi, G. and Feliziani, E. (2014) *Botrytis cinerea* (gray mold). In: *Postharvest Decay*. Elsevier, pp. 131–146.

References

Albertini, C., Thebaud, G., Fournier, E. and Leroux, P. (2002) Eburicol 14a-demethylase gene (CYP51) polymorphism and speciation in *Botrytis cinerea*. *Mycological Research* 106(10), 1171–1178.

Alfonso, C., Raposo, R. and Melgarejo, P. (2000) Genetic diversity in *Botrytis cinerea* populations on vegetable crops in greenhouses in south-eastern Spain. *Plant Pathology* 49(2), 243–251.

Avenot, H.F., Morgan, D.P., Quattrini, J. and Michailides, T.J. (2020) Phenotypic and molecular characterization of fenhexamid resistance in *Botrytis cinerea* isolates collected from pistachio orchards and grape vineyards in California. *Crop Protection* 133, 105133.

Azami-Sardooei, Z., França, S.C., Vleesschauwer, D. and Höfte, M. (2010) Riboflavin induces resistance against *Botrytis cinerea* in bean, but not in tomato, by priming for a hydrogen peroxide-fueled resistance response. *Physiological and Molecular Plant Pathology* 75(1–2), 23–29.

Aziz, A., Poinssot, B., Daire, X., Adrian, M., Bézier, A. *et al.* (2003) Laminarin elicits defense responses in grapevine and induces protection against *Botrytis cinerea* and *Plasmopara viticola*. *Molecular Plant-Microbe Interactions* 16(12), 1118–1128.

Behmann, J., Steinrücken, J. and Plümer, L. (2014) Detection of early plant stress responses in hyperspectral images. *ISPRS Journal of Photogrammetry and Remote Sensing* 93, 98–111.

Boff, P., Kastelein, P., Kraker, J., Gerlagh, M. and Köhl, J. (2001) Epidemiology of grey mould in annual waiting-bed production of strawberry. *European Journal of Plant Pathology* 107, 615–624.

Bonaterra, A., Badosa, E., Daranas, N., Francés, J., Roselló, G. *et al.* (2022) Bacteria as biological control agents of plant diseases. *Microorganisms* 10(9), 1759.

Brauna-Morževska, E., Bankina, B. and Kan, eps, J. (2019) *Botrytis* genus fungi as causal agents of legume diseases: A review. *Research for Rural Development* 2, 63–69. DOI: 10.22616/rrd.25.2019.050.

Bulger, M.A., Ellis, M.A. and Madden, L.V. (1987) Influence of temperature and wetness duration on infection of strawberry flowers by *Botrytis cinerea* and disease incidence of fruit originating from infected flowers. *Phytopathology* 77(8), 1225–1230.

Caproni, C.M., Rodrigues, F.A., Rezende, R.A.L.S., Pomella, A.W.V. and Pasqual, M. (2021) Application of biological products in strawberry cultivars to control *Botrytis cinerea*. *Research, Society and Development* 10(5), e8710514655–e8710514655.

Chen, P.H., Chen, R.Y. and Chou, J.Y. (2018) Screening and evaluation of yeast antagonists for biological control of *Botrytis cinerea* on strawberry fruits. *Mycobiology* 46(1), 33–46. DOI: 10.1080/12298093.2018.1454013.

Chen, N., Yang, S., You, D., Shen, J., Ruan, B. *et al.* (2023) Systematic genetic modifications of cell wall biosynthesis enhanced the secretion and surface-display of polysaccharide degrading enzymes in *Saccharomyces cerevisiae*. *Metabolic Engineering* 77, 273–282.

Chinta, Y.D., Eguchi, Y., Widiastuti, A., Shinohara, M. and Sato, T. (2015) Organic hydroponics induces systemic resistance against the air-borne pathogen, *Botrytis cinerea* (gray mould). *Journal of Plant Interactions* 10(1), 243–251.

Ciliberti, N., Fermaud, M., Roudet, J. and Rossi, V. (2015) Environmental conditions affect *Botrytis cinerea* infection of mature grape berries more than the strain or transposon genotype. *Phytopathology* 105(8), 1090–1096.

Collinge, D.B., Jensen, D.F., Rabiey, M., Sarrocco, S., Shaw, M.W. *et al.* (2022) Biological control of plant diseases – What has been achieved and what is the direction? *Plant Pathology* 71(5), 1024–1047. DOI: 10.1111/ppa.13555.

Cortes, M.V.C.B., Barbosa, E.T., Oliveira, M.I.S., Maciel, L.H.R., Lobo Junior, M. *et al.* (2023) Trichoderma harzianum marker-free strain construction based on efficient CRISPR/Cas9 recyclable system: A helpful tool for the study of biological control agents. *Biological Control* 184, 105281. DOI: 10.1016/jbiocontrol.2023.105281.

Dean, R., Van Kan, J.A.L., Pretorius, Z.A., Hammond-Kosack, K.E., Di Pietro, A. *et al.* (2012) The top 10 fungal pathogens in molecular plant pathology. *Molecular Plant Pathology* 13(4), 414–430.

del Fabbro, R., Crivelli, L., Lacertosa, G., Digeronimo, G., Calari, A. *et al.* (2016) Activity of Aureobasidium Pullulans (Botector) Against Grey Mold on Grape, Strawberry and Tomato.

D'ercole, N. (1986) Maladies Fongiques et Bacteriennes du Fraisier.

Diao, Y., Larsen, M.M., Kamvar, Z.N., Zhang, C., Li, S. *et al.* (2020) Genetic differentiation and clonal expansion of Chinese *Botrytis cinerea* populations from tomato and other crops in China. *Phytopathology* 110(2), 428–439. DOI: 10.1094/PHYTO-09-18-0347-R.

Droby, S. and Lichter, A. (2007) Post-harvest botrytis infection: Etiology, development and management. In: *Botrytis: Biology, Pathology and Control*. Springer, pp. 349–367.

Droby, S., Wisniewski, M., Macarisin, D. and Wilson, C. (2009) Twenty years of postharvest biocontrol research: Is it time for a new paradigm? *Postharvest Biology and Technology* 52(2), 137–145.

Elad, Y. and Pertot, I. (2014) Climate change impacts on plant pathogens and plant diseases. *Journal of Crop Improvement* 28(1), 99–139.

Elad, Y. and Shtienberg, D. (1995) *Botrytis cinerea* in greenhouse vegetables: Chemical, cultural, physiological and biological controls and their integration. *Integrated Pest Management Reviews* 1, 15–29.

Elad, Y., Williamson, B., Tudzynski, P. and Delen, N. (2007) *Botrytis: Biology, Pathology and Control*. Springer, Netherlands, pp. 1–8.

Elmer, P.A.G. and Reglinski, T. (2006) Biosuppression of *Botrytis cinerea* in grapes. *Plant Pathology* 55(2), 155–177.

Faretra, F. and Pollastro, S. (1991) Genetic basis of resistance to benzimidazole and dicarboximide fungicides in Botryotinia fuckeliana (*Botrytis cinerea*). *Mycological Research* 95(8), 943–951.

Feliziani, E. and Romanazzi, G. (2016) Postharvest decay of strawberry fruit: Etiology, epidemiology and disease management. *Journal of Berry Research* 6(1), 47–63.

Fernández-Ortuño, D., Pérez-García, A., Chamorro, M., Peña, E., Vicente, A. *et al.* (2017) Resistance to the SDHI fungicides boscalid, fluopyram, fluxapyroxad and penthiopyrad in *Botrytis cinerea* from commercial strawberry fields in Spain. *Plant Disease* 101(7), 1306–1313.

Fillinger, S. and Elad, Y. (2015) Botrytis-the Fungus, the Pathogen and Its Management in Agricultural Systems 1–486. DOI: 10.1007/978-3-319-23371-0.

Fillinger, S. and Elad, Y. (2016) *Botrytis-the Fungus, the Pathogen and Its Management in Agricultural Systems*, Vol. 486. Springer, Cham, Switzerland.

Fira, D., Dimki, I., Beric, T., Lozo, J. and Stankovic, S. (2018) Biological control of plant pathogens by Bacillus species. *Journal of Biotechnology* 285, 44–55. DOI: 10.1016/j.jbiotec.2018.07.044.

Fournier, E., Levis, C., Fortini, D., Leroux, P., Giraud, T. *et al.* (2003) Characterization of Bc-hch, the *Botrytis cinerea* homolog of the Neurospora crassa het-c vegetative incompatibility locus, and its use as a population marker. *Mycologia* 95(2), 251–261.

Geng, L., Fu, Y., Peng, X., Yang, Z., Zhang, M. *et al.* (2022) Biocontrol potential of trichoderma harzianum against *Botrytis cinerea* in tomato plants. *Biological Control* 174. DOI: 10.1016/j.biocontrol.2022.105019.

Hua, L., Yong, C., Zhanquan, Z., Boqiang, L., Guozheng, Q. *et al.* (2018) Pathogenic mechanisms and control strategies of *Botrytis cinerea* causing post-harvest decay in fruits and vegetables. *Food Quality and Safety* 2(3), 111–119.

Huss, P. and Raman, S. (2020) Engineered bacteriophages as programmable biocontrol agents. *Current Opinion in Biotechnology* 61, 116–121. DOI: 10.1016/j.copbio.2019.11.013.

Jalali, F., Zafari, D. and Salari, H. (2016) Efficacy of the volatile organic compounds of *Trichoderma asperellum* in control of *Botrytis cinerea*, the causal agent of strawberry gray mold.

Köhl, J., Kolnaar, R. and Ravensberg, W.J. (2019) Mode of action of microbial biological control agents against plant diseases: Relevance beyond efficacy. *Frontiers in Plant Science* 10, 845.

Koike, S.T., Gladders, P. and Paulus, A.O. (2007) *Vegetable Diseases: A Color Handbook*. Gulf Professional Publishing, Houston, TX.

Kułek, B. and Floryszak-Wieczorek, J. (2002) Local and systemic protection of poinsettia (*Euphorbia pulcherrima* Willd.) against *Botrytis cinerea* Pers. induced by benzothiadiazole. *Acta Physiologiae Plantarum* 24, 273–278.

Lahlali, R., Ezrari, S., Radouane, N., Kenfaoui, J., Esmaeel, Q. *et al.* (2022) Biological control of plant pathogens: A global perspective. *Microorganisms* 10(3), 596. DOI: 10.3390/microorganisms10030596.

Lenneman, B.R., Fernbach, J., Loessner, M.J., Lu, T.K. and Kilcher, S. (2021) Enhancing phage therapy through synthetic biology and genome engineering. *Current Opinion in Biotechnology* 68, 151–159.

Leroux, P. (2007) Chemical control of Botrytis and its resistance to chemical fungicides. In: *Botrytis: Biology, Pathology and Control*. Springer, Dordrecht, Netherlands, pp. 195–222.

Lumînare, C.M. (2022) The main fungal diseases in strawberries crop-review. *Scientific Papers Series A. Agronomy. LXV* 2(2), 464477.

Maung, C.E.H., Baek, W.S., Choi, T.G. and Kim, K.Y. (2021) Control of grey mould disease on strawberry using the effective agent, Bacillus amyloliquefaciens Y1. *Biocontrol Science and Technology* 31(5), 468–482.

Mehari, Z.H., Elad, Y., Rav-David, D., Graber, E.R. and Meller Harel, Y. (2015) Induced systemic resistance in tomato (*Solanum lycopersicum*) against *Botrytis cinerea* by biochar amendment involves jasmonic acid signaling. *Plant and Soil* 395, 31–44.

Mercier, J., Roussel, D., Charles, M.-T. and Arul, J. (2000) Systemic and local responses associated with UV-and pathogen-induced resistance to *Botrytis cinerea* in stored carrot. *Phytopathology* 90(9), 981–986.

Miura, I., Kamakura, T., Maeno, S., Hayashi, S. and Yamaguchi, I. (1994) Inhibition of enzyme secretion in plant pathogens by mepanipyrim, a novel fungicide. *Pesticide Biochemistry and Physiology* 48(3), 222–228.

Notte, A.-M., Plaza, V., Marambio-Alvarado, B., Olivares-Urbina, L., Poblete-Morales, M. *et al.* (2021) Molecular identification and characterization of *Botrytis cinerea* associated to the endemic flora of semi-desert climate in Chile. *Current Research in Microbial Sciences* 2, 100049.

Paz, I.C.P., Santin, R.C.M., Guimarães, A.M., Rosa, O.P.P., Quecine, M.C. *et al.* (2018) Biocontrol of *Botrytis cinerea* and *Calonectria gracilis* by eucalypts growth promoters Bacillus spp. *Microbial Pathogenesis* 121, 106–109. DOI: 10.1016/j.micpath.2018.05.026.

Petrasch, S., Knapp, S.J., Kan, J.A.L. and Blanco-Ulate, B. (2019) Grey mould of strawberry, a devastating disease caused by the ubiquitous necrotrophic fungal pathogen *Botrytis cinerea*. *Molecular Plant Pathology* 20(6), 877–892.

Poveda, J. (2020) Use of plant-defense hormones against pathogen-diseases of postharvest fresh produce. *Physiological and Molecular Plant Pathology* 111, 101521.

Poveda, J., Barquero, M. and González-Andrés, F. (2020) Insight into the microbiological control strategies against *Botrytis cinerea* using systemic plant resistance activation 10. DOI: 10.3390/agronomy10111822.

Rahman, M.U., Liu, X., Wang, X. and Fan, B. (2024) Grapevine gray mold disease: Infection, defense and management. *Horticulture Research* 11(9), uhae182.

Rhouma, A., Hajji-Hedfi, L., Ben Othmen, S., Kumari Shah, K., Matrood, A.A.A. *et al.* (2022) Strawberry grey mould, a devastating disease caused by the airborne fungal pathogen *Botrytis cinerea*. *Egyptian Journal of Phytopathology* 50(2), 44–50. DOI: 10.21608/ejp.2022.161763.1070.

Robinson-Boyer, L., Jeger, M.J., Xu, X.-M. and Jeffries, P. (2009) Management of strawberry grey mould using mixtures of biocontrol agents with different mechanisms of action. *Biocontrol Science and Technology* 19(10), 1051–1065.

Rosslenbroich, H.-J. and Stuebler, D. (2000) *Botrytis cinerea*—history of chemical control and novel fungicides for its management. *Crop Protection* 19(8–10), 557–561.

Savas, N.G., Yildiz, M., Eltem, R. and Ozkale, E. (2021) Determination of antifungal efficiency of some fungicides and secondary metabolites of trichoderma species against *Botrytis cinerea*. *Journal of Environmental Biology* 42(3), 705–713. DOI: 10.22438/JEB/42/3/MRN-1589.

Segarra, G., Elena, G. and Trillas, I. (2013) Systemic resistance against *Botrytis cinerea* in Arabidopsis triggered by an olive marc compost substrate requires functional SA signalling. *Physiological and Molecular Plant Pathology* 82, 46–50.

Sharma, K.K. (2018) Trichoderma in agriculture: An overview of global scenario on research and its application. *International Journal of Current Microbiology and Applied Sciences* 7, 1922–1933.

Sood, M., Kapoor, D., Kumar, V., Sheteiwy, M.S., Ramakrishnan, M. *et al.* (2020) Trichoderma: The "secrets" of a multitalented biocontrol agent. *Plants* 9(6), 762. DOI: 10.3390/plants9060762.

Spada, M., Pugliesi, C., Fambrini, M., Palpacelli, D. and Pecchia, S. (2023) Knockdown of Bmp1 and Pls1 virulence genes by exogenous application of rnai-inducing dsRNA in *Botrytis cinerea*. *International Journal of Molecular Sciences* 24(5), 1–22. DOI: 10.3390/ijms24054869.

Staats, M., Baarlen, P. and Kan, J.A.L. (2005) Molecular phylogeny of the plant pathogenic genus *Botrytis* and the evolution of host specificity. *Molecular Biology and Evolution* 22(2), 333–346.

Thambugala, K.M., Daranagama, D.A., Phillips, A.J.L., Kannangara, S.D. and Promputtha, I. (2020) Fungi vs. fungi in biocontrol: An overview of fungal antagonists applied against fungal plant pathogens. *Frontiers in Cellular and Infection. Microbiology* 10, 604923.

Tomas-Grau, R.H., Requena-Serra, F.J., Hael-Conrad, V., Martínez-Zamora, M.G., Guerrero-Molina, M.F. *et al.* (2018) Soft mechanical stimulation induces a defense response against *Botrytis cinerea* in strawberry. *Plant Cell Reports* 37, 239–250.

Trotel-Aziz, P., Couderchet, M., Vernet, G. and Aziz, A. (2006) Chitosan stimulates defense reactions in grapevine leaves and inhibits development of *Botrytis cinerea*. *European Journal of Plant Pathology* 114, 405–413.

Vicedo, B., Flors, V., Leyva, O.M., Finiti, I., Kravchuk, Z. *et al.* (2009) Hexanoic acid-induced resistance against *Botrytis cinerea* in tomato plants. *Molecular Plant-Microbe Interactions* 22(11), 1455–1465.

Wafaa, M.H. and Mostafa, A.E.S. (2012) Production and optimization of *Pseudomonas fluorescens* biomass and metabolites for biocontrol of strawberry grey mould. *American Journal of Plant Sciences* 3(7), 20690.

Wang, K., Jin, P., Han, L., Shang, H., Tang, S. *et al.* (2014) Methyl jasmonate induces resistance against *Penicillium citrinum* in chinese bayberry by priming of defense responses. *Postharvest Biology and Technology* 98, 90–97.

Wang, X., Glawe, D.A., Kramer, E., Weller, D. and Okubara, P.A. (2018) Biological control of *Botrytis cinerea*: Interactions with native vineyard yeasts from Washington state. *Phytopathology* 108(6), 691–701. DOI: 10.1094/PHYTO-09-17-0306-R.

Wang, C., Cao, Y., Wang, Y., Sun, L. and Song, H. (2019) Enhancing surfactin production by using systematic CRISPRi repression to screen amino acid biosynthesis genes in *Bacillus subtilis*. *Microbial Cell Factories* 18(1), 1–13. DOI: 10.1186/s12934-019-1139-4.

Widiastuti, A., Yoshino, M., Saito, H., Maejima, K., Zhou, S. *et al.* (2011) Induction of disease resistance against *Botrytis cinerea* by heat shock treatment in melon (*Cucumis melo* L.). *Physiological and Molecular Plant Pathology* 75(4), 157–162.

Williamson, B., Tudzynski, B., Tudzynski, P. and Van Kan, J.A.L. (2007) *Botrytis cinerea*: The cause of grey mould disease. *Molecular Plant Pathology* 8(5), 561–580.

Wilkinson, S.W., Pastor, V., Paplauskas, S., Pétriacq, P. and Luna, E. (2018) Long-lasting P-aminobutyric acid-induced resistance protects tomato fruit against *Botrytis cinerea*. *Plant Pathology* 67(1), 30–41.

Zhang, Z., Qin, G., Li, B. and Tian, S. (2014) Infection assays of tomato and apple fruit by the fungal pathogen *Botrytis cinerea*. *Bio-Protocol* 4(23), e1311–e1311.

Zhang, Z., Qin, G., Li, B. and Tian, S. (2015) Effect of cinnamic acid for controlling gray mold on table grape and its possible mechanisms of action. *Current Microbiology* 71, 396–402.

Zimmerli, L., Métraux, J.-P. and Mauch-Mani, B. (2001) p-Aminobutyric acid-induced protection of Arabidopsis against the necrotrophic fungus *Botrytis cinerea*. *Plant Pathology* 126(2), 517–523.

21 Basal Stem Rot Disease of Oil Palm in Indonesia

IRDA SAFNI AND LISNAWITA

Department of Agrotechnology, Faculty of Agriculture, Universitas Sumatera Utara, Medan, Indonesia

Introduction

Basal stem rot (BSR) disease, primarily caused by *Ganoderma boninense*, poses the greatest threat to sustainable oil palm production in South-east Asia, particularly in Indonesia. Early symptoms of BSR are often subtle and easily overlooked, beginning with slight discoloration of older leaves, which gradually fade from green to yellowish, starting at the base of the frond midrib. Over time, plant growth becomes stunted, with young leaves appearing shorter than normal and fewer midribs developing. The plant then exhibits wilting symptoms and irregular patches of chlorosis on the leaf surface, which later progress to necrosis. Older leaves begin to dry from tip to base but do not detach immediately, remaining attached to the stem and forming a distinctive "skirt" around the base.

As the disease progresses, more distinctive signs emerge, including unopened spear leaves (Fig. 21.1A), basal stem decay (Fig. 21.1B), soft and blackish-brown stem tissue, and eventually the development of basidiocarps at the base of the trunk (Fig. 21.1C) or around old frond bases, often concealed by vegetation or leaf litter. Basidiocarp formation occurs more frequently during wet seasons or under high humidity, which conditions that favors fungal growth and sporulation. Infection typically begins at the roots, where the fungus penetrates and colonizes vascular tissues, gradually disrupting water and nutrient flow. This systemic invasion causes internal stem decay and structural weakening, ultimately resulting in palm collapse (Fig. 21.1D), which marks the final stage of *G. boninense* infection. The most severe BSR-related losses have been recorded in Indonesia and Malaysia, with economic damage becoming significant when the infection affects more than 10% of the plantation.

Fig. 21.1. Field symptoms of basal stem rot in oil palm, (a) unopened multiple spear leaves; (b) trunk decay at the palm base; (c) basidiocarp of Ganoderma; and (d) palm collapse.

History of Oil Palm and BSR Disease's Occurrence in Indonesia and Its Economic Impact

Oil palms were first introduced to present-day Indonesia as ornamental plants for botanical gardens. However, by the early 20th century, their commercial benefits became evident, leading to widespread agricultural cultivation, particularly on the islands of Sumatra and Borneo. In 2006, Indonesia became the world's leading palm oil producer and continues to hold this position. The palm oil sector is a vital component of the Indonesian economy, contributing significantly to the renewable energy, food, and oleochemical industries. Indonesian palm oil represents approximately 25% of global vegetable oil production and 59% of the world's total palm oil output. In 2023, Indonesia produced 47.08 million tons of palm oil, of which 10.2 million tons were allocated for domestic food needs, 2.3 million tons for the oleochemical industry, 10.6 million tons for biodiesel, and 23.98 million tons for exports. Currently, palm oil accounts for nearly one-third (75.17 million tons) of the world's total vegetable oils. Despite these achievements, the industry faces serious productivity challenges due to BSR disease.

BSR caused by *Ganoderma* was first reported by Thompson in 1931 in the Asia-Oceania region on older palms following replanting in Malaysia. In Indonesia, the disease was first documented by PT Socfin Indonesia, a private oil palm company, in Serdang Badagai, North Sumatra, in 1954. This may represent the earliest recorded occurrence of BSR in Indonesia or the broader South-east Asian region. Initially, the Ganoderma species associated with oil palm was identified as *G. lucidum*. Over the next five decades, around 15 other species names were recorded, including *G. boninense G. chalceum, G. miniactocinctum, G. tornatum,* and *G. zonatum*. While several additional species have suggested as potential pathogens, *G. boninense* is currently recognized as the principle causal agent of BSR in South Asia and Oceania, including Indonesia.

Although various diseases threaten the oil palm industry, BSR caused by *G. boninense* has posed the most serious challenge to sustainable cultivation in South-east Asia for over 80 years. It remains the most destructive disease affecting the oil palm industry in Indonesia. While Africa, Papua New Guinea, and Thailand experience comparatively lower incidences of BSR, Indonesia and Malaysia suffer the most severe losses. In Indonesia, BSR disease can reduce oil palm yield per hectare significantly, with plant mortality rates exceeding 50% in heavily infested plantations.

The fungus infects and decomposes roots, bole, and trunk tissues, thereby preventing upward transport of water and nutrients. External symptoms include yellowing and wilting of fronds, unopened spear leaves, and eventual plant collapse. In severely impacted plantations, widespread infections can result in palm death and up to 68% reduction in oil palm productivity.

Although the disease has long been recognized in coastal regions, a survey of 13-year-old palms in both inland and peat soils revealed a typical disease incidence of approximately 30%. Similar issues are reported across Indonesian plantations. In certain inland areas of North Sumatra, BSR has killed up to 67% of oil palm trees within 25 years of planting. More severe palm losses – up to 25% within seven years – have been observed in fields where oil palm stumps were left in the ground during replanting. Meanwhile, in a peatland oil palm farm in Sumatra, BSR has destroyed between 30% and 54% of the palm stand, with an estimated yield reduction of 0.5 to 0.7 t/ha/yr over 14 years.

When the disease affects more than 10% of the stand, financial losses begin to accrue. Yield reduction in infected palms may occur either through from two main factors : the total loss of dues to palm death (direct loss), and through decreased weight of fresh fruit bunches (FFB) in still-living but infected palms (indirect loss). For every palm lost, FFB production decreases by an average of 0.16 t/ha. Once the stand is reduced by 50%, the average FFB yield declines by 35%. By the midpoint of the palms' economic lifespan, BSR can kill up to 80% of a plantation, resulting in cumulative FFB losses ranging from 0.04 t to 4.34 t/ha, depending on the plantation's age (10–22 years after planting). Recent probabilistic modeling estimates that a single plantation may suffer economic losses due to BSR in Indonesia, potential losses due to BSR are estimated at more than US$256 million for every 1% increase in disease incidence, with an annual regional impact of approximately US$500 million across South-east Asia. Among Indonesia's major oil palm-growing regions, North Sumatra has the highest infection level, followed by Riau and Jambi provinces, also located in Sumatra (Hendra *et al.*, 2024).

Environmental factors related to disease incidence are critical to the development of effective strategies. Poor soil nutrients can weaken plant resistance, making them more susceptible to infection. Susanto *et al.* (2013) reported that the incidence and infection rate of *Ganoderma* were higher in sandy-textured soils than in clay-textured soil. This is attributed to the physical properties of loose sandy soils, which have high porosity, fewer microorganisms, and more efficient water circulation, all of which contribute to a higher infection rate compared to clay soil.

Furthermore, Hendra *et al.* (2024) found that the level of BSR disease infection in various soil types in Sumatra was highest in peat soils (24.79 trees/ha), followed by alluvial (13.41 trees/ha), volcanic (7.25 trees/ha), and inland soils (0.07 trees/ha). The abundance of organic matter and plant debris in peat soil creates a nutrient-rich and conducive environment for *Ganoderma* development. The ideal humidity range for the growth of *G. boninense* mycelial growth is between 50% and 60%, with an optimal temperature range of 25°C to 32°C being optimal (Peng *et al.*, 2019). The high infection rate in peat soil is associated with the higher planting density typically used in these areas—150 to 180 oil palms/ha—compared to mineral soil, which have 136 to 148 oil palms/ha. This denser planting results in shorter distances between trees, increasing root contact and thereby the potential for infection. Dolmat (2005) reported that in deep peatland, a planting density of 200 oil palm trees/ha resulted in an infection rate that was 70 and 36% higher than those densities of 160 and 120 trees/ha, respectively.

Characteristics of *Ganoderma boninense*, the Causal Pathogen that Infects Oil Palm in Indonesia

Ganoderma boninense is a parasitic fungus with complex life cycle. Although not fully elucidated, its reproductive cycle resembles that of other Basidiomycetes and includes both sexual and vegetative reproduction. It produces sexual spores (basidiospores) and vegetative mycelia. Many researchers consider the basidiospores produced by fruiting bodies to be the primary inoculum, as they are easily dispersed by wind, water, or animals.

BSR infection typically originates from an existing inoculum source, such as infected trees, stumps, or trunks left in the field. The fungus forms a fruiting body that releases basidiospores, which may spread via air, insects, or wind. Fruiting bodies have diverse morphological characters, as reported by Gulo *et al.* (2024) (Table 21.1). Under suitable environmental conditions, basidiospores germinate to produce monokaryotic mycelium (n). When two compatible monokaryotic mycelia meet, they fuse to form dikaryotic mycelium (2n), which is more aggressive and pathogenic.

The mycelium then grows and spreads through the soil, facilitating contact between the roots of healthy oil palm roots and infected oil palm trees. Once the mycelium reaches the roots of healthy plants, it invades the root tissue, degrading lignin, cellulose, and hemicellulose through extracellular enzymes. This leads to internal decay of root and stem structures, disrupting water and nutrient transport and contributing directly to the physiological symptoms observed in infected palms.

At this stage, external symptoms remain invisible, but internal damage has already begun, rendering the plant physiologically unstable. As the infection advances, external symptoms emerge, beginning with foliar discoloration, stunted growth, and unopened spear leaves – signs of impaired vascular function due to fungal colonization. This is followed by drying of the spear leaves, wilting of leaf sheaths, and the eventual appearance of fungal fruiting bodies on the lower stem surface (Elkhateeb and Daba, 2022). Infected palms undergo rapid health deterioration, culminating in death. These dead palms subsequently become new inoculum, enabling the fungus to spread further through plant tissue or soil, thereby perpetuating the disease cycle repeats (Fig. 21.2).

Management Strategies and Control Methods

As a soil-borne pathogen, *Ganoderma* employs a survival strategy by forming compact melanin-rich mycelium on infected roots or at the base of the stem. This structure protects the pathogen from chemical and biological agents, allowing it to persist for extended periods. Consequently, the potential for damage from BSR increases annually, presenting ongoing challenges for effective disease control. Several management strategies can be employed to mitigate the impact of *Ganoderma*.

Monitoring and Early Detection System

Infection by *G. boninense* infection often remains undetected in its early stages because symptoms develop internally. Therefore, early detection

Table 21.1. Fruiting bodies morphological of Ganoderma spp. at the Adolina Seed Center, Indonesia (Gulo et al., 2024).

		Fruiting Bodies Morphologycal of Ganoderma spp.		
No	Name, codeSamples, location	Upper	Lower	Description
1	39–19/G-3187-1(Blok AD-04)	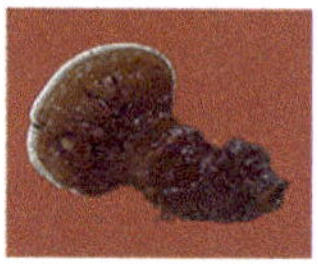	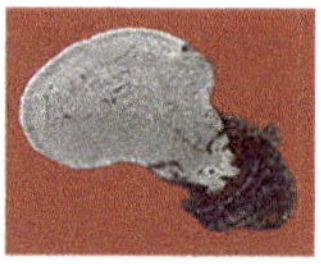	Upper: bronze-orange, fleshy, substipitate; Lower: white; the fruit body stalk is flabelliform and porous.
2	26–6/G-3187-2(Blok AD-04)			Upper: orange, fleshy, sessile; Lower: grayish white; the fruit body stalk is flabelliform and porous
3	40–20/G-3187-3(Blok AD-04)			Upper: orange-blond, fleshy, sessile; Lower: grayish white; the fruit body stalk is flabelliform and porous
4	7–2/G-3187-4(Blok AD-02)			Upper: dark brown-blond, hard, substipitate; Lower: grayish white; the fruit body stalk is flabelliform and porous
5	10–7/G-3187-5(Blok AD-02)			Upper: orange-blond, fleshy, substipitate; Lower: grayish white; the fruit body stalk is flabelliform and porous

Continued

Table 21.1. Continued

		Fruiting Bodies Morphologycal of Ganoderma spp.		
No	Name, codeSamples, location	Upper	Lower	Description
6	6–2/G-3187-6(Blok AD-02)		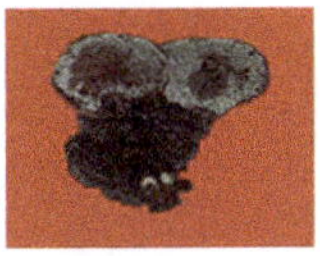	Upper: dark brown, hard, substipitate; Lower: grayish white; the fruit body stalk is flabelliform and porous
7	9–2/G-3187-7(Blok AD-02)			Upper: dark brown, hard, sessile; Lower: grayish white; the fruit body stalk is flabelliform and porous

and systematic monitoring are critical to preventing disease and preserving plantation productivity.

Initial detection methods include visual observation of early signs, such as changes in leaf coloration, midrib drying, and failure of spear leaves to open. However, since these symptoms typically appear only after the infection is well advanced, more sensitive techniques are required.

Molecular-based diagnostic methods, such as conventional PCR and quantitative PCR (qPCR), enable the detection of *Ganoderma* DNA in root, stem, or soil samples. Additionally, electrochemical and biosensor-based techniques can be used for early identification. Remote sensing tools, such as drone-based aerial imagery, are also effective in detecting changes in canopy structure. Furthermore, metabolite analysis using liquid chromatography-mass spectrometry (LC-MS) provides an advanced approach to identifying biochemical changes associated with infection.

By implementing these early detection and monitoring strategies consistently, disease progression can be curtailed, and control measures can be applied more effectively and promptly (Kurihara *et al.*, 2022; Karunarathna *et al.*, 2024; Cahyadi *et al.*, 2025).

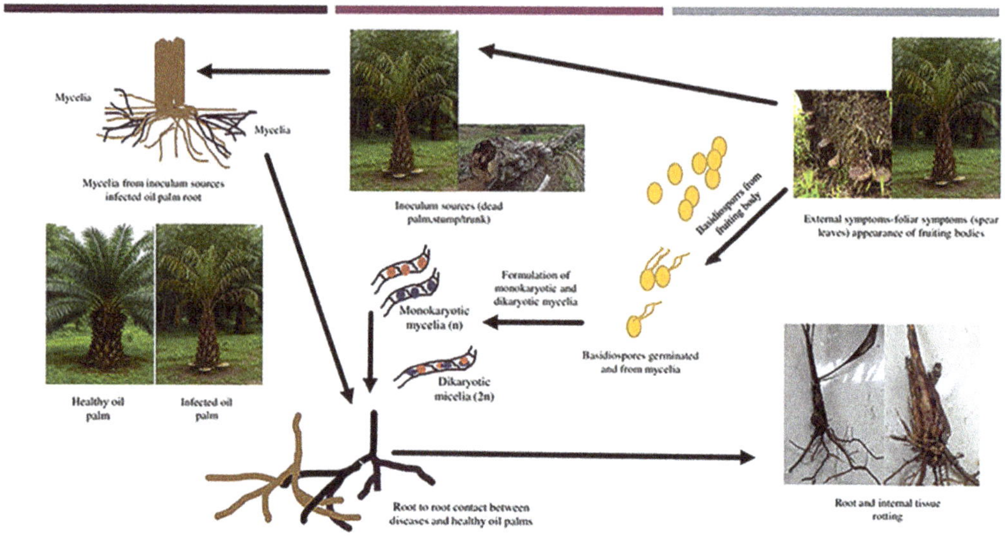

Fig. 21.2. Disease cycle of basal stem rot of oil palm (Adapted with permission from Bharudin *et al.* (2022); Karunarathna *et al.* (2024)).

Use of Disease-Free Seedlings

The use of pathogen-free planting materials is a crucial first step in preventing the spread of inoculum in the field. Seedlings produced through tissue culture and certified as disease-free have a significantly lower risk of being infected by pathogens. Tissue culture in oil palm is an important technique for vegetative propagation in oil palm, offering several advantages over conventional methods. One key advantage is the potential to propagate plants with resistance to multiple diseases simultaneously. For instance, plants resistant to both *Fusarium* and *Ganoderma* can be selected by initially screening several crosses of clonal plants for resistance to Fusarium wilt. These selected clones are then planted in areas endemic to *Ganoderma* to further identify those resistant to BSR. Tissue culture techniques hold promise for producing pathogen-free planting materials; however, continuous efforts are required to optimize protocols and maintain the necessary systems, infrastructure, and laboratory standards.

Technical Culture Control (Short-Term)

The control of BSR through technical culture practices aims to reduce the rate of infection. In established oil palm plantations, implementing sound culture practices is often recommended as a short-term disease management strategy. These practices are designed to halt disease progression, eliminate or reduce the sources of pathogen inoculum, and provide temporary recovery to affected oil palm plants. Cultural methods include land clearing (clean clearance), sanitation, stem chipping, deboling, soil cultivation, digging large planting holes, and constructing isolation trenches. While these techniques may not completely eradicate the disease, their combined application can reduce the risk of infection and minimize economic losses caused by *Ganoderma*.

Biological Control

Biological control agents (BCAs) are gaining increasing attention as environmentally friendly alternatives to chemical fungicides in the management of BSR in oil palm plantations. Over recent decades, numerous BCAs have been investigated for their ability to suppress *Ganoderma,* with most studies conducted *in vitro*. These BCAs include actinomycetes, arbuscular mycorrhizal fungi, *Bacillus* spp., *Diaporthe* spp., mycoparasites, *Penicillium* spp., *Streptomyces* spp., and *Trichoderma* spp. However, *in vitro* results must be validated through nursery and field trials to confirm the agents' practical efficacy.

Lisnawita *et al.* (2025) evaluated six Trichoderma isolates obtained from the rhizosphere of healthy oil palm plants in the Bukit Kijang plantation. *In vitro* tests showed that four isolates effectively inhibited *G. boninense,* with inhibition zone percentages ranging from 64.93 to 78.78%, while the remaining two isolates demonstrated inhibition below 50%. Further testing of these four isolates on 3-month-old Dura × Pisifera oil palm seedlings revealed that two isolates—*T. asperellum* and *T. virens* (as identified through molecular techniques)—significantly suppressed *Ganoderma* development and promoted vegetative growth after 7 months of inoculation. These two isolates are currently undergoing limited-scale field trials on immature oil palm trees. Commercial BCA products already available in the market include Marfu-P and Greemi-G (produced by PPKS), among others.

In addition to BCAs, another promising biotechnological approach is the development of genetically resistant oil palm varieties. Several Indonesian plantations have reported success in breeding *Ganoderma*-resistant palm through extensive research. Notable examples include DxP Socfindo Moderate Resistant Gano (MTG) (PT. Socfin Indonesia), DxP 540 NG (PPKS), DxP Spring MR Gano and DxP Themba MR Gano (PT. ASD Bakrie Oil Palm Seed), and DxP Dami Mas IGR (PT. Dami Mas Sejahtera). Offspring from crosses such as Cameroon x Cameroon, Cameroon x AVROS, and Nigeria x Nigeria have shown tolerance to BSR. Furthermore, tolerance genes have been identified in Zaire x Cameroon cultures, while partial resistance genes have been located in Congo x Cameroon cultures.

Plant Specialized Metabolites as Bioprotectants

In response to many environmental variables, plants produce some primary and secondary metabolites. Plant and other organisms produce and utilize metabolites in their metabolism. Certain specialized metabolites play an important role in the plant defense system against fungi, bacteria, and other pests, depending on their specific metabolite profile. Saragih and Lisnawita (2024) reported that administering *Asystasia gangetica* leaf extract at a concentration of 8 mg/ml to 2-month-old Dura × Pisifera oil palm seedlings infected with *Ganoderma* stimulated vegetative growth, suppressed disease incidence, and reduced disease severity in 8 months after inoculation. These results are supported by research by Saragih *et al.* (2021), who discovered that oil palm seedlings infected with *Ganoderma* contained isothiocyanate compounds (N=C=S), whereas this compound was absent in seedlings treated with specialized metabolites from *Diplazium esculentum* and *Rivina humilis*, both of which possess antifungals properties. Thus, the administration of specialized metabolites from these weed species may enhance the defense of oil palm seedlings against *Ganoderma* attacks and serve as an effective bioprotectant (Collinge *et al.*, 2022).

Chemical Control

Chemical control measures, such as the use of fungicides, have long been employed to manage *Ganoderma*-infected oil palms. For instance, organomercury and potassium hydroxyquinoline sulfate have been injected into diseased trees. More recently, the application of hexaconazole via pressure injection, stem injection, soil drenching, or soil injection has shown effectiveness in reducing the spread of BSR in infected palms. Pyraclostrobin has also proven successful in mitigating BSR in seedlings and has demonstrated beneficial physiological effects, including increased plant height, stem diameter, root mass, photosynthetic rate, PSII quantum efficiency, and relative leaf chlorophyll content. However, its efficacy in mature palms remains underexplored. Despite the availability, several effective fungicides, their use raises concerns about environmental contamination and the development of fungicide resistance.

In addition to fungicides, minerals also play an important role in enhancing plant resistance to pathogens through the induction of spesialized metabolites. Disease severity can be reduced through appropriate minerals fertilization, making it a viable strategy for managing calcium has been shown to reduce BSR severity and suppress *G. boninense* growth in infected seedlings. These minerals may also reinforce cell wall and middle lamella, thereby limiting pathogen penetration.

Biotechnological Approaches

Biotechnological innovations are increasingly being explored to complement conventional strategies for BSR management. One promising approach is the genetic transformation of oil palm, although this remains technically challenging due to the crop's long life cycle and the complexities of tissue culture propagation. Nonetheless, transformation has been achieved using *Agrobacterium tumefaciens*-mediated and biolistic methods. These enable the introduction of resistance-related genes, such as those encoding pathogenesis-related (PR) proteins, chitinases, and antifungal peptides.

Another promising strategy is Host-Induced Gene Silencing (HIGS), which involves engineering oil palms to produce double-stranded RNAs (dsRNAs) that silence specific essential genes in *Ganoderma*. Advances in fungal genomics and transcriptomics have facilitated the identification of virulence-associated genes – particularly those related to lignin degradation and effector secretion – that serve as potential HIGS targets. The integration of HIGS with genomics-based target identification offers a species-specific and environmentally benign approach to suppress fungal pathogenicity.

Integrated *Ganoderma* Management (IGM)

The individual control methods described above have not yet provided fully satisfactory outcomes in limiting the spread of *Ganoderma* infection in the field. Consequently, an integrated control strategy is necessary, combining multiple approaches. While no single method offers completely disease control, integration of these methods can minimize infection risk and reduce economic losses caused by *Ganoderma*. Figure 21.2 presents an integrated management framework tailored to different growth stages of oil palm: immature palms, mature palms, and replanting areas. The process begins with a Ganoderma census and symptom identification, followed by management interventions based on the age of the palms and severity of infection. The Disease Severity Index (DSI) is employed to assess infection levels and guide treatment decisions.

Currently no single control method has been effective for managing BSR disease. Therefore, it is recommended to apply more than one method by either combining the above methods or applying integrated control.

Furthermore, collaborative efforts involving scientists, breeders, and farmers are essential for developing and implementing sustainable strategies against this destructive plant pathogen. Through rigorous scientific investigation and evidence-based practices, it is possible to safeguard global plant health and mitigate the severe economic consequences of *Ganoderma* infections (Fig. 21.3).

Future Research Areas to Improve BSR Management

Emerging microbiome engineering approaches focus on utilizing Plant Growth Promoting Microorganisms (PGPMs)—including biofertilizers, biostimulants, bioprotectants, and biological control agents—to preserve microbial diversity for oil palm sustainability. These efforts must be aligned with ongoing climate change research.

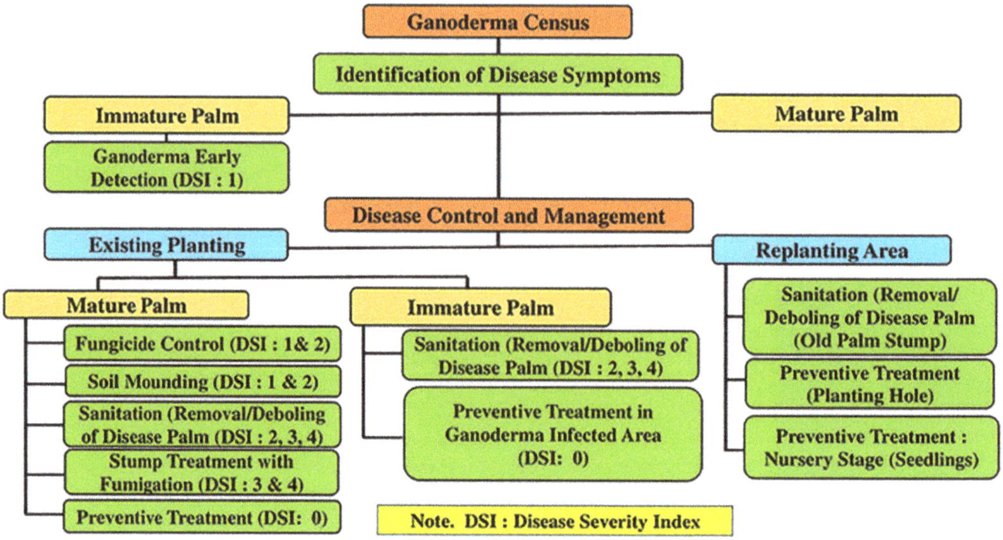

Fig. 21.3. Integrated *Ganoderma* Management (IGM) framework for oil palm plantations (adapted with permission from MPOB in Pascal Biotech, 2025). DSI 0: No disease symptoms; the tree looks healthy. DSI 1: Early symptoms, such as discoloration at the base of the trunk or slight signs of damage. DSI 2: Moderate symptoms, with more pronounced damage at the trunk base and early signs appearing on other parts of the tree. DSI 3: Severe symptoms, with clear damage at the trunk base and visible symptoms affecting most parts of the tree. DSI 4: The tree is dead or near death, with extensive damage throughout all parts of the tree.

Innovative genetic and detection systems are needed to enhance disease resistance and improve management precision. This includes the application of genome editing technologies, the identification and tracking of *Ganoderma* resistance genes, and the integration of molecular diagnostics into precision agriculture frameworks.

Spray-Induced Gene Silencing (SIGS) presents a non-transgenic, RNA-based approach to silencing key pathogenicity genes in *G. boninense*. SIGS involves the external application of synthetic double-stranded RNAs (dsRNAs), and its effectiveness can be significantly enhanced through nanocarrier-based delivery systems that improve dsRNA stability and uptake.

CRISPR/Cas genome editing offers a promising avenue for engineering disease-resistant oil palm varieties and modulating plant-microbe interactions. Functional genomics and transcriptomic analyses of both host and pathogen will be critical to identifying novel molecular targets, thereby enabling the development of durable, precise, and environmentally sustainable BSR management strategies.

Discussion Points

1. How was the oil palm initially introduced to Indonesia?
2. Explain the impact of *Ganoderma* infection on the oil palm.
3. Describe the integrated *Ganoderma* management.
4. Why have the various control methods described above not yet provided satisfactory results in managing *Ganoderma* in the field?
5. Research Task: Write an essay on the prospects for BSR disease control using one of the future research areas outlined above.

Further Reading

Govender, N., Mui-Yun, W. and Paterson, R.R.M. (2020) Opportunities for new-generation *Ganoderma boninense* biotechnology. Grand Challenges in Fungal Biotechnology, 477–500.

Keerthana, J., Amrutha Lakshmi, M., Aditya, S., Ajesh, B.R. and Manyam, P. (2023) Detection and management of basal stem rot of oil palm: Classical to modern approaches. In: *In: Detection, Diagnosis and Management of Soil-Borne Phytopathogens*. Springer Nature Singapore, Singapore, pp. 225–267.

Khoo, Y.W. and Chong, K.P. (2023) *Ganoderma boninense*: General characteristics of pathogenicity and methods of control. *Frontiers in Plant Science* 14, 1156869.

Rupaedah, B., Prasetyo, A.E., Hidayat, F., Asiani, N., Wahid, A. *et al.* (2024) Evaluation of microbial biocontrol agents for Ganoderma boninense management in oil palm nurseries. *Journal of the Saudi Society of Agricultural Sciences* 23(3), 236–244.

Siddiqui, Y., Surendran, A., Paterson, R.R.M., Ali, A. and Ahmad, K. (2021) Current strategies and perspectives in detection and control of basal stem rot of oil palm. *Saudi Journal of Biological Sciences* 28(5), 2840–2849.

References

Bharudin, I., Wahab, A.F.F.A., Samad, M.A.A., Yie, N.X., Zairun, M.A. *et al.* (2022) Review update on the life cycle, plant–microbe interaction,genomics, detection and control strategies of the oil palm pathogen *Ganoderma boninense*. *Biology* 11, 251. DOI: 10.1088/1755-1315/1413/1/012032.

Cahyadi, M.N., Syariz, M.A., Taufany, F., Lisnawita,Wismaroh, S.S. *et al.* (2025) Integration of low-cost gnss and multispectral camera to increase oil palm position accuracy and health monitoring. *Civil Engineering Journal* 11(3), 988–1010. DOI: 10.28991/CEJ-2025-011-03-010.

Collinge, D.B., Jensen, D.F., Rabiey, M., Sarrocco, S., Shaw, M.W. *et al.* (2022) Biological control of plant diseases – what has been achieved and what is the direction? *Plant Pathology* 71(5), 1021–1228. DOI: 10.1111/ppa.13555.

Dolmat, M.T. (2005) *Technologies for Planting Oil Palm on Peat*. Malaysian Palm Oil Board, Malaysia, pp. 12–30.

Elkhateeb, W.A. and Daba, G.M. (2022) The precious *Ganoderma* mushroom and plant diseases. *Journal of Pharmacy and Drug Development* 1(1), 1–7. DOI: 10.58489/2836-2322/005.

Gulo, G.K., Lisnawita and Safni, I. (2024) Characterization of basal stem rot pathogen in oil palm trees in Adolina Seed Center, North Sumatra, to support sustainable food security. *IOP Conference Series: Earth and Environmental Science* 1413, 012032. DOI: 10.1088/1755-1315/1413/1/012032.

Hendra, H., Haryadi, D., Sidhu, M., Samosir, Y. and Tan, J.S. (2024) Comparative incidence of Ganoderma basal stem rot and upper stem rot disease in oil palm in relation to soil type andgeographical distribution. *IOP Conference Series: Earth and Environmental Science* 1308(1), 012016. DOI: 10.1088/1755-1315/1308/1/012016.

Karunarathna, S.C., Patabendige, N.M., Lu, W., Asad, S. and Hapuarachchi, K.K. (2024) An in-depth studyof phytopathogenic *Ganoderma*: Pathogenicity, advanced detection techniques, control strategies, andsustainable management. *Journal of Fungi* 10(414). DOI: 10.3390/jof10060414.

Kurihara, J., Koo, V.-C., Guey, C.W., Lee, Y.P. and Abidin, H. (2022) Early detection of basal stem rot diseasein oil palm tree using unmanned aerial vehicle-based hyperspectral imaging. *Remote Sensing* 14, 799. DOI: 10.3390/rs14030799.

Lisnawita, Safni, I., Hanum, H., Effendy, M., Sitepu, S.F. *et al.* (2025) Indigenous Trichoderma isolates from Bukit Kijang, North Sumatra, Indonesia, are biocontrol agents against *Ganoderma boninense* in infected oil palm seedlings. *Ecological Frontiers* 45(1), 175–184. DOI: 10.1016/j.ecofro.2024.10.006.

Malaysian Palm Oil Board (MPOB) in Pascal Biotech (2025) Integrated Ganoderma Management. Available at: https://www.pascal-biotech.com (accessed 15 February 2025).

Peng, S.H.T., Yap, C.K., Ren, P.F. and Chai, E.W. (2019) Effects of environment and nutritional conditions on mycelial growth of *Ganoderma boninense*. *International Journal of Oil Palm* 2(3), 95–107. DOI: 10.35876/ijop.v2i3.

Saragih, W.S. and Lisnawita. (2024) Aktivitas antifungi ekstrak daun *Asystasia gangetica* (L.) T. Anderson terhadap bibit kelapa sawit (*Elaeis guineensis* Jacq.) terinfeksi *Ganoderma boninense*. *Jurnal Agrotek Tropika* 12(3), 632–641. DOI: 10.23960/jat.v12i3.7239.

Saragih, W.S., Purba, E., Lisnawita and Basyuni, M. (2021) The Fourier transform infrared spectroscopy from *Diplazium esculentum* and *Rivina humilis* analysis to reveals the existence of necessary components in oil palm plantations of *Ganoderma boninense* control. *Biodiversitas* 22(9), 3645–3651. DOI: 10.13057/biodiv/d220902.

Susanto, A., Prasetyo, A.E. and Wening, S. (2013) Laju infeksi Ganoderma pada empat kelas tekstur tanah. *Jurnal Fitopatologi Indonesia* 9(2), 39–39. DOI: 10.14692/jfi.9.2.39.

22 Resistant Grapevine Varieties to Downy Mildew

I. Salotti, T. Caffi, G. Fedele and V. Rossi

Department of Sustainable Crop Production (DI.PRO.VES.), Università Cattolica del Sacro Cuore, Via E. Parmense 84, 29122, Piacenza, Italy

Introduction

Downy mildew (DM) is likely the most important disease of grapevine in grape-growing areas with frequent rains. The causal agent of DM is the obligate, biotrophic oomycete *Plasmopara viticola* (Berk et Curtis) Berlese et De Toni, which originated from North America and was accidentally introduced in Europe at the end of the 19th century (Millardet, 1881). After the first report of the presence of DM in European vineyards in 1878, the pathogen rapidly spread throughout Europe causing periodic, severe epidemics (Gessler *et al.*, 2011). Control of DM largely relies on fungicides that, however, increase crop protection costs and can have a negative impact on human health and the environment (Peressotti *et al.*, 2010). The search for alternative methods to control *P. viticola* is of paramount importance, and grapevine varieties showing partial resistance to DM represent an innovation that may boost sustainability in viticulture (Töpfer *et al.*, 2011).

While the cultivated Eurasian grapevine *V. vinifera* L. is generally susceptible to DM, a long adaptation and coevolution history of *P. viticola* with wild American and Asian genotypes of the genera *Vitis* and *Muscadinia* led to the development of mechanisms of defence against the pathogen in some resistant genotypes (Kortekamp *et al.*, 1997; Bellin *et al.*, 2009; Gessler *et al.*, 2011). Resistance is conferred by quantitative trait loci (QTLs) named *Rpv* (meaning: "Resistance to *P. viticola*"); QTLs are regions of DNA associated with a specific phenotypic trait, which influence its variation through genetic interactions with each other and the environment. Due to the complexity of these interactions, there is a high variability in the level of resistance to DM among grapevine species (Boso and Kassemeyer, 2008; Bove and Rossi, 2020). Resistant accessions in North American *Vitis* species allow *P. viticola* to complete its life cycle, though one or more stages of the disease cycle are disrupted, including spore germination, penetration, colonization of the inner host tissue, and sporangial production. Other genotypes as some

Table 22.1. List of most exploited genes or resistance to *Plasmopara viticola* (*Rpv*) and their origins.

Rpv	Chromosome	Origin of resistance	References
Rpv1	12	*Muscadinia rotundifolia*	Merdinoglu *et al.* (2003)
Rpv2	18	*M. rotundifolia*	Wiedemann-Merdinoglu *et al.* (2006)
Rpv3	18	*Vitis rupestris*, *V. linecumii*, *V. labrusca*, *V. riparia*	Bellin *et al.* (2009); Di Gaspero *et al.* (2012); Zyprian *et al.* (2017)
Rpv5	9	*V. riparia*	Marguerit *et al.* (2009)
Rpv6	12	*V. riparia*	Marguerit *et al.* (2009)
Rpv8	14	*V. amurensis*	Blasi *et al.* (2011)
Rpv9	7	*V. riparia*	Moreira *et al.* (2011)
Rpv10	9	*V. amurensis*	Schwander *et al.* (2012)
Rpv12	14	*V. amurensis*	Venuti *et al.* (2013)
Rpv13	12	*V. riparia*	Moreira *et al.* (2011)
Rpv14	5	*V. cinerea*	Ochssner *et al.* (2016)

Muscadinia species or some exotic Asian grapes are able to prevent stomata penetration or halt hyphal growth during host tissue colonization, avoiding the establishment of infection, symptoms' onset, or sporulation (Jürges *et al.*, 2009). Depending on the host genotype and the *Rpvs* involved in the resistance response, different mechanisms may be activated by the plant to stop or alter *P. viticola* development, such as a hypersensitive response (Bellin *et al.*, 2009; Venuti *et al.*, 2013; Zyprian *et al.*, 2017), callose and lignin accumulation (Kortekamp *et al.*, 1997; Gindro *et al.*, 2003), synthesis of stilbene phytoalexins (Gindro *et al.*, 2003), cell necrosis (Boso and Kassemeyer, 2008; Bellin *et al.*, 2009; Zini *et al.*, 2015), induction of peroxidase activity (Kortekamp *et al.*, 1998), and accumulation of phenolic compounds in the plant tissues surrounding the infection sites (Alonso-Villaverde *et al.*, 2011). To date, 24 *Rpvs* have been identified and described for their chromosome position (Available at: www.vivic.de, accessed 18 December 2023); a list of the most exploited *Rpv* and their origins is reported in Table 22.1.

Partially resistant varieties originate from the introgression of *Rpv* genes from wild grapevines. Since the discovery of *Rpv*, many breeding programmes have been developed to obtain new varieties (hybrids) that combine the resistance to DM with the desirable agronomic and grape-quality traits carried by *V. vinifera* cultivars (Töpfer *et al.*, 2011; Merdinoglu *et al.*, 2018; Töpfer and Trapp, 2022). Anderson (2013) reported that resistant hybrids for wine production are mainly spread in America and Eastern Europe, specifically in Brazil (about 41,046 ha, 83% of the national viticultural surface), USA (about 11,980 ha, 5%), Moldova (about 11,656 ha, 13%), Russia (about 9430 ha, 37%), Hungary (about 7450 ha, 11%), Ukraine (about 3251 ha, 6%), and Canada (about 2680 ha, 27%). Limited diffusion of resistant varieties has been linked to restrictive regulations, especially in the European Union (EU), where the Regulation 1308/2013/EC stated that resistant varieties could be grown to produce Table and PGI (Protected Geographic Indication) wines, but not PDO (Protected Denomination of Origin) wines (which admit only the varieties in which the whole genome is from *V. vinifera*). This regulation has been recently modified (Regulation 1308/2013/EC updated on 08/12/2023), so that PDO wines can be obtained from vine varieties belonging to *V. vinifera* or a cross between the *V. vinifera* and other species of the genus *Vitis*. This may increase the commercial use of partially resistant varieties and then their plantation acreage.

Brief History of Breeding for Resistant Varieties

After the introduction of *P. viticola* in Europe, breeding programmes started in several European countries between the end of the 19th and the beginning of the 20th century (Töpfer *et al.*, 2011; Merdinoglu *et al.*, 2018; Töpfer and Trapp, 2022).

Nevertheless, the origins of grapevine breeding preceded efforts to obtain resistant varieties in Europe by about one hundred years. First breeding activities arose in America to marry the agronomical and quality traits of *V. vinifera* with American native vines and obtain hybrids adapted to local climates and with high-quality berry characteristics for winemaking or fresh consumption (Hedrick, 1908). The first variety successfully bred was Alexander (J. Alexander, around 1800), a hybrid originated from *V. labrusca*. Several hybrids followed, including Cunningham [*V. aestivalis* ×*V. vinifera*; J. Cunningham, 1812], Isabella [*V. labrusca* × *V. vinifera*; I. Gibbs, 1816], Catawba [*V. labrusca* × *V. vinifera*; J. Scholl, 1819], Norton [*V. aestivalis* ×*V. vinifera*; D. N. Norton, 1830], and Concord [*V. labrusca* × *V. vinifera*; E. W. Bull, 1849]. These variates are the funders of the so-called 'American hybrids' (Töpfer *et al.*, 2011; Di Gaspero *et al.*, 2012; Available at: www.vivc.de, accessed 18 December 2023).

In Europe, the first breeding activities started in France by both public institutions and private viticulturists as A. Seibel (1844–1936) and B. Seyve (1864–1939), resulting in thousands of varieties, the so-called 'French hybrids'; these hybrids were characterized by great variability in grape quality and resistance expression. They were primarily crosses between *V. vinifera* varieties and interspecific hybrids or crossed progeny so referring to them with the term 'hybrid' is improper and should be avoided at least in a scientific context (Töpfer and Trapp, 2022). Most popular French varieties included Leon Millot [Millardet et Grasset 101 O.P. (*V. vinifera* × *V. riparia*) × Goldriesling (*V. vinifera*); E. Kuhlmann, 1911], Seyval Blanc [Seibel × Rayon D'Or, two varieties derived from several interspecific crossing, with resistance-bearing ancestor *V. riparia*; B. Seyve, 1919], Chambourcin [Joannes Seyve 11369 × Plantet, two varieties derived from several interspecific crossing, with resistance-bearing ancestor *V. rupestris*; J. Seyve, 1960] (Gessler *et al.*, 2011; Töpfer *et al.*, 2011; Available at: www.vivc.de, accessed 18 December 2023). In 1929, French varieties were grown on about 250,000 ha, with an increasing trend until a peak in 1958 when they accounted for 500,000 ha (Töpfer *et al.*, 2011). Later on, French government policies and advances in crop protection products dropped the plantation acreage, although the cultivation of some varieties survived in limited areas, like Chambourcin in Loire Valley. The political ban was mainly due to poor quality linked to varieties showing off-flavours recognizable at the juice level, such as the 'foxy' taste (e.g. due to furaneol, methyl anthranilate, or 2-aminoacetophenon) inherited from *V. labrusca*, or developed after the fermentation process for wine-making (Rapp and Versini, 2015; Töpfer and Trapp, 2022). Also, the effective use of copper against DM and sulphur against powdery mildew reduced the need for cultivars with resistance to these pathogens (Gessler *et al.*, 2011; Töpfer *et al.*, 2011).

Although breeding programmes stopped in France and in other grape-growing European areas, breeding activities continued in other countries like Germany and Hungary during the 20th century. For instance, privately-funded breeding programmes by e.g. as E. Bauer, in which valuable French varieties were exploited for the development of German varieties that matched grape quality and DM resistance like Regent [Diana x Chambourcin; G. Alleweldt, 1967] (Gessler *et al.*, 2011; Töpfer *et al.*, 2011; Available at: www.vivc.de, accessed 18 December 2023), originally conceived for the German market and currently grown on 1618 ha (Deutsches Weininstitut, 2023).

Nowadays, breeding programmes exploiting resistance from both American and Asian wild grapes are running in several institutions, e.g., in the USA at Cornell University, UC Davis, and Florida AM University; in China at more than 20 universities and research institutes; in Australia at CSIRO; in Germany at JKI Gelweilerhof and WBI Freiburg i.Br; in France at INRA; in Italy at the University of Udine, Fondazione Edmund Mach, CREA-VE, and VCR research centre (Available at: www.vivairauscedo.com/en/vcr-research-center/, accessed 20 December 2023).

Effect of Resistance on *Plasmopara viticola* Disease Cycle: The Components of Partial Resistance

DM epidemics start with the germination of a relatively small number of *P. viticola* oospores, which were formed at the end of the previous season and overwintered in the leaf litter or in soil. Primary infections are caused by zoospores released by the sporangia formed by germinating oospores and splashed to plant surfaces by rainfall. A massive clonal multiplication involving secondary infections carried on by air-borne sporangia containing zoospores developed from lesions follows primary infections (Blaeser and Weltzien, 1979; Fig. 22.1). Splash-deposited or sporangia-released zoospores go through different processes to cause infection, which simply swim (through flagella) to the stomata (i.e., the penetration sites), germinate (i.e., for a germ tube) and penetrate into the host tissue, form intercellular hyphae that invade the host tissue and form haustoria to uptake nutrients, produce sporangiophores and sporangia on the surface of the affected tissue (i.e., the DM lesion).

Partial resistance conferred by *Rpvs* affects several stages of the disease cycle (Fig. 22.2), including penetration, colonization of the inner tissue, the

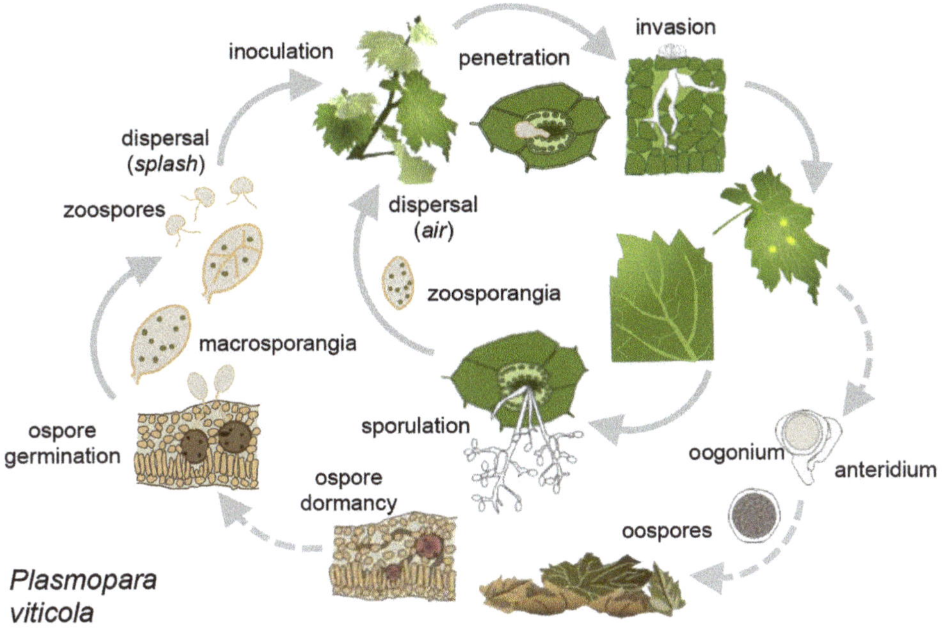

Fig. 22.1. Life cycle of *Plasmopara viticola*, the causal agent of grapevine downy mildew (from Muccinelli and Rossi, 2022).

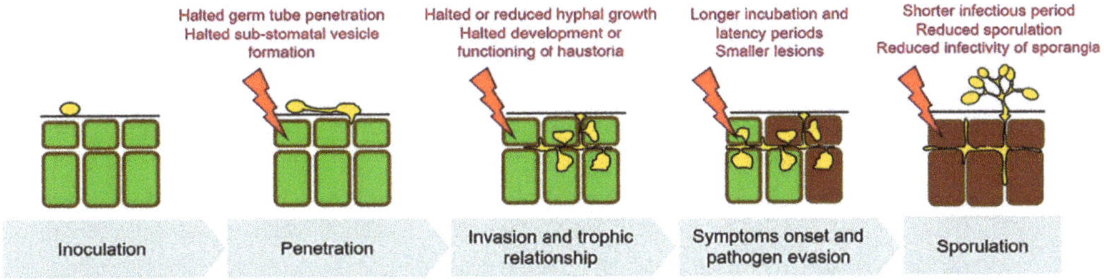

Fig. 22.2. Effect of components of partial resistance on the main steps of *Plasmopara viticola* infection cycle.

duration of the latent period (i.e., the period between infection and pathogen evasion), the length of the infectious period (i.e., the period in which lesions support the continuous production of sporangia), and sporulation (Van der Plank, 1963). These processes are called "components of partial resistance"; in fact, the resistance shown in each process is a component of the phenotype exhibited by the plant, i.e., the overall resistance that regulates the progress of an epidemic resulting from a chain of disease cycles during the host-growing season (Parlevliet, 1979; Willocquet *et al.*, 2017). The components of resistance have been studied in several pathosystems, with particular attention to reduced infection efficiency, a longer latent period, a shorter infectious period, reduced sporulation, and smaller lesion size. Recently, Bove and Rossi (2020) have investigated the components' partial resistance to *P. viticola* on several grapevine genotypes carrying one or more *Rpvs* from *V. rupestris* or *V. amurensis*, in comparison to a sensitive variety (Merlot). This study reported great variability in the expression of resistance, depending on the genotype and the component of partial resistance examined, leading to low to high resistance phenotypes.

Infection efficiency or frequency are defined as the number of DM lesions generated by each zoospore deposited on the leaf surface or by the proportion of inoculated sites, respectively (Lalancette *et al.*, 1987; Bove and Rossi, 2020). Bove and Rossi (2020) reported a reduced infection frequency in partially resistant varieties ranging from 6 to 43%, in comparison to the sensitive variety. Such a reduction is the consequence of the ability of the plant to prevent or halt some steps of the infection process, like penetration of the germ tube through stomata, formation of the sub-stomatal vesicles, invasion and development or functioning of haustoria. This ability is connected to defence mechanisms activated by the plant before or soon after *P. viticola* infection; common defence mechanisms are callose deposition in stomata, presence of inner cuticular rims, and hypersensitive response (HR). Gindro *et al.* (2003) observed in the resistant variety Solaris [Merzling × Geisenheim 6493], which carries *Rpv3* and *Rpv10*, callose deposition around the germinating zoospores starting from 7 hour after inoculation, and zoospores were enveloped into stomata secretions within 24 hour. Furthermore, stomata in the vicinity of the infection sites deposited callose in and around the stomatal opening. Jürges *et al.* (2009) and Alonso-Villaverde *et al.* (2011) detected, in several North American and Siberian *Vitis* species and *M. rotundifolia*, the presence of an inner cuticular rim at the transition between the sub-stomatal cavity and the stoma, which suppress the hyphal growth at early stages. An HR has been reported in several varieties (Gindro *et al.*, 2003; Bellin *et al.*, 2009). The HR consists of the rapid and localized plant cell death induced after infection to prevent pathogen nutrition and growth. The HR involves the massive production and accumulation of reactive oxygen species (e.g., hydrogen peroxide) that can modulate localized plant cell death of several cells around the infection site, other than the cells invaded by or having direct contact with the pathogen. The HR is generally recognized by the presence of brown, dead cells at the infection site and a visible brown lesion may develop if sufficient cells die (Fig. 22.3).

Longer latent periods are related to halted or reduced hyphal growth in the mesophyll of the host plant due to mesophyll morphology or reduced availability of nutrients from host cells. For instance, larger intercellular spaces observed in resistant varieties like Solaris imply that the hyphae of *P. viticola* have to cover a longer distance to target the cells and develop haustoria (Alonso-Villaverde *et al.*, 2011). Also, the HR causes cell death and reduces the number or the functioning of haustoria and, therefore, the uptake of nutrients from host cells by the pathogen (Bellin *et al.*, 2009). Bove and Rossi (2020) observed in the resistant varieties Bronner [Merzling x Geisenheim 6494], which carries *Rpv3* and *Rpv10*, and Johanniter [Riesling Weiss × Freiburg 589–54], which carries *Rpv3*, an increase in the length of the latent period of 1 or 2 days compared to the sensitive variety Merlot.

Leaf anatomy and lack of nutrients are also related to shorter infectious periods, reduced sporulation, and smaller lesions. Mesophyll thickness as well

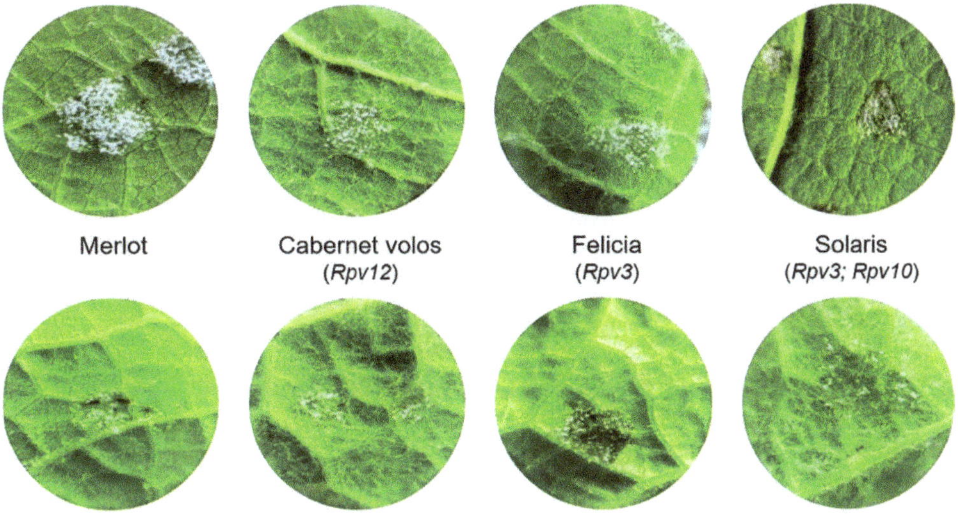

Merlot | Cabernet volos (*Rpv12*) | Felicia (*Rpv3*) | Solaris (*Rpv3; Rpv10*)

Regent (*Rpv3*) | Johanniter (*Rpv3*) | Fleurtai (*Rpv12*) | Bronner (*Rpv3; Rpv10*)

Fig. 22.3. Foliar discs showing sporulation of *Plasmopara viticola* on the sensitive control Merlot and grapevine varieties carrying one or more genes of resistance (*Rpvs*). Hypersensitive response (black, dotted necrosis) is clearly visible in Solaris and Bronner beneath sporulation.

as the accumulation of resveratrol, flavonoids, peroxidase compounds, and lignin in mesophyll cells (Alonso-Villaverde *et al.*, 2011) all limit the growth of the pathogen in the host and result in smaller lesions. In the variety Bianca [Bouvier × Villard Blanc], which carries *Rpv3*, an HR at the infection sites within 2 days post inoculation did not halt mycelial growth but reduced pathogen performances, including reduced sporangiophore density and reduced numbers of sporangia produced per unit of leaf area (Bellin *et al.*, 2009). The reduction in the number of sporangia produced per lesion in partially resistant varieties observed by Bove and Rossi (2020) ranged from 44 to 99% compared to the sensitive variety (Fig. 22.3). In some of the resistant varieties the infectivity of sporangia produced on DM lesions (i.e. their ability to cause new infections) was also affected; for instance, in Bronner, the sporangial infectivity was reduced to about 22% compared to the sensitive variety.

Effect of Partial Resistance on Epidemic Development

A polycyclic epidemic results from the concatenation of infection cycles during the plant-growing season (Zadoks and Schein, 1979). As detailed in the previous paragraph, components of partial resistance act on both the current and the next disease cycles, and therefore slow down the progress of DM epidemics because of a reduced affected area (in terms of both the number of lesions and their size), longer latency periods, and reduced sporulation (as reduced infectious period, sporangiophore density, and number of sporangia). From an epidemiological point of view, components of partial resistance contribute to the reduction of the apparent infection rate ® of the epidemic, i.e., the progress rate of the disease (Van der Plank, 1963; Parlevliet, 1979). For this reason, partial resistance has been also called *r*-reducing resistance (Parlevliet, 1979).

To explain the effect of partial resistance on DM epidemic development, Bove *et al.* (2021) proposed a simulation model that includes the following four components of partial resistance: (1) infection efficiency; (2) latency period; (3) sporulation; and (4) infectious period. Components of partial resistance were investigated, alone or combined, for their effects on epidemic development on leaves and clusters.

When components were studied alone, Bove *et al.* (2021) observed that reduced infection efficiency had the greatest effect on the slowdown of the epidemic, followed by sporulation, infectious period, and latency period. This component, in fact, reduces primary infection originating from the overwintering oospores on foliage in spring and suppresses secondary infections on both leaves and bunches during the growing season. Increase in the length of the latent period (with later starting of sporulation on DM lesions) and reduction of both the length of the infectious period (with the consequent reduction of the time over which the infectious, sporulating lesions contribute to the epidemic) and sporulation rate in the resistant varieties had an effect on the abundance of secondary inoculum and then in the number of new infections.

Epidemics developed in plants with low resistance levels (i.e., low reduction of infection efficiency, short latent periods, long infectious periods, or low reduction of sporulation) originated a disease progress curve with a logistic shape on clusters and a logistic shape followed by a decline due to leaf senescence on foliage. These curves were similar to the S-shaped curves observed in sensitive varieties, although final levels of disease on leaves and clusters were lower. On the contrary, epidemics developed in plants with high resistance present reduced rates of disease progress, leading to slowly growing, flattened curves and further lower final disease levels (Fig. 22.4). Varying levels of disease on foliage and clusters were observed combining low, medium, and high resistance in the four considered components, with negligible disease on foliage when all the components expressed high resistance (Fig. 22.5).

Risk of Overcoming Resistance to Downy Mildew

Most of the genetic factors conferring resistance to DM and currently exploited in breeding programmes are located in genomic regions rich in NBS-LRR-like resistance genes that translate into nucleotide-binding-site/leucine-rich-repeat (NBS–LRR) protein receptors (Merdinoglu *et al.*, 2018). These genes operate and translate the flow of events from pathogen perception to the defence responses and eventually hypersensitive plant cell death and are potentially involved in a gene-for-gene interaction with the pathogen (Peressotti *et al.*, 2010; Merdinoglu *et al.*, 2018). The gene-for-gene model implies the presence of a resistance gene ® in the plant and its corresponding avirulence (*Avr*) gene in the pathogen, and their signal-mediated interaction

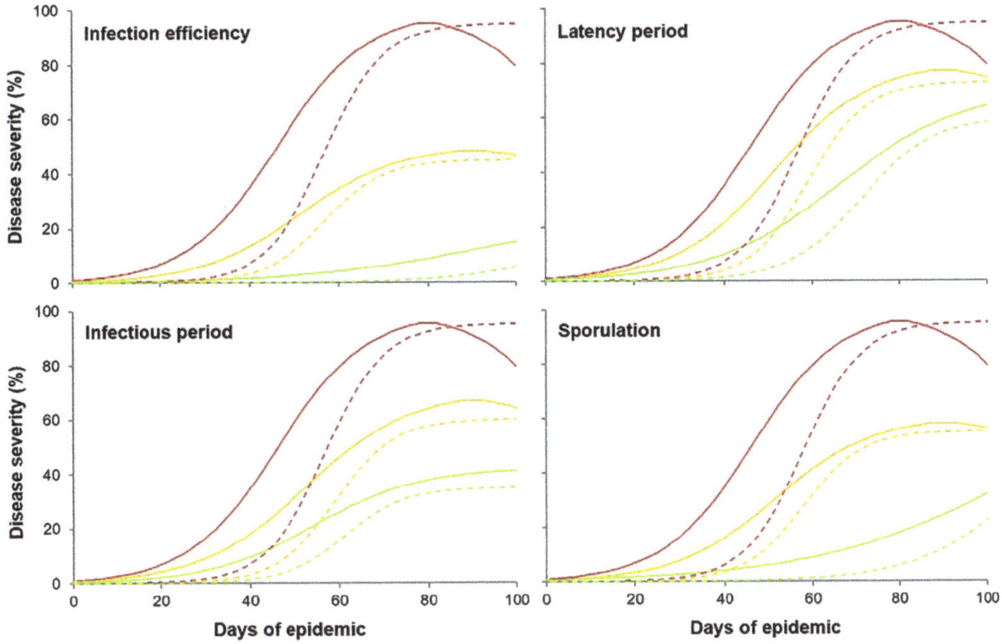

Fig. 22.4. Individual effects of four components of resistance on the progress of downy mildew on the foliage (solid lines) and on clusters (dashed lines) for sensitive (red), moderate resistant (orange), and highly resistant (green) grapevine varieties.

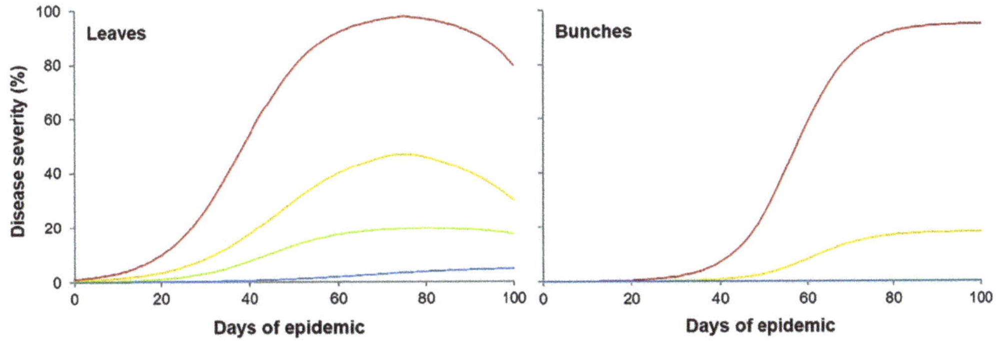

Fig. 22.5. Combined effect of resistance components on leaves and clusters for a sensitive grapevine variety (red) and varieties simultaneously expressing two (longer latent period and reduced sporulation; orange), three (reduced infection efficiency and sporulation, longer latent period; green), and four (reduced infection efficiency and sporulation, longer latent period, short infectious period; blue) components of resistance.

leads to the activation of plant defence and consequently to resistance. Resistance, however, is often not an stable trait and the protection given by resistance genes can be overcome by a virulent strain of the pathogen, due to the recognition of a protein encoded by the pathogen avirulence gene by the product of the corresponding plant resistance gene (Merdinoglu *et al.*, 2018).

Breakdown of resistance has already been observed in *P. viticola*-grapevine pathosystem, directly on some partially resistant varieties (Peressotti *et al.*, 2010; Delmotte *et al.*, 2014) or through the identification of*P. viticola* genomic region associated with the resistance breakdown (Paineau *et al.*, 2023). Peressotti *et al.* (2010) described the emergence of a *P. viticola* isolate (isolate SL) that specifically overcame *Rpv3*, the major resistance gene carried by the variety Bianca at chromosome 18. Delmotte *et al.* (2014) reported the erosion of partially resistance in the variety Regent, which carries *Rpv3*, supporting the evidence of an adaptative pattern of *P. viticola* to partial resistance components; for instance, a high level of sporulation was observed in Regent. Further studies on the adaption of the pathogen to partial resistance strengthen the hypothesis of the evolution of DM population particularly in traits linked to pathogenicity and reported that isolates from resistant varieties were more aggressive than isolates from susceptible hosts, as they had a shorter latency period and higher levels of spore production (Delmas *et al.*, 2016).

The concept of resistance durability is therefore fundamental for disease control, especially in a perennial crop like grapevine, which is meant to stay in the field for 30 years or more (Peressotti *et al.*, 2010; Töpfer and Trapp, 2022). Furthermore, the introgression of a resistance gene in grapevine genome is a long and costly process that can last 25 years (Töpfer and Trapp, 2022) and the number of genes of resistance is still limited despite the attempts to find new sources (Merdinoglu *et al.*, 2018). Breeding programmes, therefore, need to make sure that the resistance introduced in a variety is not only effective but also persists over time, despite the constant evolution of pathogen (Peressotti *et al.*, 2010).

To ensure resistance durability, operative approaches should be undertaken at plant, pathogen, and agronomic levels. Thanks to advantages in biotechnologies like the use of marker-assisted selection (MAS), it is possible to identify genotypes that combine the desired resistance factors and, thus, to design varieties with very high resistance by combining multiple genetic factors conferring partial resistance (Merdinoglu *et al.*, 2018). This strategy based on the introgression of more than one resistance gene in the plant genome is called 'pyramiding resistance genes'. Although plants with pyramided genes have been observed to present a significantly higher resistance than single-*Rpv* genomes (Bove and Rossi, 2020; Salotti *et al.*, 2022), the overcoming of partial resistance conferred by pyramided *Rpv3* and *Rpv12* has been reported (Wingerter *et al.*, 2021).

Along with new breeding programmes, investigations on *P. viticola* evolutionary potential, partial resistance adaption, and mating system may help to assess the risk of resistance breakdown and guide strategies for long-lasting resistance (Delmotte *et al.*, 2014; Delmas *et al.*, 2016). Durability also depends on environmental conditions and cultural practices, because of their influence on pathogen biology, the number of offspring produced in a season, and epidemic progress. Appropriate crop protection strategies to control DM and manage pathogen resistance to fungicides would reduce the selection pressure and thus avoid a forced, fast evolution of *P. viticola* populations. This means that partially resistant variety need proper plant protection interventions, as it is discussed in the section "Disease control" of this Case Study.

Sharing Life with Many Fungal Pathogens

Grapevine cultivation is threatened by many pathogens other than *P. viticola* for which disease control is needed, including *Erysiphe necator* (synonym: *Uncinula necator*, causal agent of powdery mildew, PM), *Phyllosticta ampelicida* (synonym: *Guignardia bidwellii*, causal agent of black-rot, BR), and *Botrytis cinerea* (causal agent of grey mould, GM). Genes of resistance have been identified to control PM and BR (Available at: www.vivc.de, accessed 18 December 2023) and have been used in breeding programmes with particular emphasis on PM (Zini *et al.*, 2015). These genes are known as *Ren* or *Run* (acronyms of Resistance to *Erysiphe necator* and *Uncinula necator*, respectively) and have been successfully introgressed into several partially resistant varieties carrying also *Rpvs*. It is a long and complicated process to develop varieties that incorporate *Rpvs*, *Ren/Run* along with resistance to BR and resilience to BR, and match the other main breeding goals of grape quality, yield, and abiotic stress tolerance (Töpfer and Trapp, 2022).

In previous research, resistance to mildews was largely evaluated by performing monocycle, *in vitro* bioassays involving the artificial inoculation of leaf discs (Boso and Kassemeyer, 2008; Bellin *et al.*, 2009) but little was done in field conditions, where the co-occurrence of mildews and other diseases during a growing season is common despite the differences in pathogen life cycles and environmental requirements. Under these conditions, the different pathogens compete for the same infection sites. If we consider the crop as a finite collection of infection sites (or carrying capacity) with two mutually exclusive states, healthy or infected (Savary *et al.*, 2018), and we assume that a site can be occupied by only one pathogen, then the occurrence of a disease limits the occurrence of the others. Knowing the behaviour of current resistant varieties when facing multiple diseases may help support disease management in the field and informing further breeding strategies.

In a recent work, the concept of 'resistance pattern' was introduced (Salotti *et al.*, 2022). This term is commonly used in bacteriology and refers to the description of antibiotic resistance of a single isolate to multiple antibiotics. Similarly, in this context a resistance pattern is defined as the description of the resistance level of a variety to all the diseases occurring in a field experiment. Salotti *et al.* (2022) tested in a 4-year field trial, 15 varieties, most of them carrying *Rpvs* and *Ren/Run*, in comparison with the sensitive variety Merlot, the resistance level to DM, PM, and BR on both leaves and bunches. Based on field observations, authors clustered the varieties into four groups (Fig. 22.6). The first group was composed by the sensitive control which had low resistance to mildews and intermediate resistance to BR. The second group included only a native Georgian variety (Rkatsitelii) that showed intermediate resistance to DM (despite the absence of *Rpv* loci in its genome) and to BR, but susceptibility to PM. The third group was composed by varieties showing high resistance to mildews, primarily thanks to their genome including varying combinations of *Rpv3*, *Rpv10*, *Rpv12*, *Ren3*, *Ren8*, and *Ren9*, and good resistance to BR. The last group was constituted of two varieties, one carrying *Rpv3* and the other *Rpv12*, that showed high resistance to DM, intermediate to PM, and low to BR.

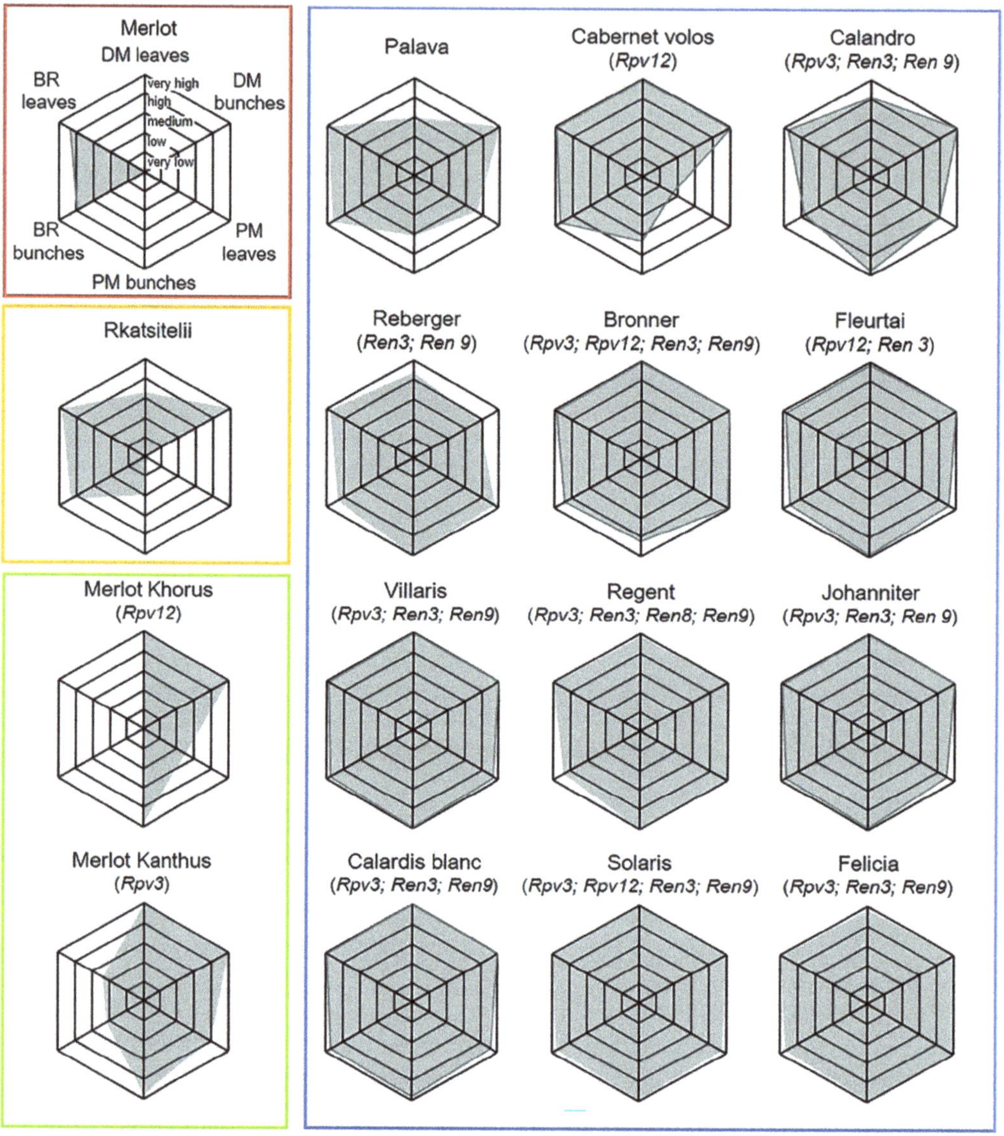

Fig. 22.6. Spider graphs representing the resistance pattern of grapevine varieties with regard to downy mildew (DM), powdery mildew (PM), and black rot (BR) on leaves and bunches. Varieties are listed by name and resistant genes (*Rpv*–resistance to *Plasmopara viticola*; and *Ren*–resistance to *Erysiphe necator*) in brackets. In each spider graph, the black net ranges from very low to very high resistance. Varieties were evaluated in a 4-year field experiment in comparison with the sensitive control Merlot. Coloured boxes group the varieties based on their resistance pattern: Low resistance to mildews and intermediate resistance to BR (red); intermediate resistance to DM and to BR, but susceptibility to PM (orange); high resistance to DM, intermediate to PM, and low to BR (green); high resistance to mildews and good resistance to BR (blue) (modified from Salotti *et al*., 2022).

Given the varying responses to diseases, there is the need to develop a proper strategy for each resistance pattern to manage multiple diseases in vineyards. The resistance pattern of a variety, in fact, indicates which pathogens may develop and require light or intensive control and provide useful information to address crop protection decisions.

Disease Control

Although some studies reported that resistant grapevine varieties can reduce the application of fungicides by 100% (Wingerter *et al.*, 2021), the complete omission of chemical protection on resistant varieties is discouraged to prevent any possible negative effects on grape yield and wine quality and, even more, to increase resistance durability (Töpfer *et al.*, 2011; Delmas *et al.*, 2016; Wingerter *et al.*, 2022). Nevertheless, considering that more than 70% of the fungicides used in Europe is applied to vineyards (Muthmann and Nadin, 2007), the use of resistant varieties promises to reduce fungicide applications and face challenges for viticulture sustainability, including loss of active fungicidal compounds, regulatory framework (e.g., in Europe: Directive 2009/128/EC, European Green Deal 2050, and Farm to Fork Strategy), climate change, and extreme weather events effects on biotic and abiotic stresses (Töpfer and Trapp, 2022).

To protect partially resistant varieties and guarantee resistance durability, it is generally recommended to apply 2 or 3 treatments between flowering and bunch closure (Töpfer and Trapp, 2022). The reduction in fungicide applications against the mildews, however, may result in the emergence of other 'minor' pathogens, such as BR, that would need specific control (Salotti *et al.*, 2022; Töpfer and Trapp, 2022).

Information on the resistance patterns provides a more nuanced and useful perspective than that provided by single-pathogen laboratory studies on the resistance of varieties under field conditions, facing multiple diseases. This information could support both the strategic decisions (i.e., decisions involving several years at the crop level like the varietal choice) and tactical decisions (i.e., decisions made day-by-day in response to what is happening to the crop like the actions required to control a disease outbreak) of grape disease management (Rossi *et al.*, 2012, 2019).

At the strategic level, resistance patterns may help growers select the variety to be planted in a specific environment. For example, in a location with frequent spring rains and long leaf wetness, i.e., a location where DM is favoured, a variety showing intermediate to high resistance to *P. viticola* would be preferred. On the other hand, the planting of sensitive or varieties with low resistance should be carefully considered in locations with a history of severe PM and BR, especially in organic farms in which the use of synthetic fungicides is banned.

At the tactical level, if a variety expresses high resistance to DM and PM but high susceptibility to BR, intensive control of *P. ampelicida* but light control of *P. viticola* and *E. necator* would be required. For such a variety, tactical DM and PM management can require few fungicide applications, which can be scheduled using an infection risk-based approach. For this purpose, information on resistance patterns could be used to modify mathematical models that predict the risk of infection based on the weather conditions (Rossi *et al.*, 2008; Caffi *et al.*, 2011; Brischetto *et al.*, 2021) so that they provide variety-specific predictions of disease risk. In a recent study, mathematical equations relating *P. viticola* infection severity to hydrothermal time (a physiological time accounting for the combined effect of temperature and wetness duration) were developed and used to define variety-specific critical thresholds for predicting DM infection in several *Vitis* accessions carrying *Rpvs* (Salotti *et al.*, 2023). Application of these thresholds in a simulation study for scheduling fungicide sprays under three contrasting climates (mild and moist, warm and dry, and warm and moist) demonstrated that the number of fungicide applications can be reduced by 48% on average (Fig. 22.7). Therefore, the implementation of resistance patterns into mathematical models would provide more precise information to growers on whether and when a fungicide treatment would be needed for any grape variety.

In addition to the use of resistant varieties, innovative approaches for reducing the use of synthetic fungicides in disease control include the exploitation of natural control mechanisms and specifically biocontrol microorganisms and microbicidal molecules of natural origin, both of which are expected to have a low impact on human health and the environment, with no or limited side effects on non-target organisms. A recent meta-analysis on the use of biocontrol agents to control grape diseases showed great variability in their efficacy but also promising results. For instance, the control of PM under field conditions had 44.8–52.3% efficacy, depending on the microorganism and the environmental conditions (Altieri, 2024). Similar efficacies were observed for biocontrol agents or botanicals against *Botrytis cinerea* (Fedele *et al.*, 2020), or resistant inducers against DM (Taibi *et al.*, 2023). Alternative products, however, have been primarily applied against diseases susceptible to *V. vinifera* varieties, and their use in combination with partially resistant varieties needs further research and fine-tuning. Possible additive effects resulting from the integration of genetic resistance with bio-based products may open new horizons for innovative, less chemical-dependent viticultural systems.

Group Discussion Questions

1. Which are opportunities and challenges of partially resistant varieties to support the viticultural transition towards sustainability?
2. Which strategies should be implemented in practical viticulture for ensuring durability of resistance towards mildews?
3. How to select the most suited resistant variety (or varieties) in a specific context (i.e., environmental conditions/oenological objectives)?
4. Figure 22.8 shows the information on individual effects of four components of resistance on the progress of DM for two varieties. Define the overall resistance level of each variety by assigning one of the following classes of resistance: Very low, low, medium, high, and very high.

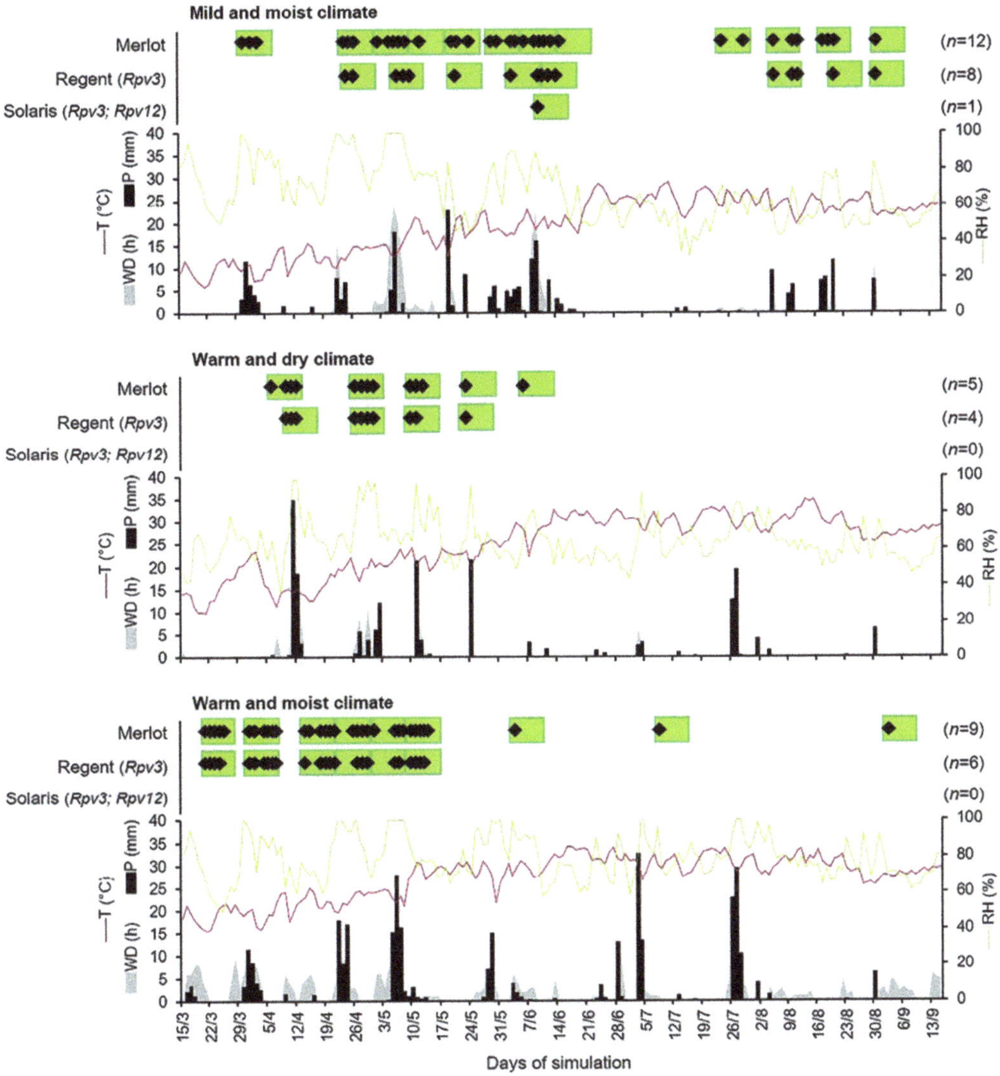

Fig. 22.7. Risk of infection by *Plasmopara viticola* and fungicide applications under contrasting climate scenarios for three representative varieties, i.e., the sensitive control Merlot, Regnet and Solaris, carrying one and two *Rpvs*, respectively. Weather data are reported for each climate type: Air temperature (T, °C, red line); relative humidity (RH, %, green line); precipitation (P, mm, black bars); and wetness duration (WD, hours, grey area). Risk of infection (represented by black diamonds) are predicted using the model of Rossi *et al.* (2008) for primary infections and Brischetto *et al.* (2021) for secondary infections. Copper-based fungicides are applied before a predicted infection risk and are assumed to remain effective for 7 days (green rectangles; number of treatments *n* in brackets).

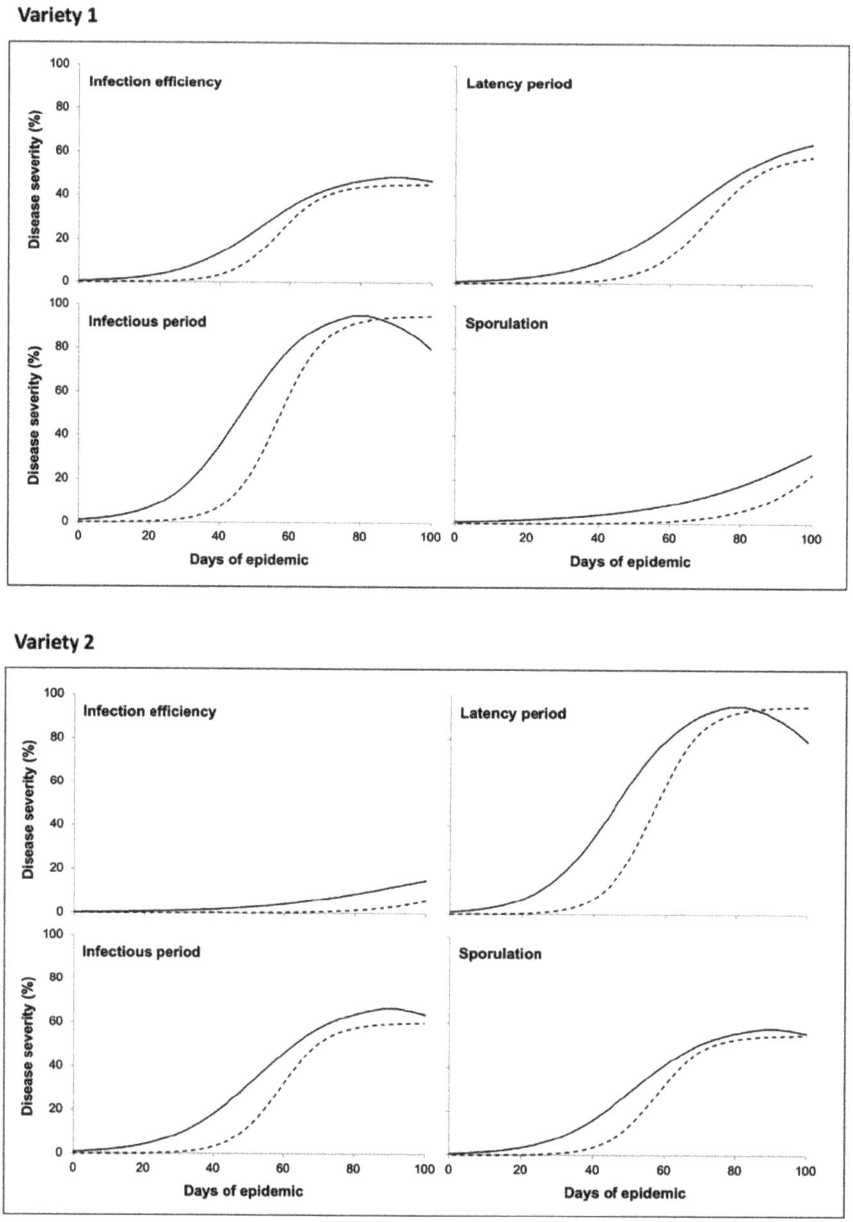

Fig. 22.8. Individual effects of four components of resistance on the progress of downy mildew on the foliage (solid lines) and on clusters (dashed lines) for variety 1 (upper panel) and for variety 2 (lower panel).

Funding statement

Irene Salotti researcher position is funded by the 'National Research Centre for Agricultural Technologies (Agritech)' CN00000022. PNRR-M4C2, Investment 1.4– Fundedbythe'European Union–NextGenerationEU'–CUP: J33C22001160007.

Further Readings

Basler, P. and Pfenninger, H. (2003) Disease-resistant cultivars as a solution for organic viticulture. *Acta Horticulturae* 603, 681–685. DOI: 10.17660/ActaHortic.2003.603.94.

Bavaresco, L. (2019) Impact of grapevine breeding for disease resistance on the global wine industry. *Acta Horticulturae* 1248, 7–14. DOI: 10.17660/actahortic.2019.1248.2.

Pertot, I., Caffi, T., Rossi, V., Mugnai, L., Hoffmann, C. *et al.* (2017) A critical review of plant protection tools for reducing pesticide use on grapevine and new perspectives for the implementation of IPM in viticulture. *Crop Protection* 97, 70–84.

References

Alonso-Villaverde, V., Voinesco, F., Viret, O., Spring, J.L. and Gindro, K. (2011) The effectiveness of stilbenes in resistant Vitaceae: Ultrastructural and biochemical events during *Plasmopara viticola* infection process. *Plant Physiology and Biochemistry* 49, 265–274.

Altieri, V. (2024) New approaches to the biocontrol of *Botrytis cinerea* in grapevine. In: *XXXVI Ciclo, Lezioni a.a. 2022/23*. Università Cattolica del Sacro Cuore, Piacenza.

Anderson, K. (2013) *Which Winegrape Varieties are Grown Where?* University of Adelaide Press, Adelaide, Australia. DOI: 10.20851/winegrapes.

Bellin, D., Peressotti, E., Merdinoglu, D., Wiedemann-Merdinoglu, S., Adam-Blondon, A.F. *et al.* (2009) Resistance to *Plasmopara viticola* in grapevine 'Bianca'is controlled by a major dominant gene causing localised necrosis at the infection site. *Theoretical and Applied Genetics* 120, 163–176.

Blaeser, M. and Weltzien, H.C. (1979) Epidemiologische Studien an Plasmopara viticola zur Verbesserung der Spritzterminbestimmung/Epidemiological studies to improve the control of grapevine downy mildew (*Plasmopara viticola*). *Zeitschrift Für Pflanzenkrankheiten Und Pflanzenschutz/Journal of Plant Diseases and Protection* 489–498.

Blasi, P., Blanc, S., Wiedemann-Merdinoglu, S., Prado, E., Rühl, E.H. *et al.* (2011) Construction of a reference linkage map of *Vitis amurensis* and genetic mapping of Rpv8, a locus conferring resistance to grapevine downy mildew. *Theoretical and Applied Genetics* 123, 43–53.

Boso, A.S. and Kassemeyer, H.H. (2008) Different susceptibility of European grapevine cultivars for downy mildew. *Vitis* 47, 39–49. DOI: 10.5073/vitis.2008.47.39-49.

Bove, F. and Rossi, V. (2020) Components of partial resistance to *Plasmopara viticola* enable complete phenotypic characterization of grapevine varieties. *Scientific Reports* 10, 585. DOI: 10.1038/s41598-020-57482-0.

Bove, F., Savary, S., Willocquet, L. and Rossi, V. (2021) Modelling the effect of partial resistance on epidemics of downy mildew of grapevine. *European Journal of Plant Pathology* 161, 847–864.

Brischetto, C., Bove, F., Fedele, G. and Rossi, V. (2021) A weather-driven model for predicting infections of grapevines by sporangia of *Plasmopara viticola*. *Frontiers in Plant Science* 12, 636607. DOI: 10.3389/fpls.2021.636607.

Caffi, T., Rossi, V., Legler, S.E. and Bugiani, R. (2011) A mechanistic model simulating ascosporic infections by Erysiphe necator, the powdery mildew fungus of grapevine. *Plant Pathology* 60, 522–531.

Delmas, C.E., Fabre, F., Jolivet, J., Mazet, I.D., Richart Cervera, S. *et al.* (2016) Adaptation of a plant pathogen to partial host resistance: Selection for greater aggressiveness in grapevine downy mildew. *Evolutionary Applications* 9, 709–725.

Delmotte, F., Mestre, P., Schneider, C., Kassemeyer, H.H., Kozma, P. *et al.* (2014) Rapid and multiregional adaptation to host partial resistance in a plant pathogenic oomycete: Evidence from european populations of *Plasmopara viticola*, the causal agent of grapevine downy mildew. *Infection, Genetics and Evolution* 27, 500–508.

Deutsches Weininstitut (2023). Available at: www.winesofgermany.com (accessed 15 January 2024).

Di Gaspero, G., Copetti, D., Coleman, C., Castellarin, S.D., Eibach, R. *et al.* (2012) Selective sweep at the Rpv3 locus during grapevine breeding for downy mildew resistance. *Theoretical and Applied Genetics* 124, 277–286.

Fedele, G., González-Domínguez, E., Si Ammour, M., Languasco, L. and Rossi, V. (2020) Reduction of *Botrytis cinerea* colonization of and sporulation on bunch trash. *Plant Disease* 104, 808–816.

Gessler, C., Pertot, I. and Perazzolli, M. (2011) Plasmopara viticola: A review of knowledge on downy mildew of grapevine and effective disease management. *Phytopathologia Mediterranea* 50, 3–44.

Gindro, K., Pezet, R. and Viret, O. (2003) Histological study of the responses of two *Vitis vinifera* cultivars (resistant and susceptible) to *Plasmopara viticola* infections. *Plant Physiology and Biochemistry* 41, 846–853.

Hedrick, U.P. (1908) *The Grapes of New York*. J.B. Lyon Co, Albany.

Jürges, G., Kassemeyer, H.H., Dürrenberger, M., Düggelin, M. and Nick, P. (2009) The mode of interaction between *Vitis* and *Plasmopara viticola* Berk. and Curt. Ex de Bary depends on the host species. *Plant Biology* 11, 886–898.

Kortekamp, A., Wind, R. and Zvprian, E. (1997) The role of callose deposits during infection of two downy mildew-tolerant and two-susceptible *Vitis* cultivars. *Vitis-Geilweilerhof* 36, 103–104.

Kortekamp, A., Wind, R. and Zyprian, E. (1998) Investigation of the interaction of *Plasmopara viticola* with susceptible and resistant grapevine cultivars. *Journal of Plant Diseases and Protection* 105, 475–488.

Lalancette, N., Ellis, M.A. and Madden, L.V. (1987) Estimating infection efficiency of *Plasmopara viticola* on grape. *Plant Disease* 71, 981–983.

Marguerit, E., Boury, C., Manicki, A., Donnart, M., Butterlin, G. *et al.* (2009) Genetic dissection of sex determinism, inflorescence morphology and downy mildew resistance in grapevine. *Theoretical and Applied Genetics* 118, 1261–1278.

Merdinoglu, D., Wiedeman-Merdinoglu, S., Coste, P., Dumas, V., Haetty, S. *et al.* (2003) Genetic analysis of downy mildew resistance derived from *Muscadinia rotundifolia*. *Acta Horticulturae* 603, 451–456. DOI: 10.17660/ActaHortic.2003.603.57.

Merdinoglu, D., Schneider, C., Prado, E., Wiedemann-Merdinoglu, S. and Mestre, P. (2018) Breeding for durable resistance to downy and powdery mildew in grapevine. *OENO One* 52, 203–209.

Millardet, A. (1881) *Notes Sur Les Vignes Américaines Et Opuscules Divers Sur Le Même Sujet*. Ferret, Bordeaux, France.

Moreira, F.M., Madini, A., Marino, R., Zulini, L., Stefanini, M. *et al.* (2011) Genetic linkage maps of two interspecific grape crosses (*Vitis* spp.) used to localize quantitative trait loci for downy mildew resistance. *Tree Genetics and Genomes* 7, 153–167.

Muccinelli, M. and Rossi, V. (2022) Le avversità e il prontuario dei prodotti fitosanitari. Vite Il nuovo Muccinelli, Edagricole, Italy.

Muthmann, R. and Nadin, P. (2007) The use of plant protection products in the European union. In: *Eurostat Statistical Books*. European Commission, Luxembourg.

Ochssner, I., Hausmann, L. and Töpfer, R. (2016) Rpv14, a new genetic source for *Plasmopara viticola* resistance conferred by *Vitis cinerea*. *Vitis* 55, 79–81.

Paineau, M., Minio, A., Mestre, P., Fabre, F., Mazet, I.D. *et al.* (2023) An effector deletion leads to the breakdown of partial grapevine resistance to downy mildew bioRxiv, 1–12. DOI: 10.1101/2023.08.17.553663.

Parlevliet, J.E. (1979) Components of resistance that reduce the rate of epidemic development. *Annual Review of Phytopathology* 17, 203–222.

Peressotti, E., Wiedemann-Merdinoglu, S., Delmotte, F., Bellin, D., Gaspero, G. *et al.* (2010) Breakdown of resistance to grapevine downy mildew upon limited deployment of a resistant variety. *BMC Plant Biology* 10, 1–11.

Rapp, A. and Versini, G.G. (2015) Methylanthranilate ("foxy taint") concentrations of hybrid and vitis vinifera wines. *Vitis* 35, 215–216.

Rossi, V., Caffi, T., Giosuè, S. and Bugiani, R. (2008) A mechanistic model simulating primary infections of downy mildew in grapevine. *Ecological Modelling* 212, 480–491.

Rossi, V., Caffi, T. and Salinari, F. (2012) Helping farmers face the increasing complexity of decision-making for crop protection. *Phytopathologia Mediterranea* 51, 457–479.

Rossi, V., Sperandio, G., Caffi, T., Simonetto, A. and Gilioli, G. (2019) Critical success factors for the adoption of decision tools in IPM. *Agronomy* 9, 710.

Salotti, I., Bove, F., Ji, T. and Rossi, V. (2022) Information on disease resistance patterns of grape varieties may improve disease management. *Frontiers in Plant Science* 13, 1017658. DOI: 10.3389/fpls.2022.1017658.

Salotti, I., Bove, F., Taibi, O. and Rossi, V. (2023) Predict downy mildew infection in partially resistant grapevines. In: *Abstracts Book of IOBC-WPRS Integrated Protection in Viticolture, 3–5 October 2023*, IOBC-WPRS, Logroño, Spain.

Savary, S., Nelson, A.D., Djurle, A., Esker, P.D., Sparks, A. *et al.* (2018) Concepts, approaches, and avenues for modelling crop health and crop losses. *European Journal of Agronomy* 100, 4–18.

Schwander, F., Eibach, R., Fechter, I., Hausmann, L., Zyprian, E. *et al.* (2012) Rpv10: A new locus from the Asian *Vitis* gene pool for pyramiding downy mildew resistance loci in grapevine. *Theoretical and Applied Genetics* 124, 163–176.

Taibi, O., Salotti, I. and Rossi, V. (2023) Plant resistance inducers affect multiple epidemiological components of *Plasmopara viticola* on grapevine leaves. *Plants* 12, 2938.

Töpfer, R. and Trapp, O. (2022) A cool climate perspective on grapevine breeding: Climate change and sustainability are driving forces for changing varieties in a traditional market. *Theoretical and Applied Genetics* 135, 3947–3960.

Töpfer, R., Hausmann, L., Harst, M., Maul, E., Zyprian, E. *et al.* (2011) New horizons for grapevine breeding. *Fruit, Vegetable and Cereal Science and Biotechnology* 5, 79–100.

Van der Plank, J.E. (1963) *Plant Diseases: Epidemics and Control*. Academic Press Inc, New York.

Venuti, S., Copetti, D., Foria, S., Falginella, L., Hoffmann, S. *et al.* (2013) Historical introgression of the downy mildew resistance gene Rpv12 from the Asian species *Vitis amurensis* into grapevine varieties. *PLoS ONE* 8, e61228. DOI: 10.1371/journal.pone.0061228.

Wiedemann-Merdinoglu, S., Prado, E., Schneider, C., Coste, P., Onimus, C. *et al.* (2006) Resistance to downy mildew derived from *Muscadinia rotunfifolia*: Genetic analysis and use of molecular markers for breeding. In: *Proceeding 5th International Workshop on Grapevine Downy Mildew and Powdery Mildew*, Istituto Agrario di San Michele all'Adige, San Michele all'Adige, Italy, p. 28.

Willocquet, L., Savary, S. and Yuen, J. (2017) Multiscale phenotyping and decision strategies in breeding for resistance. *Trends in Plant Science* 22, 420–432.

Wingerter, C., Eisenmann, B., Weber, P., Dry, I. and Bogs, J. (2021) Grapevine Rpv3-, Rpv10- and Rpv12-mediated defense responses against *Plasmopara viticola* and the impact of their deployment on fungicide use in viticulture. *BMC Plant Biology* 21, 1–17.

Wingerter, C., Eisenmann, B., Kortekamp, A. and Bogs, J. (2022) Resistance properties of new fungus-resistant grapevine cultivars against *Plasmopara viticola* and the impact of their deployment on fungicide use in viticulture. In: *BIO Web of Conferences*, EDP Sciences, p. 02006 (Vol. 50).

Zadoks, J.C. and Schein, R.D. (1979) *Epidemiology and Plant Disease Management*. Oxford University Press, UK.

Zini, E., Raffeiner, M., Di Gaspero, G., Eibach, R., Grando, M.S. *et al*. (2015) Applying a defined set of molecular markers to improve selection of resistant grapevine accessions. *Acta Horticulturae* 7, 3–78. DOI: 10.17660/ActaHortic.2015.1082.9.

Zyprian, E., Ochßner, I., Schwander, F., Šimon, S., Hausmann, L. *et al*. (2017) Quantitative trait loci affecting pathogen resistance and ripening of grapevines. *Molecular Genetics and Genomics* 29, 1573–1594. DOI: 10.1007/s00438-016-1200-5.

23 *Neoscytalidium dimidiatum*: An Emerging Threat in Pineapple

Zaiton Sapak

Plant Pathology Laboratory, Faculty of Plantation and Agrotechnology, Universiti Teknologi MARA, Jasin Campus, Merlimau 77300, Melaka, Malaysia

Introduction

Fungus *Neoscytalidium dimidiatum* is known for its highly adaptive and opportunistic traits. With the acceleration of climate change, the emergence of new diseases caused by this fungal pathogen has become a recent concern in many countries including Malaysia. There have been numerous first reports on the fungal pathogen causing diseases in different crops and countries from 2023 to 2024 alone. For example, the fungus was discovered to cause stem canker in pitahaya in Brazil, India, Malaysia and Ecuador, fruit rot of pears and blight and root rot of chickpea in Türkiye, branch dieback of pagoda tree, fruit rot of olive and leaf spot of Aloe vera in China and leaf blight of lily in Iran.

In Malaysia, this pathogen was reported to cause stem end rot in pineapple (Kuruppu *et al.*, 2021), and our recent study found that it also attacks pineapple leaves and causes leaf spot disease (Afifah *et al.*, 2024). The disease incidence ranging from 10% to 15% observed in several pineapple farms across Peninsular Malaysia. The disease symptoms initially appear as small brown spots, which gradually coalesce to form large brown to creamy-white lesions covering the entire leaf surface. Severe infections eventually lead to plant death over time.

Other plants affected by this pathogen include pitahaya, guava, lilies, and mother-in-law's tongue plant. In September 2018, several cases of MD2 pineapple fruits exhibiting internal black rot symptoms were notified by pineapple growers in Johor and Kedah. Prior knowledge indicates that this disease is caused by the same pathogen of fruitlet core rot (*Penicillium funiculosum*) that is commonly found in Malaysia (Lim and Rohrbach, 1980). Fruitlet core rot disease causes black sunken lesions on fruit skin, in which the fungal pathogen gets into the fruit via wounded skin tissues possibly caused by sucking insect pests, sunburn, and mechanical injuries. In contrast, internal fruit rot has no visible disease symptoms on pineapple fruit skin and the disease can only be observed on flesh after it is cut for serving. Moreover, there is no evidence of visible penetration by the pathogen from the fruit skin. This problem has become a major concern among pineapple growers in this country because the infected pineapples cannot be identified and sorted out before being packed for local or international markets.

Economic Importance of Pineapple

Pineapple (*Ananas comosus*) holds significant economic importance in the global tropical fruit market. As of 2024, pineapples accounted for approximately 37% of the export quantities of major tropical fruits, with global exports reaching around 3.3 million tonnes (FAO, 2023). Pineapple contributes to the economies of many tropical and subtropical countries, generating employment and income for growers, traders, and industrial workers in processing industries. Pineapple is used for both fresh consumption and processed products such as juice, canned fruit, sauce, jam, and dried snacks, which make this tropical fruit a valuable commodity in the food industry. Its economic value is also increased by its by-products, which include fibre and animal feed. The global demand for pineapple continues to rise due to its nutritional benefits, versatility, and consumer preference, strengthening its position as a key agricultural product in international markets.

Common Diseases in Pineapple

Like other crops, pineapple is also susceptible to several diseases caused by pathogenic microorganisms such as fungi, bacteria, viruses, and nematodes, which can significantly reduce fruit quantity and quality. Additionally, these pathogens can attack all parts of pineapple plants, including roots, butt, stem, leaves, flowers, fruitlets and fruits. Some of the economically important diseases of pineapple are summarized in Table 23.1.

These diseases have been reported to cause yield losses in pineapple production, for instance, bacterial heart rot and fruit collapse can reduce yield productivity by up to 40%. The causal pathogen attacks not only the central whorl of the pineapple plant but also the fruits. Other pathogens like the fungus *Thielaviopsis paradoxa*, also attack two parts of the pineapple plant, namely the butt and fruit. It is hardly possible to control disease caused by mealybug wilt-associated virus subgroup I (PMWaV-1), subgroup II (PMWaV-2), subgroup III (PMWaV-3) and two putative of PMWaV-4 and PMWaV-5, which have currently become a main threat to pineapple farms. Other diseases caused by pathogenic fungi, yeast and nematodes are contributing to pineapple yield losses if disease management is not properly handled.

Table 23.1. Major diseases of pineapple caused by several types of pathogenic microorganisms.

Disease	Pathogen	Symptoms
Bacterial heart rot	*Dickeya zeae* syn. *Erwinia chrysanthemi*	Water-soaked leaf base lesions, brown discolouration, foul odour, leaf detachment
Fruit collapse	*Dickeya zeae* syn. *Erwinia chrysanthemi*	Juice leakage, rotten-smelling gas bubbles, internal fruit decay
Butt rot	*Thielaviopsis paradoxa*	Black, decayed tissues at the sucker base, crown, and slip stem
Black rot	*Thielaviopsis paradoxa*	Water-soaked, black patches on post-harvest fruit
Fruitlet core rot	*Penicillium funiculosum*, *Fusarium* spp.	Internal browning of fruitlets, dry rot, blackened sunken spots
Fusariosis	*Fusarium guttiforme*, *F. ananatum*, *F. fujikuroi* and many other Fusarium species.	Brown discolouration, gum exudation, malformed fruits, dry rot
Pineapple wilt disease	*Pineapple mealybug wilt-associated viruses* (PMWaV-1 to 5)	Leaf discolouration, loss of turgidity, dieback, plant death
Stem-end rot & internal black spot	*Neoscytalidium dimidiatum*	Internal fruit decay, black lesions at the stem end, no visible skin damage
Leaf spot disease	*Exserohilum rostratum*, *Curvularia clavata*, *N. dimidiatum*, *Fusarium* spp.	Grey, yellow, or brown spots on leaves, necrosis
Yeasty disease	*Candida* spp., *Hanseniaspora valbyensis*, *Saccharomyces* spp.	Gas-filled, spongy fruit, fermentation, fruit collapse

Neoscytalidium dimidiatum and Host Ranges

Neoscytalidium dimidiatum (Penz.) Crous & Slippers is an opportunistic fungal pathogen that belongs to the family of *Botryosphaeriaceae* and has a wide range of hosts. The pathogen has been reported as the main cause of numerous diseases on economically important fruit crops such as dragon fruit, plum, fig, pineapple, tomato, and lemon. This fungal pathogen can cause numerous disease symptoms such as canker, rot, dieback, blight, and wilt that can attack all parts of a plant. Numerous reports on *N. dimidiatum*-related diseases in plants have been published (see Table 23.2). Mainly, this pathogen has been reported on dragon fruit (*Hylocereus* spp.) to cause canker on stem and fruit, which has contributed to huge post-harvest losses of up to

Table 23.2. Some of the economically important plants attacked by *Neoscytalidium dimidiatum*.

Host	Disease	Country	Authors
Pitaya (*Hylocereus undatus*, *H. polyrhizus*, *Selenicereus megalanthus*)	Stem and Fruit Canker	China, Thailand, Malaysia, Israel, The United States, Taiwan	Lan *et al*. (2012); Ezra *et al*. (2013); Mohd *et al*. (2013); Sanahuja *et al*. (2016); Hong *et al*. (2020); Dy *et al*. (2022), Khoo *et al*. (2023)
Citrus (*Citrus sinensis* (L.) Osbeck cv. Tarocco Scirè)	Shoot Blight, Canker, and Gummosis	Italy	Polizzi *et al*. (2009)
Almond (*Prunus dulcis*)	Stem Canker, Branch Dieback, Shoot Blight and Fruit Rot	United States, Turkey	Nouri *et al*. (2018); Ören *et al*. (2020)
Fig (*Ficus carica*)	Branch Wilt, Dieback, Canker, Gummosis, Limb Dieback	Australia, the United States, Egypt	Ray *et al*. (2010); Al-Bedak *et al*. (2018); Gusella *et al*. (2021)
Mango (*Mangifera indica*)	Dieback and Canker	Australia	Ray *et al*. (2010)
Tomato (*Solanum lycopersicum* L.)	Foliar Blight and Root Rot	Turkey	Türkölmez *et al*. (2019)
Physic Nut (*Jatropha curcas*)	Collar And Root Rot	Brazil	Machado *et al*. (2012)
Grapes (*Vitis vinifera*)	Canker and Dieback	Turkey	Oksal *et al*. (2019)
Pistachio (*Pistacia vera*)	Canker, Shoot Blight, and Root Rot	Turkey	Derviş *et al*. (2019)
Lemon balm (*Melissa officinalis*)	Blight	Turkey	Özer et al. (2022)
Olive (*Olea europaea*)	Canker and Leaf Scorch	Turkey	Güney *et al*. (2022)
Pineapple (*Ananas comosus*)	Stem End Rot	Malaysia	Kuruppu *et al*. (2021)
Apple (*Malus domestica*)	Branch Dieback and Canker	Turkey	Ören *et al*. (2022)

60–80%. There have also recent reports of its effect on olives, apples, pineapple, guava, lemon, and fig. Knowledge of the characteristics and pathogenicity of the pathogen, disease symptoms, host range, and epidemiology of the disease is needed for alarming the emerging new disease on fruit crops caused by *N. dimidiatum* and disease distribution in many countries.

The spread of *N. dimidiatum* is facilitated by airborne conidia, infected plant material, and water splashes. The pathogen thrives in environments with high humidity and warm temperatures, conditions that favour

Fig. 23.1. Healthy MD2 pineapple fruit based on external appearance with (a) no evidence of pathogen penetration via fruit skin but (b) the appearance of internal black spot disease on fruit flesh after cut.

sporulation and infection. Wounded or stressed plants are more susceptible, with factors such as mechanical injuries, insect feeding, and environmental stress enhancing disease severity.

New Disease Caused by *N. dimidiatum* in Pineapple Fruit

In September 2018, MD2 pineapple fruit displaying internal black rot was detected on pineapple farms located in Johor, south of Peninsular Malaysia. The disease symptom was not visible on pineapple fruit skin and was only observed on the flesh after it has been cut for serving (see Fig. 23.1). Moreover, there is no evidence of visible penetration by the pathogen through the fruit skin. To address the issue, experiments were conducted involving isolation, pathogenicity testing, and identification of the primary causal pathogen from 20 samples of infected fruit collected from pineapple farms in Johor. Infected areas of fruit flesh were cut into small pieces (0.5 cm × 0.5 cm), and surface sterilized with 1% sodium hypochlorite (NaOCl) for 30 s, followed by 70% ethanol for 30 s, and rinsed thrice with sterilized water before placing on Potato Dextrose Agar (PDA). The PDA plates were incubated at room temperature (28 ± 2°C) in natural light. After 5 days of incubation, the potential causal pathogen was purified using a single conidial isolation technique for morphological and molecular characterizations. All 25 isolates displayed similar phenotypes. Based on morphological observation on PDA, the colonies were initially white, consisting of aerial mycelia but gradually darkening as the culture aged (see Fig. 23.2). Microscopic features of the

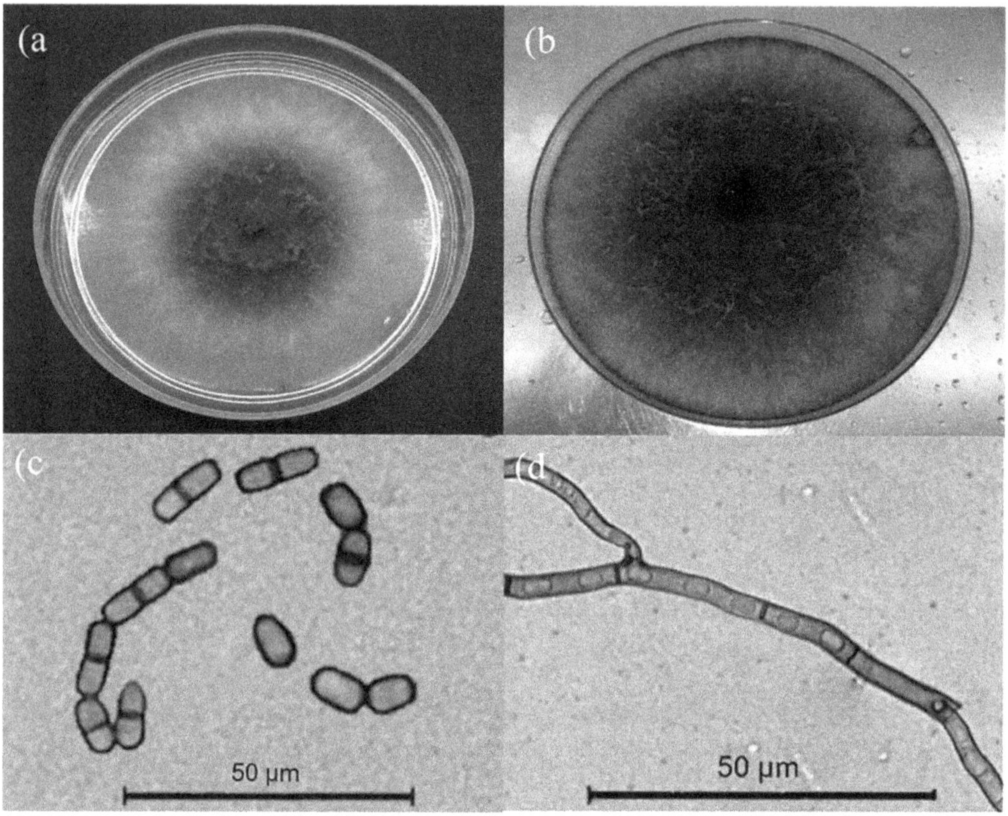

Fig. 23.2. Initial growth of *Neoscytalidium dimidiatum* on PDA with (a) white aerial mycelia and (b) gradually darkened as the culture aged. (c) Arthroconidia with various shapes, thick-walled, hyaline-to-brown and occurring singly or in chains and (d) mycelia were branched and disarticulated into 0- to 1-septate.

14-day-old fungal culture showed that the mycelia were branched with 0–1 septa, pigmented, and brown. Arthroconidia were ellipsoid to ovoid or round shaped, hyaline, with rounded apex, truncate base, and occurring singly or in chains averaging $9 \pm 3 \times 5 \pm 2$ µm (n=20). Based on the morphological characteristics, the fungal isolates were tentatively identified as *Neoscytalidium* species. The potential pathogen was further identified through PCR amplification of the Internal Transcribed Spacer (ITS) region using ITS1 and ITS4 primers and BLAST homology search as *Neoscytalidium dimidiatum* (Penz.) Crous & Slippers based on 100% similarity to a reference sequence (accession no. KU204558.1). The sequence was deposited in Gen Bank (accession no. OR366479).

The pathogenicity test was performed on mature pineapple fruits (index 1) using a pricking method. Sterile toothpicks were placed on PDA containing the pathogen and incubated at 28 ± 2°C until the toothpicks were fully colonized by the mycelia. Each fruit was wounded with three colonized toothpicks inserted into healthy tissue and left intact until the end of the

Fig. 23.3. The pathogenicity test of the MD2 pineapple fruits with the pathogen displayed: (a) disease symptom of rot and (b) the control fruit remains healthy.

experiment. Fruits in the control treatment were implanted with sterile toothpicks without mycelia. Three replicates were used for each treatment and the experiment was repeated twice. After 7 days, the fruits were cut vertically, and disease symptom consisting of brown lesions was observed on fruits treated with the pathogen-impregnated toothpicks (see Fig. 23.3a) while the fruits in the control group were symptomless (see Fig. 23.3b). The pathogen was successfully reisolated from fruit rot symptoms in which the cultural and morphological characteristics were identical to those of the original cultures.

This study confirmed that internal black rot in pineapple fruit is caused by *N. dimidiatum*. Besides end stem rot disease, this pathogen can get into pineapple fruit without causing any damage to the external fruit skin. In our previous study, this pathogen was also found to cause leaf spot disease on pineapple leaves, with disease incidence ranging from 10% to 15% observed in several pineapple farms across Peninsular Malaysia. The disease symptoms initially appear as small brown spots, which gradually coalesce to form large brown to creamy-white lesions covering the entire leaf surface (see Fig. 23.4). Severe infections eventually lead to plant death over time.

Control Methods for Managing Disease Caused by *N. dimidiatum*

Several control methods have been proposed for controlling crop diseases caused by *N. dimidiatum*, including cultural practices, chemical and

Fig. 23.4. Brown leaf spot disease on pineapple leaves. (a) Coalescing spots form a large creamy-white lesion on the adaxial surface, while a corresponding black lesion is visible on the abaxial surface of the leaf (b).

biological controls, genetic resistance planting materials and integrated disease management.

Cultural practices play an important role in preventing the spread and development of the disease. Sanitation, crop rotation, proper Irrigation system, improved air circulation through adequate planting distance, plant canopy management, and disinfection of tools for pruning are the most common practices that could be implemented.

Furthermore, several studies have suggested applying selected chemical fungicides, for instance, azoxystrobin + difenoconazole, floupyram + tebuconazole, and tebuconazole as the most effective fungicides for controlling *N. dimidiatum* in apricot and Fluazinam in pitaya. However, due to the potential for resistance development, they should be applied strategically. A combination of systemic and contact chemical fungicides is advisable to enhance efficacy and delay resistance.

Recent studies show some biological control agents such as *Trichoderma asperellum*, *Streptomyces* spp., *Bacillus subtilis* and plant-based biopesticides such as clove essential oil have the potential to be an alternative method for controlling the diseases caused by *N. dimidiatum*.

Moreover, breeding-resistant cultivars are a long-term and sustainable strategy for disease management that could be achieved through genetic engineering and grafting techniques. Recent advances in the identification of resistant genotypes, molecular breeding, and tissue culture techniques have significantly improved the efficiency of developing disease-resistant varieties. Marker-assisted selection and genetic mapping are increasingly employed to accelerate the breeding process. Although progress in pineapple breeding is generally slower compared to some other crops due to its biological constraints, it remains an effective method for managing major pineapple diseases.

A holistic disease management approach by integrating all available control measures could maximize effectiveness and sustainability in controlling the diseases caused by *N. dimidiatum*. At present, there are no studies on the effective control for the infection *N. dimidiatum* in pineapple fruits.

Future Research to Understand and Manage the Disease in Pineapple Fruit

This disease could be one of the most challenging diseases to control as the symptoms cannot be notified from external fruit skin. Therefore, more studies on this disease are needed such as:

1. Identification of the phenological stage and entry route of internal black spot disease pathogen to get into pineapple fruits.
2. Understand the mechanisms of browning or blackening that lead to internal black rot symptoms on pineapple fruit flesh.
3. Exploration of potential control methods for disease management, including the use of biological control agents, biofilm for fruit coating in farms, and plant-based bio fungicides.

Conclusion

The emergence of *N. dimidiatum* as a new causal pathogen of internal black rot in pineapples represents a significant threat to pineapple production and export. The ability of the fungal pathogen to infect without visible external symptoms complicates early detection and post-harvest quality assurance, posing economic risks for growers and exporters. Future research into how the pathogen enters pineapple fruits, the mechanisms of infection, and the development of black rot symptoms is essential to effectively address this disease and identify the most effective control strategies.

Discussion Points

1. Discuss strategies for early detection of this disease and methods to prevent the pathogen from entering pineapple fruits.
2. Explain how climate change contributes to the emergence of new diseases in agriculture and how farmers can adapt to these challenges.
3. Suggest farming practices that farmers can adopt to control the disease and prevent its spread.

Further Reading

EFSA Panel on Plant Health (PLH), Bragard, C., Baptista, P., Chatzivassiliou, E., Serio, F. *et al.* (2023) Pest categorisation of *Neoscytalidium dimidiatum. EFSA Journal* 21(5), e08001.

da Silva, J.H.B., Araújo, D.B., de Andrade, F.H.A., Toledo, L.C.M., Júnior, G.T.S. *et al.* (2023) The main diseases in the culture of pineapple: A review. *Scientific Electronic Archives* 16(8).

Dervis, S. and Özer, G. (2023) Plant-associated *Neoscytalidium dimidiatum*—taxonomy, host range, epidemiology, virulence, and management strategies: A comprehensive review. *Journal of Fungi* 9(11), 1048.

Sapak, Z. and Nusaibah, S.A. (2024) Common diseases in pineapple and their management. In: *Advances in Tropical Crop Protection*. Springer Nature Switzerland, Cham, pp. 85–104.

References

Afifah, M., Sapak, Z., Noor, N.M., Ab Wahab, M.Z., Anuar, I.S.M. *et al.* (2024) First report of leaf spot of pineapple (*Ananas comosus* (L.) Merr caused by *Neoscytalidium dimidiatum* in Malaysia. *Plant Disease* 108(4), 1100.

Al-Bedak, O.A., Mohamed, R.A. and Seddek, N.H. (2018) First detection of *Neoscytalidium dimidiatum* associated with canker disease in Egyptian ficus trees. *Forest Pathology* 48(2), e12411.

Derviş, S.S., Türkölmez, Ş., Çiftçi, O., Ulubaş Serçe, Ç. and Dikilitas, M. (2019) First report of *Neoscytalidium dimidiatum* causing canker, shoot blight, and root rot of pistachio in Turkey. *Plant Disease* 103(6), 1411.

Dy, K.S., Wonglom, P., Pornsuriya, C. and Sunpapao, A. (2022) Morphological, molecular identification and pathogenicity of *Neoscytalidium dimidiatum* causing stem canker of *Hylocereus polyrhizus* in southern Thailand. *Plants* 11(4), 504.

Ezra, D., Liarzi, O., Gat, T., Hershcovich, M. and Dudai, M. (2013) First report of internal black rot caused by *Neoscytalidium dimidiatum* on *Hylocereus undatus* (Pitahaya) fruit in Israel. *Plant Disease* 97(11), 1513.

FAO (2024) *Major Tropical Fruits Market Review. Preliminary Results 2024*. Food and Agriculture Organization of the United Nations.

Güney, ˙I.G., Özer, G., Türkölmez, S.¸ and Dervis¸, S. (2022) Canker and leaf scorch on olive (*Olea europaea* L.) caused by *Neoscytalidium dimidiatum* in Turkey. *Crop Protection* 157, 105985.

Gusella, G., Morgan, D.P. and Michailides, T.J. (2021) Further investigation on limb dieback of fig (*Ficus carica*) caused by *Neoscytalidium dimidiatum* in California. *Plant Disease* 105(2), 324–330.

Hong, C.F., Gazis, R., Crane, J.H. and Zhang, S. (2020) Prevalence and epidemics of *Neoscytalidium* stem and fruit canker on Pitahaya (*Hylocereus* spp.) in South Florida. *Plant Disease* 104(5), 1433–1438.

Khoo, Y.W., Tan, H.T., Khaw, Y.S., Li, S.F. and Chong, K.P. (2023) First report of *Neoscytalidium dimidiatum* causing stem canker on *Selenicereus megalanthus* in Malaysia. *Plant Disease* 107(1), 222.

Kuruppu, M., Siddiqui, Y., Kong, L.L., Ahmed, K. and Ali, A. (2021) First report of postharvest stem end rot disease on MD2 pineapple fruits caused by *Neoscytalidium dimidiatum* in Malaysia. *Plant Disease* 105(5), 1564.

Lan, G.B., He, Z.F., Xi, P.G. and Jiang, Z.D. (2012) First report of brown spot disease caused by *Neoscytalidium dimidiatum* on *Hylocereus undatus* in Guangdong, Chinese mainland. *Plant Disease* 96(11), 1702.

Lim, T.K. and Rohrbach, K.G. (1980) Role of *Penicillium funiculosum* strains in the development of pineapple fruit diseases. *Phytopathology* 70(7), 663–665.

Machado, A.R., Pinho, D.B., Dutra, D.C. and Pereira, O.L. (2012) First report of collar and root rot of physic nut (*Jatropha curcas*) caused by *Neoscytalidium dimidiatum* in Brazil. *Plant Disease* 96(11), 1697.

Mohd, M.H., Salleh, B. and Zakaria, L. (2013) Identification and molecular characterizations of *Neoscytalidium dimidiatum* causing stem canker of red-fleshed dragon fruit (*Hylocereus polyrhizus*) in Malaysia. *Phytopathology* 161(11–12), 841–849.

Nouri, M.T., Lawrence, D.P., Yaghmour, M.A., Michailides, T.J. and Trouillas, F.P. (2018) *Neoscytalidium dimidiatum* causing canker, shoot blight and fruit rot of almond in California. *Plant Disease* 102(8), 1638–1647.

Oksal, E., Çelik, Y. and Özer, G. (2019) *Neoscytalidium dimidiatum* causes canker and dieback on grapevine in Turkey. *Australasian Plant Disease Notes* 14(1), 33.

Ören, E., Koca, G., Gencer, R. and Bayraktar, H. (2020) First report of *Neoscytalidium novaehollandiae* associated with stem canker and branch dieback of almond trees. *Australasian Plant Disease Notes* 15(1), 17.

Ören, E., Palacıoglu, G., Koca, G., Ozan, G.N. and Bayraktar, H. (2022) First report of *Neoscytalidium dimidiatum* causing branch dieback and canker on apple in Turkey. *Plant Pathology* 104(1), 429.

Polizzi, G., Aiello, D., Castello, I., Viale, A., Groenwald, J.Z. *et al.* (2009) Occurrence, molecular characterisation, and pathogenicity of *Neoscytalidium dimidiatum* on citrus in Italy. In: Gentile, A. and La Malfa, S. (eds) *Proceedings of the International Symposium on Citrus Biotechnology*, pp. 237–243.

Ray, J.D., Burgess, T. and Lanoiselet, V.M. (2010) First record of *Neoscytalidium dimidiatum* and *N. novaehollandiae* on *Mangifera indica* and *N. dimidiatum* on *Ficus carica* in Australia. *Australasian Plant Disease Notes* 5(1), 48–50.

Sanahuja, G., Lopez, P. and Palmateer, A.J. (2016) First report of *Neoscytalidium dimidiatum* causing stem and fruit canker of *Hylocereus undatus* in Florida. *Plant Disease* 100(7), 1499.

Türkölmez, S., Dervi, S.S., Çiftçi, Ç.U. and Dikilitas, M. (2019) New disease caused by *Neoscytalidium dimidiatum* devastates tomatoes (*Solanum lycopersicum*) in Turkey. *Crop Protection* 118, 21–30.

Conclusion: Looking Forward—Prediction, Detection, and Management in Plant Health

Trisna Tungadi and Francis Wamonje

Detection and Identification in the Modern Era

Early detection and accurate identification remain among the most challenging aspects of disease management. Symptoms often resemble one another across pathogen groups. Yellowing, wilting, stunting, and mosaic patterns occur in bacterial, viral, and fungal infections alike. Infection may be hidden in roots or seeds, invisible on exposed plant parts. Environmental factors further complicate matters: drought stress or mineral deficiencies can exacerbate symptoms, making identification based on visual assessment alone unreliable. The initial step of sampling and preserving plant material until it reaches the laboratory is often the most difficult, as knowing which part of the plant to sample and when matters considerably. Poor sampling timing or location can produce false negatives or false positives, undermining the entire diagnostic process.

Molecular and serological methods

Molecular methods remain the most accurate means of identifying plant pathogens, though cost, time, and the requirement for specialized equipment and skilled personnel constrain their deployment. Polymerase chain reaction (PCR) has been the standard for amplifying pathogen genomic DNA. Its workflow requires DNA extraction, access to laboratory equipment, and expertise to interpret results. Loop-mediated isothermal amplification (LAMP) offers an increasingly popular alternative, operating on similar principles as PCR but running isothermally on simple heating blocks, handheld devices, or battery-powered units. This enables on-site testing by growers, agronomists, and plant health inspectors without laboratory access, facilitating rapid decision-making during routine surveys, border checks, and seed certification. Sample processing is straightforward and inexpensive; readout is simple, involving color change or lateral flow dipsticks that non-specialists can interpret. The trade-off lies in complex initial primer design and susceptibility to false positives due to high sensitivity.

ELISA remains widely used for detecting plant pathogens through antibody–antigen binding, with the enzyme-linked antibody producing a color change indicating pathogen presence. ELISA's appeal stems from its sensitivity, rapidity, cost-effectiveness, and high-throughput capacity, allowing screening of large sample numbers, particularly for viruses, bacteria, and fungi.

All molecular and serological methods share a fundamental limitation: they require prior knowledge of the pathogen being sought to enable proper primer or antibody design. Plant pathogens evolve continuously under selection pressure. For viruses, emerging strains persistently challenge the reliability of existing detection methods. Knowledge-sharing and reporting across stakeholders are critical to ensure that emerging strains are monitored and that detection methods are updated accordingly.

Digital diagnostics: The rise of smartphone apps and artificial intelligence

Digital identification through smartphone apps represents a significant shift in diagnostic accessibility. Many of these applications utilize image analysis, artificial intelligence, and extensive disease databases to identify plant pathogens directly from field or greenhouse photos (Minhans *et al.*, 2025). The widespread adoption of these technologies indicates that they have surpassed the experimental stage. Plantix, possibly the most widely used example, expanded from covering 30 crop–disease combinations at launch to nearly 800 by late 2025, with its user base growing from 10 million downloads in 2021 to approximately 34 million installations 4 years later. Field trials in India, Kenya, and Vietnam have shown diagnostic accuracy rates of 88% for key crops such as tomato, maize, and rice, with case studies reporting yield increases of 25% when timely interventions are made based on app-guided diagnoses. In Malawi, where farmers once waited 10–14 days for expert analysis, instant diagnoses via mobile apps now reduce this to seconds, enabling interventions during critical periods that might otherwise be missed.

The process is straightforward: users photograph symptomatic plants, and the app compares them with known patterns using machine-learning models trained on extensive data sets. This technology proves especially helpful where traditional extension services are lacking. In Tamil Nadu, Plantix helped smallholder farmers differentiate between tomato leaf curl virus and nutrient deficiency, which appear similar but require different management approaches.

Looking ahead, integration is likely to replace standalone apps. New systems will link mobile diagnostics with drone cameras and sensors to automate data collection across fields. Augmented reality tools under development will let users see disease overlays directly on crops in real time. Satellite systems now detect plant stress days or weeks before visible symptoms appear, with multispectral data highlighting shifts in chlorophyll, water stress, and tissue damage that signal disease risk. These technologies point to a future where diagnosis starts with automated alerts, not farmer observation, moving toward genuinely proactive plant health management.

The drawbacks remain significant. Technology adoption varies, and language barriers persist since many apps primarily support English or other major languages. Local context is often lacking: farmers in Myanmar using less-known local bean varieties found limited data in Plantix's database, which restricted its diagnostic usefulness. These systems work best on crops and conditions well-represented in their training data, and they face difficulties with regional specialties or new pathogen strains not yet widely documented. Additionally, while apps are increasingly reliable at identifying visible symptoms, they cannot detect root diseases, latent systemic infections, or pathogens that may be present before symptoms appear. These limitations are likely to continue until diagnostics expand beyond image analysis to include biochemical or molecular methods integrated with mobile technology.

Predicting Disease in a Changing World

Climate as a primary driver

Plant disease dynamics and outbreaks are closely linked to weather and climate conditions, with epidemics often triggered by specific weather patterns that favor pathogen establishment on susceptible hosts. Integrating fundamental knowledge of how plant viruses, fungi, phytoplasmas, or bacterial pathogens interact with their host plants and respond to heat, cold, and humidity is essential for translating this understanding into better risk prediction. Virus diseases and their insect vectors provide some of the best examples.

Warmer winters, for instance, allow aphid populations to persist longer and emerge earlier. Studies across Europe documented that the first spring flight of key species such as *Myzus persicae* advanced by 2 weeks for every 1°C increase in mean January–February temperature (Zhou *et al.*, 1995). In China, a 60-year analysis revealed that winter warming led to earlier hatching of overwintering eggs and extended migration periods for three major aphid species (Wu *et al.*, 2020). These shifts increase the likelihood of aphid outbreaks and cause earlier spring migrations, giving populations greater opportunity to build to damaging levels and prolonging virus infection periods.

In another study conducted over 10 years, barley yellow dwarf virus (BYDV) damage to winter cereals and the population dynamics of the aphid vector, *Rhopalosiphum padi* were monitored across different fields in northern France. Data showed that warmer January-to-August periods were associated with a higher percentage of viruliferous aphids in autumn, probably because higher temperatures boost aphid population growth and accelerate virus transmission (Fabre *et al.*, 2003).

Weather-based epidemiological modeling

Epidemiological modeling uses weather data and forecasts to predict how plant diseases spread by linking environmental conditions to pathogen activity. Temperature, humidity, rainfall, and wind patterns strongly influence survival, reproduction, and movement of bacterial and fungal pathogens, as well as the insect vectors that transmit phytoplasmas and viruses. By combining field observations with weather-based models, researchers can estimate when infection is most likely to occur, how quickly the disease will progress, and where it may spread next.

These models have evolved considerably over the past 50 years, advancing from simple multiple regression approaches to more sophisticated process-based mechanistic models that incorporate biological knowledge and real-time environmental data (González-Domínguez *et al.*, 2023). Recent machine-learning applications in wheat disease forecasting achieved high predictive accuracy, with artificial neural networks demonstrating R-squared values of 0.96 for yellow rust and 0.98 for powdery mildew during calibration, based on the analysis of optimal weather conditions (Singh *et al.*, 2024). These models support growers by identifying high-risk periods, helping to optimize monitoring, spraying, and other management decisions, ultimately reducing crop losses and improving the timing of control strategies.

Operational forecasting systems

Several operational early warning systems now integrate weather forecasting with disease surveillance to provide near-real-time risk assessments. The wheat rust early warning system developed for Ethiopia combines field and mobile phone surveillance data with spore dispersion and disease environmental suitability forecasting, distributing fortnightly advisories to policy makers and SMS alerts directly to smallholder farmers (Allen-Sader *et al.*, 2019). This system averted disaster in 2021 when early field surveys detected a new strain of yellow rust under weather conditions described as "super optimal" for rust development. The early warning enabled rapid government deployment of fungicides, resulting in a bumper wheat harvest rather than widespread crop failure.

Model-based forecasting that integrates airborne inoculum monitoring, environmental data, and satellite-based monitoring offers particular promise for managing crop diseases at regional scales (Ojiambo *et al.*, 2017). Several forecasting systems have been deployed in developing countries despite infrastructure challenges. Models like FARM-D and CLIMEX integrate weather data, crop models, and historical disease records to serve as predictive tools for disease outbreaks, model pathogen distribution changes, and new strains based on climate data while also incorporating vector activity data to predict disease risks.

Such systems represent a paradigm shift from reactive to anticipatory disease management, particularly valuable in developing countries where access to traditional extension services remains limited. The integration of information and communication technologies, including Internet of Things (IoT) sensors, remote sensing, and mobile applications, has transformed the feasibility of real-time disease monitoring and response, enabling rapid dissemination of forecasts directly to farming communities.

Socio-economic factors as drivers

Beyond climate, socio-economic factors play a crucial role in disease emergence and spread. Trade globalization involving international agreements over free trade or tariff reductions has opened new pathways for large-scale transfer of crop produce from one continent to another, while more efficient methods of rapid air and sea transport, combined with loosened plant quarantine regulations, have facilitated this movement.

Billions of plants consumed in North America are produced in Central America, and Europe receives large numbers from Africa and Asia, with favorable climates and labor costs providing incentives for plant production in tropical regions. The bacterium *Xylella fastidiosa* exemplifies this phenomenon: imported horticultural plants and nursery stock may remain asymptomatic in early infection stages, allowing the pathogen to enter new countries via the global trade network. Recent outbreaks and interceptions across European and Mediterranean countries demonstrate how easily infected material can escape detection.

Countries often prove reluctant to report plant disease outbreaks due to potential economic consequences, including barriers to export markets, leading to underreported disease surveillance and undue political and economic influence on nations. The convergence of these socio-economic pressures with accelerating climate change creates conditions particularly conducive to disease emergence, establishment, and transboundary movement. This underscores the need for integrated surveillance systems that account for both biological and human-mediated pathways of pathogen dispersal, combining meteorological forecasting with trade network analysis and epidemiological modeling to provide comprehensive risk assessments.

The Future of Disease Management

Gene editing and the CRISPR revolution

Breeding disease-resistant or disease-tolerant crop varieties remains one of the most effective long-term strategies for protecting yields. Resistant plants can halt infection or slow disease development, reducing the need for pesticides and providing more sustainable protection. However, pathogens can evolve new strains that overcome resistance. Additionally, breeding takes many years and must comply with strict regulatory processes. Even then, access to high-quality certified seed is not always readily available or affordable.

The emergence of CRISPR-Cas9 and related genome-editing platforms has fundamentally transformed crop improvement by enabling precise, targeted modifications to plant genomes. Unlike traditional genetic modification, which typically involves introducing foreign DNA from other species, CRISPR allows researchers to make specific changes to a plant's own genetic material, edits that could, in principle, arise through conventional breeding but would require decades of crossing and selection to achieve. The technology has been successfully deployed across a wide range of crops to enhance disease resistance through several distinct strategies: knocking out susceptibility genes that pathogens exploit for infection, introducing resistance genes at precise genomic locations, and modulating defense pathways to enhance immune responses. We shall give a few examples:

For viral diseases, where chemical control options do not exist and resistance breeding faces continuous challenges, CRISPR offers multiple intervention points. The Cas13 system, which targets RNA rather than DNA, has demonstrated effectiveness in directly degrading viral genomes within infected plant cells (Zhan *et al.*, 2019). Researchers have engineered Cas13 in potato to target and cleave the RNA of potato virus Y, preventing replication and substantially reducing infection rates. An alternative strategy focuses on disrupting host genes that viruses require for successful infection, so-called susceptibility genes. Knockouts of the eukaryotic translation initiation factor *eIF4E* in rice conferred resistance to *Rice tungro spherical virus* without compromising plant vigor (Macovei *et al.*, 2018).

In bacterial diseases, CRISPR-mediated gene editing targeting the *OsERF922* gene in rice is a useful approach to enhance blast resistance. Rice blast is one of the most destructive diseases affecting rice worldwide (Wang *et al.*, 2016). In fruit, knockout of the *CsWRKY22* gene enhanced resistance to citrus canker caused by *Xanthomonas citri* in citrus (Wang *et al.*, 2019), while in apple, targeted disruption of MdDIPM4 conferred resistance to fire blight (Pompili *et al.*, 2020).

Fungal disease resistance has been achieved through modification of the *MLO* gene, a well-characterized susceptibility factor for powdery mildew. CRISPR-mediated knockout of *MLO* homologues in wheat, tomato, and grapevine all produced plants with enhanced resistance to powdery mildew caused by various fungal and oomycete pathogens (Wang *et al.*, 2014; Nekrasov *et al.*, 2017). The approach is particularly valuable in perennial crops such as grapevine and apple, where conventional breeding cycles span many years, and genetic segregation of traits proves especially laborious (Malnoy *et al.*, 2016).

Regulatory landscapes and future prospects

The regulatory treatment of gene-edited crops varies dramatically across jurisdictions. In the United States, the Department of Agriculture determined that plants modified through gene editing without the incorporation of foreign DNA would not be subject to stringent regulations governing traditional genetically modified organisms, effectively treating them as equivalent to conventionally bred varieties. Several gene-edited products have reached the US market under this framework.

The European Union initially adopted a far more restrictive stance, classifying all gene-edited organisms as genetically modified organisms subject to full regulation regardless of whether foreign DNA was introduced. However, a fundamental shift occurred with the provisional agreement in December 2025, establishing a two-tier regulatory system. Category 1 plants, which contain only changes that could have occurred naturally or through conventional breeding, now undergo streamlined verification and are exempt from GMO labeling requirements. Category 2 plants, which involve more extensive modifications, remain subject to the original GMO framework.

Looking forward, CRISPR-based disease resistance is likely to become standard practice in plant breeding programs, integrated alongside conventional selection and marker-assisted approaches rather than replacing them. The technology offers particular advantages for introducing resistance to newly emerged pathogen strains, for stacking multiple resistance genes to achieve durable protection, and for modifying traits in elite cultivars without the genetic drag associated with conventional backcrossing. Combining CRISPR with high-throughput phenotyping, genomic selection, and pathogen surveillance networks could enable genuinely predictive breeding, identifying likely future pathogen threats and preemptively engineering resistance before outbreaks occur.

The challenge extends beyond technical capability to questions of access. Ensuring that gene-editing tools remain available to public-sector researchers and smallholder-focused breeding programs, rather than becoming concentrated in commercial hands through restrictive intellectual property frameworks, will prove critical to realizing the technology's potential for global plant health and food security.

From Historical Lessons to Future Practice

Historical understanding of plant diseases has fundamentally shaped contemporary management approaches. Across fungal, bacterial, and viral pathogens, experience accumulated over decades has emphasized prevention, system-level thinking, and integration of management approaches, commonly known as integrated pest management, rather than reliance on single control measures.

Plant viruses, from the first discovery of *Tobacco mosaic virus* in the late 18th century, have demonstrated that viruses cannot be controlled or cured by chemical sprays once plants are infected. This drove plant breeding for virus resistance or tolerance, and because the majority of plant viruses are spread by insect vectors that feed on phloem, such as aphids, whiteflies, and leafhoppers, vector monitoring and control became common practice to predict and prevent viral outbreaks. Certain plant viruses persist in propagative materials, such as seeds, tubers, and cuttings, which drives the current practice of using certified, virus-free planting materials, coupled with sanitation practices that remove infected plants commonly known as "rogueing."

For bacterial diseases, latent infections and asymptomatic spread present the primary challenges. Bacterial diseases are mainly spread through infected planting materials, growing practices, and trade. Based on historical experience, current bacterial pathogen management focuses on strict hygiene and sanitation practices, regulatory frameworks, certification, quarantine, and biosecurity controls to ensure the provision of clean planting materials. Curative options are limited, as antibiotics are largely ineffective against bacterial infections in plants. Therefore, early detection and monitoring are vital, with a firm emphasis on prevention.

For fungal pathogens, while curative options exist in fungicides, the overuse of chemicals is problematic. Modern management strategies combine the use of resistant plant cultivars, cultural methods, and managed use of fungicides to prevent chemical overuse. Fungal spores are easily spread by rain, making weather-driven outbreaks common, while simultaneously the data generated lead to better epidemiology and forecasting tools based on weather patterns.

Across all plant pathogen groups, several unifying management principles emerge: prevention, systems thinking, and modern plant breeding strategies that emphasize durable resistance. Disease management requires system-level strategies that take into account plant host diversity, environmental conditions, and human behavior via changing lifestyles, socioeconomic status, trade, and travel.

Monitoring pathogen populations for adaptation and resistance is now facilitated by the accessibility and affordability of sequencing technologies. Modern molecular and diagnostic techniques are continually being developed to be faster, more robust, lower cost, and easier to perform. This, coupled with sequencing technology, means that discovery, monitoring, and identification of plant diseases are more accessible and affordable than ever before.

For plant viruses, increased access to high-throughput sequencing has led to the rapid discovery of many new plant viruses in the last decade. However, this creates its own challenge: distinguishing presence from impact. Many newly detected and reported plant viruses cause no symptoms or economic damage. Risk assessment to determine whether a plant virus poses a genuine threat to food production, or is simply part of the plant's virome, remains difficult. Historical context, biological characterization, and epidemiological data for newly discovered viruses are needed to assess risk properly. Decisions by policy makers and regulatory bodies regarding plant health require careful prioritization to avoid overregulation while still identifying and acknowledging viruses and other plant pathogens with significant potential to affect yield, trade, and food security in the long term.

The trajectory for robust plant health is one of convergence rather than displacement. Smartphone diagnostics will not replace laboratory methods but will identify which samples require deeper analysis, thus streamlining the process. Climate forecasting will not eliminate the need for agronomic, real-time judgment but will indicate when and where surveillance should intensify. CRISPR will not render conventional breeding obsolete but will accelerate the introduction of durable resistance traits into elite cultivars. The future of plant health lies in integration: diagnostic technologies feeding surveillance networks, surveillance data informing predictive models, predictive models guiding breeding priorities, and resistant varieties reducing reliance on chemical control. This integrated approach, grounded in historical lessons on prevention and systems thinking, offers the most credible path to managing plant diseases in an era of accelerating environmental change and globalized agricultural trade. Finally, the provision of training and education for future plant health practitioners and research scientists is vital. Higher education providers, industry, and government must take a proactive role in promoting and delivering courses, apprenticeships, and specialized training to build a sustainable and resilient plant health workforce for future generations.

References

Allen-Sader, C., Thurston, W., Meyer, M., Nure, E., Bacha, N. *et al.* (2019) An early warning system to predict and mitigate wheat rust diseases in Ethiopia. *Environmental Research Letters* 14(11), 115004. DOI: 10.1088/1748-9326/ab4034.

Fabre, F., Dedryver, C.A., Leterrier, J.L. and Plantegenest, M. (2003) Aphid abundance on cereals in autumn predicts yield losses caused by *Barley Yellow Dwarf* virus. *Phytopathology* 93(10), 1217–1222. DOI: 10.1094/PHYTO.2003.93.10.1217.

González-Domínguez, E., Caffi, T., Rossi, V., Salotti, I. and Fedele, G. (2023) Plant disease models and forecasting: Changes in principles and applications over the last 50 years. *Phytopathology* 113(4), 678–693. DOI: 10.1094/PHYTO-10-22-0362-KD.

Macovei, A., Sevilla, N.R., Cantos, C., Jonson, G.B., Slamet-Loedin, I. *et al.* (2018) Novel alleles of rice eIF4G generated by CRISPR/Cas9-targeted mutagenesis confer resistance to Rice tungro spherical virus. *Plant Biotechnology Journal* 16(11), 1918–1927. DOI: 10.1111/pbi.12927.

Malnoy, M., Viola, R., Jung, M.H., Koo, O.J., Kim, S. *et al.* (2016) DNA-free genetically edited grapevine and apple protoplast using CRISPR/Cas9 ribonucleoproteins. *Frontiers in Plant Science* 7, 230280. DOI: 10.3389/fpls.2016.01904.

Minhans, K., Sharma, S., Sheikh, I., Alhewairini, S.S. and Sayyed, R. (2025) Artificial intelligence and plant disease management: An agro-innovative approach. *Journal of Phytopathology* 173(3), e70084. DOI: 10.1111/jph.70084.

Nekrasov, V., Wang, C., Win, J., Lanz, C., Weigel, D. *et al.* (2017) Rapid generation of a transgene-free powdery mildew resistant tomato by genome deletion. *Scientific Reports* 7(1), 482. DOI: 10.1038/s41598-017-00578-x.

Ojiambo, P.S., Yuen, J., van den Bosch, F. and Madden, L.V. (2017) Epidemiology: Past, present, and future impacts on understanding disease dynamics and improving plant disease management—A summary of focus issue articles. *Phytopathology* 107(10), 1092–1094. DOI: 10.1094/PHYTO-07-17-0248-FI.

Pompili, V., Costa, L.D., Piazza, S., Pindo, M. and Malnoy, M. (2020) Reduced fire blight susceptibility in apple cultivars using a high-efficiency CRISPR/Cas9-FLP/FRT-based gene editing system. *Plant Biotechnology Journal* 18(3), 845–858. DOI: 10.1111/pbi.13253.

Singh, M., Savary, S., Setiya, P., Tewari, A.K. and Nain, A.S. (2024) Multivariate analysis of the effects of weather variables on white rust epidemics and yield reduction of mustard over multiple growing seasons. *Plant Pathology* 73(4), 791–809. DOI: 10.1111/ppa.13851.

Wang, Y., Cheng, X., Shan, Q., Zhang, Y., Liu, J. *et al.* (2014) Simultaneous editing of three homoeoalleles in hexaploid bread wheat confers heritable resistance to powdery mildew. *Nature Biotechnology* 32(9), 947–951. DOI: 10.1038/nbt.2969.

Wang, F., Wang, C., Liu, P., Lei, C., Hao, W. *et al.* (2016) Enhanced rice blast resistance by CRISPR/Cas9-targeted mutagenesis of the ERF transcription factor gene OsERF922. *PLOS ONE* 11(4), e0154027. DOI: 10.1371/journal.pone.0154027.

Wang, L., Chen, S., Peng, A., Xie, Z., He, Y. *et al.* (2019) CRISPR/cas9-mediated editing of OsWRKY22 reduces susceptibility to *Xanthomonas citri* subsp. *citri* in Wanjincheng orange (*Citrus sinensis* (L.) Osbeck). *Plant Biotechnol Rep* 13, 501–510. DOI: 10.1007/s11816-019-00556-x.

Wu, Y., Li, J., Liu, H., Qiao, G. and Huang, X. (2020) Investigating the impact of climate warming on phenology of aphid pests in China using long-term historical data. *Insects* 11(3), 167. DOI: 10.3390/insects11030167.

Zhan, X., Zhang, F., Zhong, Z., Chen, R., Wang, Y. *et al.* (2019) Generation of virus-resistant potato plants by RNA genome targeting. *Plant Biotechnology Journal* 17(9), 1814–1822. DOI: 10.1111/pbi.13102.

Zhou, X., Harrington, R., Woiwod, I.P., Perry, J.N., Bale, J.S. *et al.* (1995) Effects of temperature on aphid phenology. *Global Change Biology* 1(4), 303–313. DOI: 10.1111/j.1365-2486.1995.tb00029.x.

Index

www.ingramcontent.com/pod-product-compliance
Lightning Source LLC
LaVergne TN
LVHW070441061126
851858LV00010B/140
9781836994411